Nitrogen-Rich Polycyclic-Based Energetic Materials

Guangbin Cheng · Hongwei Yang · Jie Tang

Nitrogen-Rich Polycyclic-Based Energetic Materials

上海交通大学出版社
Shanghai Jiao Tong University Press

Springer

Guangbin Cheng
School of Chemistry and Chemical Engineering
Nanjing University of Science and Technology
Nanjing, China

Hongwei Yang
School of Chemistry and Chemical Engineering
Nanjing University of Science and Technology
Nanjing, China

Jie Tang
School of Chemistry and Chemical Engineering
Nanjing University of Science and Technology
Nanjing, China

ISBN 978-981-97-3416-0 ISBN 978-981-97-4569-2 (eBook)
https://doi.org/10.1007/978-981-97-4569-2

Jointly published with Shanghai Jiao Tong University Press
The print edition is not for sale in China (Mainland). Customers from China (Mainland) please order the print book from: Shanghai Jiao Tong University Press.

This Springer imprint is published by the registered company Springer Nature Singapore Pte Ltd.
The registered company address is: 152 Beach Road, #21-01/04 Gateway East, Singapore 189721, Singapore

Acknowledgements

This work was supported by the National Natural Science Foundation of China (No. 22205110, 21875110, 22075143), the Natural Science Foundation of Jiangsu Province (BK20220958) and the Fundamental Research Funds for the Central Universities (30922010404).

Introduction

The chemistry of high-energy-density materials featuring multiple nitrogen-rich heterocyclic skeletons with energetic fragments has witnessed the enormous growth since the process of manufacturing explosives successfully was improved by Alfred Nobel. The recent surge of interest in HEDMs is particularly owing to its wide application in the military and civilian fields. This book provides a systematic introduction to developments in heterocyclic (oxadiazole, tetrazole, triazole, pyrazole, imidazole, triazine and tetrazine) nitrogen-rich energetic materials with various linkages in the form of bicyclic-, tricyclic-, tetracyclic- and multicyclic-based energetic compounds. This book summarizes the recent efforts toward new high-energy-density materials and analyzes the structure–property relationship based on a series of nitrogen-rich polycyclic-based compounds. The further study on new nitrogen-rich heterocycles may be helpful to achieve a more sustainable chemical process, which could make a positive impact of chemistry not only on energetic materials but also on other fields.

Contents

About the Authors

Professor Guangbin Cheng is a doctoral supervisor, high-level talents of Jiangsu Province "333 Project", young and middle-aged academic leader of "Qinglan Project". Famous scholars and experts in the field of energetic materials. Won Outstanding Achievement Award for Scientific Research in universities of the Ministry of Education, National Defense Innovation Award, Jiangsu Provincial Innovation Award and more than 10 national invention patents. Editorial board member of journal《含能材料》and *Energetic Material Frontiers*. Published more than 100 papers in domestic and foreign core journals (20 of which had an impact factor higher than 10). Fostered more than 60 doctoral and master students. Long-term research on the design and synthesis of energetic materials, green organic synthesis reaction and pharmaceutical intermediates synthesis and technology.

Professor Hongwei Yang is a doctoral supervisor, young and middle-aged academic leader of Jiangsu Province "Qinglan Project". Won Outstanding Achievement Award for Scientific Research in universities of the Ministry of Education, National Defense Innovation Award, Jiangsu Provincial Innovation Award and more than 10 national invention patents. Editorial board member of journal《火炸药学报》and *Energetic Material Frontiers*. Published more than 80 papers in domestic and foreign core journals (20 of which had an impact factor higher than 10). Fostered more than 40 doctoral and master students. Long-term research on the design and synthesis of energetic materials and anticancer drugs, and the research of organometallic chemistry and organic synthetic chemistry.

Associate Professor Jie Tang won Outstanding Achievement Award for Scientific Research in universities of the Ministry of Education. Published more than 10 papers in domestic and foreign core journals and 5 national invention patents. Undertaken 3 national and provincial projects. Long-term research on the design, synthesis and properties of energetic materials.

Abbreviations and Symbols

Abbreviations

1,1′-BTO	1H,1′H-[5,5′-bitetrazole]-1,1′-diol
2,2′-BTO	2H,2′H-[5,5′-bitetrazole]-2,2′-diol
5-AT	5-amino-1H-tetrazole
AAOF	4-amino-N′-hydroxy-1,2,5-oxadiazole-3-carboximidamide
ACOF	4-amino-N-hydroxy-1,2,5-oxadiazole-3-carbimidoyl chloride
AG	Diaminoglycoluril
AMOF	3-amino-4-(5-methyl-1,2,4-oxadiazol-3-yl)furazan
ANTA	5-amino-3-nitro-1,2,4-triazole
AOT	3-Amino oxime-1,2,4-triazole
AP	Ammonium perchlorate
ATCN	5-amino-1H-1,2,4-triazole-3-carbonitrile
BAFF	Bis(aminofurazano)furoxan
BAFOZ	Bisaminofurazanodioxadiazine
BDTZ	5,5′-bis(3-diazo-1,2,4-trizole)
BNFF	Bis(nitrofurazano)furoxan
BNFF^R	Bis(nitrofurazano)furazan
BNTAF	3,4-bis-(1′-nitro-1′,2′,4′-triazole-3′-yl)furoxan
BNTFO-I	4,4″-dinitro-[3,3′:4′,3″-ter(1,2,5-oxadiazole)] 2′,5,5″-trioxide
BNTFO-IV	4,4″-dinitro-[3,3′:4′,3″-ter(1,2,5-oxadiazole)] 2,2′,2″-trioxide
BT	3,6-bis(3,5-dimethylpyrazol-1-yl)-1,2,4,5-tetrazine
BTAF	3,4-bis(1′,2′,4′-triazole-3′-yl)furoxan
BTATz	3,6-bis(1H-1,2,3,4-tetrazol-5-ylamino)-1,2,4,5-tetrazine
CL-20	Hexanitrohexaazaisowurtzitane
COT	3-chloride oxime-1,2,4-triazole
DAAF	3,3′-diamino-4,4′-azoxyfurazan
DAAT	3,3′-azo-(1H-5-amino-1,2,4-triazole)
DAF	3,4-diaminofurazan
DBU	2,3,4,6,7,8,9,10-octahydro-pyrimido[1,2-a]azepine
DCM	Dichloromethane

DHT	3,6-dihydrazinyl-1,2,4,5-tetrazine
DiPEA	N,N-diisopropylethylamine
DMF	*N,N*-dimethylformamide
DMSO	Dimethyl sulfoxide
DMTNBI	1,1′-dimethyl-4,4′,5,5′-tetranitro-1H,1′H-2,2′-biimidazole
DNAT	3,3′-azo-(1H-5-nitro-1,2,4-triazole)
DNBT	3,3′-diamino-5,5′-bis(1H-1,2,4-triazole)
DNBT	3,3′-dinitro-5,5′-bis(1H-1,2,4-triazole)
DNMBT	Dinitromethyl-bis-1,2,4-triazole
DPX1	(E,E)-3,4-bis(oximomethyl)furoxan
EEIP	Ethyl 3-ethoxy-3-iminopropionate
FOF-2	4,4′-oxybis(1,2,5-oxadiazole-3-carbonitrile)
H_2bta	5,5′-bis(1H-tetrazolyl)amine
H_2BTF	3,4-di(1H-tetrazol-5-yl)-1,2,5-oxadiazole 2-oxide
HATF	3-amino-4-tetrazolefurazan
HBT	5,5′-hydrazinebistetrazole
HEDMs	High-energy-density materials
HMX	Octahydro-1,3,5,7-tetranitro-1,3,5,7-tetrazocine
HTNF	4-Nitro-3-(5-tetrazole)-furoxan
K_2DNABT	1H,1′H-[5,5′-bitetrazole]-1,1′-diylbis(nitroamide) potassium
LLM-116	3,5-dinitro-4-aminopyrazole
MNNG	1-methyl-1-nitroso-2-nitroguanidine
Na_2ZT	5,5′-azotetrazolate pentahydrate
NBS	N-bromosuccinicmide
NBS	N-bromosuccinimide
NCS	N-chlorosuccinimide
NOG	3-nitro-5-guanidino-1,2,4-oxadiazole
ONC	Octanitrocubane
PBAN	Polybutadiene acrylonitrile
PETN	Pentaerythrityl tetranitrate
RDX	1,3,5-trinitro-1,3,5-triazinane
RT	Room temperature
SDIC	Sodium dichloroisocyanurate
TADPTz	3-(1H-1,2,3,4-tetrazol-5-ylamino)-6-(3,5-dimethyl-pyrazol-1-yl)-s-tetrazine
t-BuOCl	Tert-butyl hypochlorite
TCICA	Trichloroisocyanuric acid
TEAB	Tetraethylammonium bromide
TFAA	Trifluoroacetic anhydride
THA	*O-p*-toluenesulfonyl-hydroxylamine
TKX-50	5,5′-bistetrazole-1,1′-diolate
TMHI	Trimethylhydrazinium iodide
TMS	Trimethylsilyl
TNBI	4,4′,5,5′-tetranitro-1H,1′H-2,2′-biimidazole
TNP	Trinitropyrazole

TNT	2,4,6-trinitrotoluene
$TsONH_2$	O-tosylhydroxylamine
TTP	1,1,3,3-tetra(1H-tetrazol-5-yl)propane

Symbols

T_d	Thermal decomposition temperature (°C)
T_m	Melt point temperature (°C)
$\Delta_f H$	Heat of formation (kJ mol^{-1})
D	Detonation velocity (m s^{-1})
P	Detonation pressure (GPa)
d	Density (g cm^{-3})
IS	Impact sensitivity (J)
FS	Friction sensitivity (N)
I_{sp}	Specific impulse (s)

Chapter 1
Introduction

1.1 The Development of Energetic Materials

Energetic materials are a class of metastable substances whose molecular structure or composition contains explosive groups or oxidizing components and combustible components. Energetic materials can undergo a violent redox reaction under the stimulation of little external energy, quickly release their internal energy (usually accompanied by a lot of gas and heat), and form combustion or explosion [1]. Explosion reaction has two important characteristics including exothermic and rapid. The rapid exothermic reaction causes a large number of high temperature and high-pressure gaseous products generated by the reaction to instantly expand and form a high-pressure space. Then the high-pressure space quickly releases around to form an explosion. The exothermic reaction that does not release heat fast enough will not form an explosion. The rate of decomposition of energetic material is also the standard for classifying high explosives and gunpowder. The rapid reaction of high explosives is called detonation, while the slower reaction of gunpowder is called combustion. Energetic materials are mainly classified as explosives, propellants, pyrotechnics and gas generating agents based on their use in weapon systems. Among them, explosives can be subdivided into secondary explosives, primary explosives and oxidants. The primary explosive is sensitive and can be detonated under weak external stimuli (flame, spark, impact, friction, etc.) to provide impulse to detonate secondary explosives. Primary explosives generally used in detonators, fire caps and primer. Oxidants possess high oxygen content and provide oxygen for the combustion of other fuels while generating energy from their own explosions, which are often used in solid propellants. Secondary explosives can be divided into single explosives and mixed explosives according to the composition. Single explosives usually have explosive groups such as $-NO_2$, $-NHNO_2$, $-N_3$, $-C(NO_2)_3$, etc. in their molecular structures. Mixed explosives are often a mixture of single explosives and additives or oxidants, combustible agents and additives in a certain proportion, with explosive capacity, the vast majority of the application of explosives are mixed explosives [2, 3].

G. Cheng et al., *Nitrogen-Rich Polycyclic-Based Energetic Materials*,
https://doi.org/10.1007/978-981-97-4569-2_1

With the development of chemical material, new high-energy–density materials (HEDMs) represented by nitrogen-rich polycyclic-based compounds, which plays a vital role in controllable storage systems for chemical energy, have obtained the widespread application in propellants, explosives, pyrotechnics and gas generating agents [4]. "Safer, more environmentally friendly, more powerful..." are the keywords in the process of designing HEDMs [5]. The diversified energetic compounds were synthesized to meet the increasing requirement for new energetic materials. However, it is still a challenge to achieve a fine balance between high energy and low sensitivity of HEDMs because the high energy is primarily at the expense of molecular stability [6–8].

As well known, the history of explosives can be traced back to 220 years BC, when black-powder was invented by mixing flammable charcoal and sulfur with oxidant potassium nitrate by the Chinese [9]. Subsequently, the nitroglycerine was developed in the 1860s by Alfred Nobel, the pioneer scientist. Also, the well-known explosive 2,4,6-trinitrotoluene (**TNT**) (Fig. 1.1) was first synthesized in 1863 and then applied in components of melt-cast explosives [10]. Then 1,3,5-trinitro-1,3,5-triazinane (**RDX**) with excellent energetic performance was developed during the two world wars [11]. Octahydro-1,3,5,7-tetranitro-1,3,5,7-tetrazocine (**HMX**) with improved detonation properties was also discovered in later studies [12]. In 1987, hexanitrohexaazaisowurtzitane (**CL-20**) was first reported and had been paid high attention due to its excellent energetic performance [13]. The continuous work to octanitrocubane (**ONC**) with a high density to 2.1 g cm^{-3} was considered to be a milestone achievement [14].

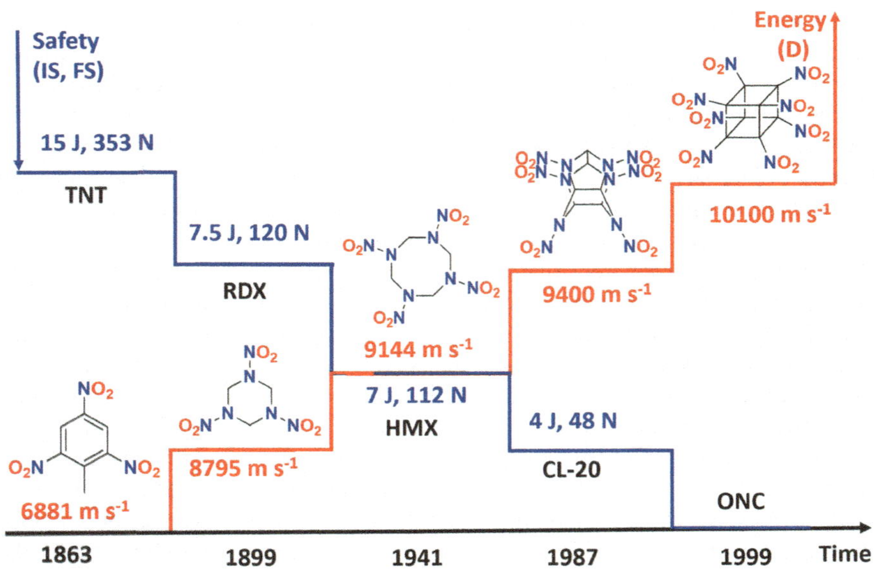

Fig. 1.1 The developmental course of traditional energetic materials

However, the incomplete oxidation of all those traditional explosives (**TNT**, **RDX**, **HMX**, etc.) during detonation leads to the significant production of soot, residue, smoke, and other incompletely oxidized toxic residues (CO, NO, etc.) due to the negative oxygen balance of these traditional explosives [15]. As for nitrogen-rich heterocycles, they often possess low sensitivity and high energy arise from large positive heats of formation due to their high nitrogen content and a large number of N–N and C–N bonds, respectively. Moreover, these heterocycles can be functionalized with energetic groups such as amino, nitro, nitramino, azido, dinitromethyl and trinitromethyl group, thereby extending the design of energetic compounds. Therefore, these heterocycles are promising compounds that fulfil many requirements in the challenging field of energetic materials research. Recently, nitrogen-rich heterocycles have become one of the most exciting scientific discoveries in chemical science. Many derivatives of heterocycle containing two or more rings with excellent performance have sprung up in the field of energetic materials. Although some related books concerning this topic have appeared, a systematic summary of nitrogen-rich polycyclic-based energetic material is rarely reported to the best of our knowledge. In this work, the up-to-date developments of nitrogen-rich polycyclic-based energetic materials including oxadiazole, tetrazole, triazole, pyrazole, imidazole, triazine and tetrazine are provided. The attention of this book is not only focused on synthesis strategies, properties, and derivatives but also the insight structure–property relationship of these heterocycles.

1.2 The Property of Energetic Materials

Judging the quality of energetic materials should be based on its performance characteristics, the main performance indicators are as follows:

(1) Density

Density is the mass of the energy-containing material per unit volume. In general, detonation performances increase proportionally with the density of energetic materials. In order to improve detonation performances in explosive mixture, it is necessary to maximize the charge density in practical application of energetic materials. The charge can be taken in several ways, such as granulation charge, casting or pressing. In fact, the appropriate charging method depends on the characteristics of the energetic material. For explosives, the maximum theoretical density is the density of single crystals.

(2) Heat of formation

Standard heat of formation is the change in enthalpy that occurs when a compound is synthesized from a stable small molecule in a standard state. The energy released when an energetic material explodes is the sum of the enthalpy of formation of the explosive product minus the enthalpy of formation of the energetic material. Heat of formation is an important parameter and a basic data to evaluate the detonation performance of energetic materials.

(3) Sensitivity

Sensitivity is the degree of ease to which an energetic material explodes under the stimulation of external energy. The sensitivity of energetic material is expressed by the minimum external energy provided by the detonation of energetic material. According to the type of stimulation, there are usually five kinds of sensitivity including impact sensitivity, friction sensitivity, shock sensitivity, electrostatic sensitivity, and thermal sensitivity. Energetic materials must be handled with care, because collisions, friction, impact, sparks or heat can cause energetic materials to explode leading to accidents. Sensitivity is one of the key properties of energetic materials, which is the criterion of safety and application reliability of energetic materials.

(4) Oxygen balance

Oxygen balance refers to the relative mass of remaining or insufficient oxygen atoms in the energetic compound when the oxidizer and the fuel are fully reacted, commonly represented by the symbol Ω. $\Omega = 0$ indicates zero oxygen balance, $\Omega > 0$ indicates positive oxygen balance, and $\Omega < 0$ indicates negative oxygen balance. Energetic materials with zero oxygen balance do not require any additional oxygen and excess fuel, and can be completely transformed into gaseous products when burned or exploded in a confined space. Zero oxygen balance is optimal for high explosives.

(5) Detonation velocity and pressure

Detonation velocity represents the stable propagation velocity of shock wave of energetic material. Detonation pressure represents the peak dynamic pressure of shock wave front during detonation of energetic material. They are important parameters to measure the energy level of energetic material.

(6) Thermal stability

The thermal stability of energetic materials is characterized by thermal decomposition temperature, which can be measured by differential scanning calorimetry (DSC). Simple DSC experiments require only a small amount of samples, which is especially important when testing dangerous energetic materials, to avoid damage to instruments and people. Thermal stability is an important property of energetic materials researcher concerned because an unacceptably decomposition temperature will limit its application.

(7) Compatibility

Compatibility refers to the ability of energetic materials to maintain stable physical properties, chemical properties and detonation properties when mixed or in contact with other materials. Compatibility includes not only the compatibility between the components of the mixed energetic material (internal compatibility), but also the compatibility between the energetic material product and the contact material (such as inert coating, elastic wall, etc.) (external compatibility). In the process of transportation, storage and use, the performance of energetic materials with poor compatibility will lead to explosion accidents. Therefore, compatibility experiments must be carried out in the design of energetic material formulation and the selection of materials in contact with energetic materials.

References

1. J.P. Agrawal, R.D. Hodgson, *Organic Chemistry of Explosives*. Wiley, Copyright 2007
2. X. Mei, H. Yang, X. Li, Y. Li, Y. Cheng, Study of some low temperature gas-generating compositions. Propellants Explos. Pyrotech. **40**, 526–530 (2015)
3. J. Tang, D. Chen, G. Zhang, H. Yang, G. Cheng, A green primary explosive: design. Synthesis Testing. Synlett **30**, 885–892 (2019)
4. M.H.V. Huynh, M.D. Coburn, T.J. Meyer, M. Wetzler, Green primary explosives: 5-nitrotetrazolato-N^2-ferrate hierarchies. Proc. Natl. Acad. Sci. U.S.A. **103**, 10322–10327 (2006)
5. M.H.V. Huynh, M.A. Hiskey, T.J. Meyer, M. Wetzler, Green primaries: environmentally friendly energetic complexes. Proc. Natl. Acad. Sci. U.S.A. **103**, 5409–5412 (2006)
6. J. Ma, G. Cheng, X. Ju, Z. Yi, S. Zhu, Z. Zhang, H. Yang, Amino-nitramino functionalized triazolotriazines: a good balance between high energy and low sensitivity. Dalton Trans. **47**, 14483–14490 (2018)
7. T.M. Klapötke, P.C. Schmid, S. Schnell, J. Stierstorfer, 3,6,7-Triamino-[1,2,4]triazolo[4,3-b][1,2,4]triazole: a non-toxic, high-performance energetic building block with excellent stability. Chem. Eur. J. **21**, 9219–9228 (2015)
8. H. Wei, C. He, J. Zhang, J.M. Shreeve, Combination of 1,2,4-oxadiazole and 1,2,5-oxadiazole moieties for the generation of high-performance energetic materials. Angew. Chem. Int. Ed. **54**, 9367–9371 (2015)
9. P. Yin, J.M. Shreeve, Chapter four-nitrogen-rich azoles as high density energy materials: reviewing the energetic footprints of heterocycles. Adv. Heterocycl. Chem. **121**, 89–131 (2017)
10. J. Wilbrand, Notiz über Trinitrotoluol. Ann. Chem. Pharm. **128**, 178–179 (1863)
11. T.L. Davis, *In The Chemistry of Powder and Explosives II* (Wiley, New York, 1943)
12. P.W. Cooper, *In Explosives Engineering* (Wiley-VCH, New York, 1996)
13. M.P. Nadler, R.A. Nissan, W.P. Norris, D.J. Vanderah, R.Y. Yee, *Polynitropolyaza Caged Explosives, Report NWC TP 7020 (Part 7)*. Naval Weapons Center (1989)
14. M. Zhang, P.E. Eaton, R. Gilardi, Hepta- and Octanitrocubanes. Angew. Chem. Int. Ed. **39**, 401–404 (2000)
15. Y. Qu, S.P. Babailov, Azo-linked high-nitrogen energetic materials. J. Mater. Chem. A **6**, 1915–1940 (2018)

Chapter 2
Bicyclic-Based Energetic Materials

Unlike the traditionally designed energetic materials, in which the fuel and oxidizing characteristics are the sole factor to determine their performance, nitrogen-rich heterocyclic compounds have some unique properties such as high heats of formation, significant gas release in the form of environmentally benign nitrogen, and commonly include moieties such as five-membered azole and six-membered azine rings. Nitrogen-rich heterocycles are at the forefront of high energy materials research and expected to achieve increased performance requirements with reasonable safety. The functionalization of nitrogen-rich heterocyclic compounds with energetic groups ($-NO_2$, $-NHNO_2$, $-N_3$, $-NH_2$, etc.) can increase nitrogen content, heats of formation and density, then improve the detonation performances. Many studies have shown that the combination of two or more nitrogen-rich heterocycles can lead to outstanding energetic properties. In this section, the synthesis, properties, and structure–property relationship of those bicyclic-based nitrogen-rich heterocyclic compounds including bicyclic oxadiazole, tetrazole, triazole, pyrazole, imidazole, tetrazine and triazine were investigated.

2.1 Bicyclic Oxadiazole-Based Energetic Compound

Among the five-membered rings, oxadiazole attracts wide interest due to its unique characteristic of an intrinsic oxygen atom, which can increase the oxygen content thus achieving oxygen balance more readily. As shown in Fig. 2.1, there are four oxadiazole isomers: 1,2,4-oxadiazole, 1,2,5-oxadiazole (furazan), 1,3,4-oxadiazole and 1,2,3-oxadiazole [1].

1,2,5-Oxadiazole (furazan) is highly energetic and has potential application in both propellant and explosive formulations owing to its highest positive heat of formation (185 kJ mol^{-1}). Moreover, 1,2,5-oxadiazole-2-oxide (furoxan) is a N-oxide derivative of furazan, which has a "latent" nitro group within one side of its

G. Cheng et al., *Nitrogen-Rich Polycyclic-Based Energetic Materials*,
https://doi.org/10.1007/978-981-97-4569-2_2

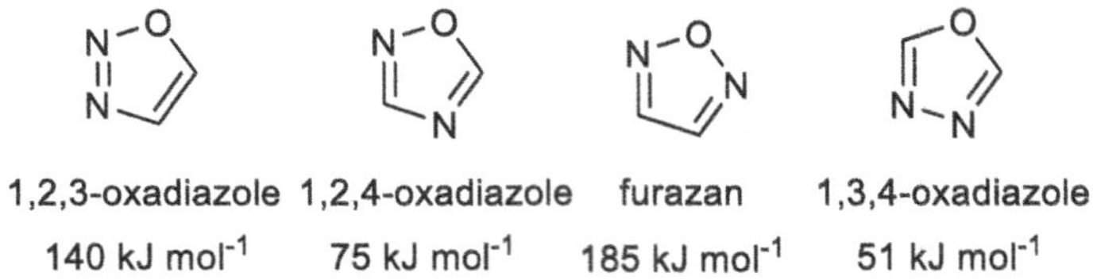

Fig. 2.1 Four isomers of oxadiazole

ring. The density and detonation velocity of an energetic molecule can be increased by ca. 0.06–0.08 g cm^{-3} and ca. 300 m s^{-1} through introducing a furoxan ring into the molecule, respectively [2]. Therefore, a large amount of bicyclic-based energetic compounds containing furoxan or furazan ring have been investigated. A wide variety of aminofurazan derivatives and aminofuroxan derivatives are known in the field of energetic materials due to their good stability and high heats of formation. As shown in Scheme 2.1, 3,3′-diamino-4,4′-bisfurazane (**2a-1**) and 3,3′-diamino-4,4′-bisfuroxane (**2a-4**) were synthesized through the cyclization reaction of tetraoximodiaminobutane in the presence of NaOH at 170 °C and bromine or chlorine at 0 °C, respectively [3, 4]. Rearrangement toward the thermodynamically more stable 4,4′-diamino-3,3′-bisfuroxane (**2a-6**) was achieved by heating **2a-4** in dioxane. In order to enhance the detonation performance, the oxidation and nitration of these aminofurazans to nitrofurazanes and nitraminoazofurazanes have also been reported [5]. In recent years, various nitraminofurazan derivatives were prepared by nitrating the corresponding aminofurazans while nitraminofuroxan derivatives have been rarely reported due to the instability of nitraminofuroxans. In continuing efforts to seek high performance, dinitraminofuroxans (**2a-5**, **2a-8**) with high density (1.91, 1.94 g cm^{-3}) and heats of formation (461.6, 430.8 kJ mol^{-1}) were synthesized through the nitration of their corresponding aminofuroxan (**2a-4**, **2a-6**) using pure nitric acid as a nitration reagent.

By using 4-amino-N′-hydroxy-1,2,5-oxadiazole-3-carboximidamide (**AAOF**) as a raw material, 3-amino-4-(4-amino-1,2,5-oxadiazol-3-yl)-1,2,5-oxadiazole-2-oxide and its isomer were synthesized [6]. The rearrangement process is similar to the above discussion about transformation between compounds **2a-4** and **2a-6**. In a typical procedure, 100% nitric acid was utilized as the nitrating reagent. Dinitramino compound **2a-8** was precipitated from the acid solution, and the corresponding energetic salts were formed by reacting these nitramines with a variety of bases in ethyl ether or CH_3CN solution [7].

Among those compounds contain nitro or nitramino group (Table 2.1, compd **2a-(2-8)**), they all exhibit the higher detonation velocity and pressure (D = 9227–9525 m s^{-1}, P = 34.7–41.7 GPa) than second generation explosives **RDX** (D = 8795 m s^{-1}, P = 34.9 GPa) and **HMX** (D = 9144 m s^{-1}, P = 39.2 GPa). However, the high sensitivity to impact and friction (IS < 3.5 J; FS < 48 N) were observed due to two nitramino or nitro groups in the molecule, which limits their application in the field of energetic materials (Table 2.1).

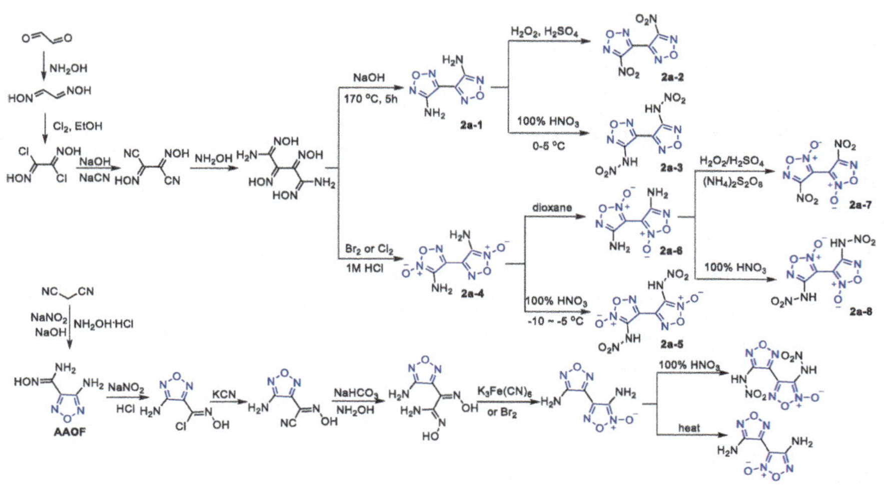

Scheme 2.1 Synthesis of C–C bonded bicyclic 1,2,5-oxadiazole derivatives

In the pursuit of the balance between energy and safety, some explosive units were introduced to oxadiazole such as azo bridge, oxy-bridge, dinitromethyl (–$CH(NO_2)_2$) and fluorodinitromethyl (–$CF(NO_2)_2$) et al. The introduction of an azo bridge to form the azofurazan moiety not only can increases the heat of formation, but also maintain the planar structure. In addition, the additional oxygen atom coordinated to the azo group was considered not only to enhance density and performance but also to stabilize the entire molecule [8]. The synthesis of 3,4-diaminofurazan (**DAF**) was first reported by Coburn in 1968 [9]. The azo-bridged compound 3,3′-diamino-4,4′-azofurazan (**2a-9**) was obtained from the coupling reaction of **DAF** with an oxidative coupling reagent (sodium hypochlorite) in a yield of 84.9%. Solodyuk's work details a number of methods to oxidize **DAF** and the 3,3′-diamino-4,4′-azoxyfurazan (**DAAF, 2a-13**) was observed as oxidative product. This best synthetic method to **DAAF** involved the use of concentrated sulfuric acid and 30% hydrogen peroxide at 16–20 °C shown in Scheme 2.2 [10]. Subsequent an optimization synthetic process of **DAAF** was designed by Francois to investigate the issues with high purity and simple manufacturability [11]. As known, the trinitroethyl group is a fascinating energetic unit in a lot of high-energy and high-density compounds. The incorporation of the trinitroethyl group into nitrogen-rich parent compound is helpful to enhance the oxygen balance and density of the energetic materials, which improves the detonation performances. In Scheme 2.2, the trinitroethyl moieties were incorporated to azofurazan and azoxyfurazan through the Mannich reaction of trinitroethanol with amino to increase the oxygen balance, respectively [12, 13]. Furthermore, the nitration reaction of secondary amine **2a-11** and **2a-15** proceeded smoothly and completed by using a mixture of 100% nitric acid and acetic anhydride as the nitration reagent [3]. The nitraminofurazanes **2a-10** and **2a-14** were synthesized by the nitration of the corresponding aminofurazanes in 100% HNO_3 at 0–5 °C in a high

Table 2.1 Energetic properties of bicyclic oxadiazole-based energetic compound

Compd	T_d[a] (°C)	$\Delta_f H$[b] (kJ mol^{-1})	D[c] (m s^{-1})	P[d] (GPa)	d[e] (g cm^{-3})	IS[f] (J)	FS[g] (N)
C–C bonded bicyclic 1,2,5-oxadiazole derivatives							
2a-2	85	414	9227	35.7	1.85	3.5	/
2a-3	80	526	9477	41.5	1.94	1.5	48
2a-5	74.5	462	9381	40.8	1.91	3.0	20
2a-7	140	449	9525	34.7	2.01	/	/
2a-8	91.3	431	9472	41.7	1.94	2	20
Azo-bridged and azoxy bridged bicyclic 1,2,5-oxadiazole derivatives							
2a-10	122	824	9255	41.4	1.91	3.0	20
2a-11	230	781	8900	36.3	1.82	7.4	111
2a-12	159	1259	9486	40.8	1.87	3.5	70
2a-14	90	772	9131	44.1	1.88	1.0	2
2a-15	190	759	9041	36.7	1.84	7.9	130
2a-16	145	961	9458	41.2	1.92	1.2	80
2a-22	166	176	8492	30.5	1.87	5	120
2a-23	140	526	8936	34.6	1.81	2	40
2a-25	119	667	10,000	/	2.00	< 1	/
Bisnitramino-methylene bridged bicyclic-1,2,5-oxadiazole derivative							
2a-28	128	299	8885	37.8	1.87	2	80
Oxy bridged bicyclic 1,2,5-oxadiazole derivatives							
2a-33	196	167	8715	34.8	1.83	2	/
2a-34	> 270	/	8497	/	1.91	14	64
C–C bonded bicyclic 1,2,5-oxadiazole-based 1,2,4-oxadiazole derivatives							
2a-37	200	295	8118	27.7	1.71	> 60	> 360
2a-38	172	413	8602	32.8	1.78	/	/
2a-39	159	356	7810	24.1	1.65	37.8	360
2a-44	89	270	8155	27.3	1.74	35	360
2a-46	176	412	8651	33.7	1.85	20	240
Bicyclic 1,2,4-oxadiazole derivatives with different linkages							
2a-47	183	− 79	8180	29.4	1.83	8.7	282
2a-50	151	− 362	8367	30.1	1.96	10	192
2a-51	124	62	8814	34.5	1.94	10	80
2a-54	166	− 273	8580	32.6	1.88	/	/
2a-55	117	174	9053	37.4	1.91	13	72
2a-56	355	334	7980	23.4	1.73	> 40	/
2a-57	123	− 81	7363	20.7	1.70	> 40	/
2a-58	140	487	9190	37.5	1.90	2	10

(continued)

Table 2.1 (continued)

Compd	T_d[a] (°C)	$\Delta_f H$[b] (kJ mol^{-1})	D[c] (m s^{-1})	P[d] (GPa)	d[e] (g cm^{-3})	IS[f] (J)	FS[g] (N)
C–C bonded bicyclic 1,3,4-oxadiazole derivatives							
2a-59	246	185	7493	20.4	1.68	40	360
2a-60·H$_2$O	132	238	8227	28	1.77	15	120
2a-61	155	219	8102	26.9	1.75	35	360
2a-62	99	231	8764	33.2	1.90	10	160
2a-63	/	– 212	8440	29.7	1.90	/	/
RDX	204	70.3	8795	34.9	1.80	7.4	120
HMX	287	74.8	9144	39.2	1.91	7.4	120

[a]Thermal decomposition temperature (onset) under nitrogen (determined by the DSC exothermal peak, 5 °C min^{-1}). [b]Heat of formation. [c]Detonation velocity. [d]Detonation pressure. [e]Density determined by gas pycnometer at 25 °C. [f]Impact sensitivity (BAM method). [g]Friction sensitivity (BAM method)

yield. The performances of azofurazan and azoxyfurazan compounds were shown in Table 2.1, compd **2a-**(**10-16**). Compounds **2a-12** and **2a-16** possess high detonation performances (D > 9458 m s^{-1}, P > 40.8 GPa) because of many nitro groups and an azo-bridge or azoxy-bridge in the molecule, which is in agreement with the above discussion. Moreover, the oxidation of **2a-9** to 3,3′-nitro-4,4′-azoxyfurazan (**2a-17**) in the mixture of sulfuric acid and hydrogen peroxide, as well as the oxidation of **DAF** using hydrogen peroxide then coupling reaction using potassium permanganate in acid to 3,3′-nitro-4,4′-azofurazan (**2a-18**) were also reported by Zelenin [14].

Additionally, the introduction of either the fluorodinitromethyl or the dinitromethyl group into oxadiazoles has been demonstrated to be an efficient way for the construction of high energy density materials [15], since the fluorodinitromethyl or the dinitromethyl groups not only can supply additional oxygen atoms but also have a more stabilizing influence than the trinitromethyl group (–C(NO_2)$_3$) especially in thermal stability [16]. The combination of furazan backbone, an azo bridge, and a energetic unit (–CF(NO_2)$_2$ or –CH(NO_2)$_2$) into a molecule were reported [17, 18]. As shown in Scheme 2.2, 3-amino-4-cyanofurazan (**2a-19**) was synthesized through the cyanation of **AAOF** in the presence of manganese dioxide [19]. Subsequently, 4,4′-dicyano-3,3′-azofurazan (**2a-20**) was prepared by the oxidation of 3-amino-4-cyanofurazan using potassium permanganate in hydrochloric acid. Compound **2a-20** was then treated with 50% aqueous hydroxylamine in ethanol, followed by diazotization in dilute HCl, to form 4,4′-bischlorohydroximoyl-3,3′-azofurazan (**2a-21**). The nitration of **2a-21** with a mixture of 100% nitric acid and trifluoroacetic acid anhydride (CF_3CO)$_2$O resulted in the formation of intermediate and then treated with KI to give the product potassium 4,4′-bis(dinitromethyl)-3,3′-azofurazanate, which act as a new energetic 3D metal–organic framework (MOF). Several attractive properties of this potassium salt including a density of 2.039 g cm^{-3}, a decomposition temperature of 229 °C, a detonation velocity of 8138 m s^{-1}, a detonation

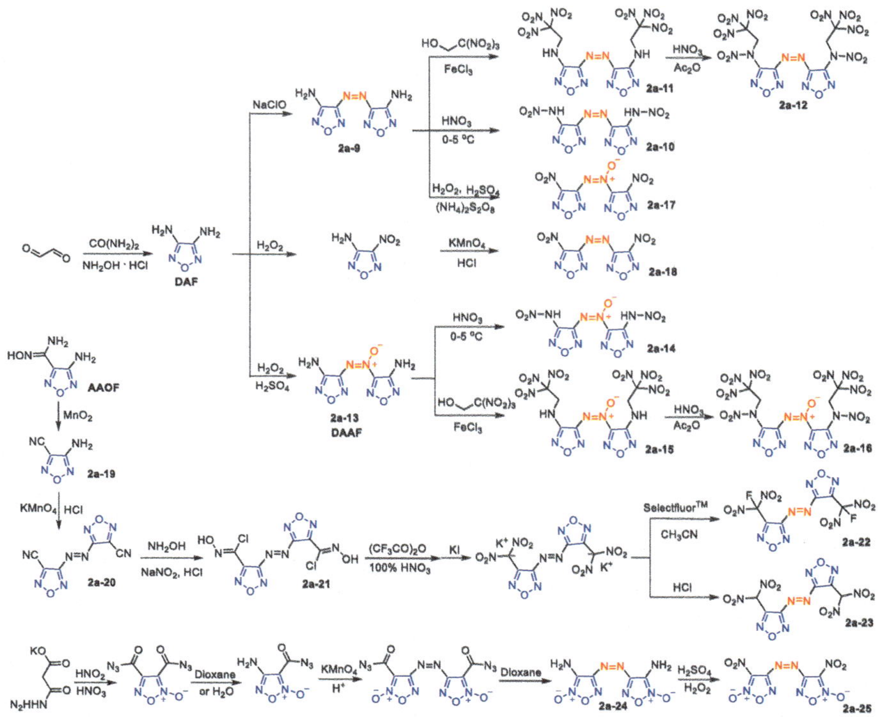

Scheme 2.2 Synthesis of azo or azoxy bridged bicyclic 1,2,5-oxadiazole derivatives

pressure of 30.1 GPa, an impact sensitivity of 2 J, and friction sensitivity of 20 N make it a good candidate as a green primary explosive. Moreover, the fluorination of these potassium salts with Selectfluor™ in anhydrous acetonitrile at 25 °C led to the formation of 4,4′-bis(fluorodinitromethyl)-3,3′-azofurazan (**2a-22**) in moderate yield. 4,4′-Bis(dinitromethyl)-3,3′-azofurazan (**2a-23**) was obtained by acidifying potassium salt with concentrated HCl in acetone. Because of the presence of the furazan rings and the dinitromethyl groups, both **2a-22** and **2a-23** exhibit high positive oxygen balances with values of 19.5% and 12.8%, respectively. Meanwhile, it is seen that the incorporation of fluorine atoms decreases the impact sensitivity (**2a-22**, 5 J cf. **2a-23**, 2 J), enhance the density (**2a-22**, 1.865 g cm^{-3} cf. **2a-23**, 1.811 g cm^{-3}) and thermal stability (**2a-22**, 166 °C cf. **2a-23**, 140 °C). However, the detonation velocity was decreased after the incorporation of fluorine atoms (**2a-22**, 8492 m s^{-1} cf. **2a-23**, 8936 m s^{-1}).

It is worth mentioning that 3,3′-diamino-4,4′-azofuroxan (**2a-24**) was first synthesized from ethyl acetoacetate by Sikderr, but the complexity of processing conditions limits its application [20]. Then, the one-pot reaction including cyclization, ammoniation and oxidation coupling was reported by J.X. He et al. from malonic acid monohydrazide monopotassium salt to obtain **2a-24** (Scheme 2.2). And **2a-24** was directly oxidized to 3,3′-dinitro-4,4′-azofuroxan (**2a-25**) by hydrogen peroxide/sulfuric acid solution, which shows a pretty high density (2.0 g cm^{-3}) and detonation velocity (10,000 m s^{-1}). However, 3,3′-dinitramino-4,4′-azofuroxan have not been synthesized due to the instability of nitramino group [21].

In addition, the combination of furazan rings functionalized with dinitromethyl and a bisnitramino-methylene bridge can impart the moieties with enhanced energy density and oxygen balance [22]. With this strategy in mind, furazans functionalized with some dinitromethyl or fluorodinitromethyl containing bisnitramino-methylene bridge were presented as HEDMS (Scheme 2.3). The chlorine oxime intermediate containing a methylene bridge was prepared from the reaction of formaldehyde and **ACOF**, which was obtained from amidoxime **AAOF** in hydrochloric acid. Then this intermediate was treated with a mixture of 100% nitric acid and trifluoroacetic anhydride (TFAA) to form **2a-26**. After compound **2a-26** was treated with KI in methanol and then acidificated with sulfuric acid, the neutral dinitromethyl compound **2a-27** was prepared. Also the fluorodinitromethyl compound **2a-28** was obtained by fluorination of the amine salt of **2a-27**, which was synthesized by acid–base neutralization reaction of **2a-26** using Selectfluor in anhydrous acetonitrile [23]. As for bisnitramino-methylene bridged compounds, the 3-amino-4-methylfurazan and 3-amino-4-nitrofurazan were condensed with formaldehyde in water with an acid catalyst to produce excellent yields of the methylene-bis-(3-aminofurazans), **2a-30** and **2a-32**, respectively [24]. Moreover, the methanediamine **2a-29** was formed through the reaction of 3-amino-4-nitro-furazan and trinitroethanol in a mixture of [emim]-[HSO4]/[bmpyrr][CF_3SO_3] (1:1) at 20 °C [25]. Compound **2a-31** can easily nitrated using a standard nitric acid-acetic anhydride procedure to yield **2a-32**. However, compound **2a-29** can not cleanly be nitrated under this condition, it was found that dinitrogen pentoxide in CH_2Cl_2 can cleanly nitrated the compound **2a-29** into the desired nitramino compound **2a-30**. Compound **2a-32** melts at 90 °C, then decomposed at 120 °C, while compound **2a-30** melted and decomposed at 99 °C. Therefore, methylene-bis-nitraminofurazans with electron-withdrawing groups are more stable to thermal than electron-donating group substituents at the 4- and 4′-position. As known, the performances of energetic compounds are closely related to their density, oxygen balance and heat of formation [26].

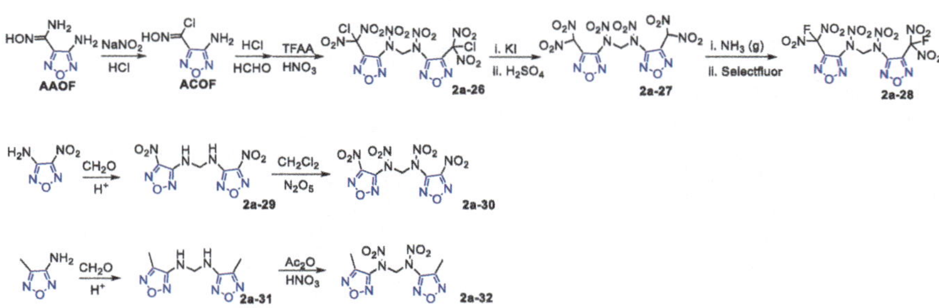

Scheme 2.3 Synthesis of bisnitramino-methylene bridged bicyclic 1,2,5-oxadiazole derivatives

Other than higher heat of formation, higher oxygen content is also required for a molecule to possess improved properties. The incorporation of high content of oxygen and nitrogen into a molecule is a new strategy to develop more powerful, less sensitivity and eco-friendly energetic materials. A series of oxy-bridged difurazanyl derivatives were synthesized due to their high stability to thermal degradation. As shown in Scheme 2.4, 3,3-bis(fluorodinitromethyl)difurazanyl ether (**2a-34**) can be synthesized through two synthetic methods [27]. The method 1 including chlorine oximation, the introduction of hydroxylamine group, oxidation, and fluorination was firstly synthesized starting from **FOF-2**, but the detailed synthetic conditions, characterization data and properties are lacking. It is difficult for the oxidation of N-hydroxy amidoxime with sodium persulfate to offer desired **2a-33**. Afterwards, a new four-step reaction including chlorine oximation, oxidation, salt formation and fluorination starting from **FOF-2** was designed (method 2). Compound **2a-34** and its intermediates were characterized thoroughly. Compared with **2a-23**, **2a-24** exhibits increase of 0.08 g cm^{-3} and 12 J in density and impact to sensitivity and decrease of 213 m s^{-1} in detonation velocity, respectively. These performance changes after incorporation of fluorine atom to dinitromethyl are in agreement with the above discussion of **2a-28**. Compared with azo-bridged difurazanyl derivatives containing fluorodinitromethyl or dinitromethyl (**2a-22**, T_{dec} = 166 °C, D = 8492 m s^{-1}, IS = 5 J; **2a-23**, T_{dec} = 140 °C, D = 8936 m s^{-1}, IS = 2 J), the oxy-bridged derivatives (**2a-34**, T_{dec} > 270 °C, D = 8497 m s^{-1}, IS = 14 J; **2a-33**, T_{dec} = 196 °C, D = 8715 m s^{-1}, IS = 2 J) exhibit higher density and decomposion temperature, lower sensitivities to impact, but lower detonation performances. This result may enrich future prospects for the design of energetic materials.

As for the difurazanyl ether moiety, 3,3-dinitrodifurazanyl ether (**2a-36**) was found to have high-energy potency. Nitro-azoxy was also introduced as a new explosophoric group through a *N*-nitration of an X-azoxy compound (X = Ac, *t*Bu) with nitronium salts [28]. As shown in Scheme 2.4, the most common method for the preparation of difurazanyl ethers has been reported through the ether bond formation from base-promoted transformation of the corresponding 3-nitro-furazans. The treatment of 3-nitro-furazan containing nitro-azoxy bridge with base, followed by flash chromatography can give product **2a-35** and by-product **2a-36** with a low yield of 8%

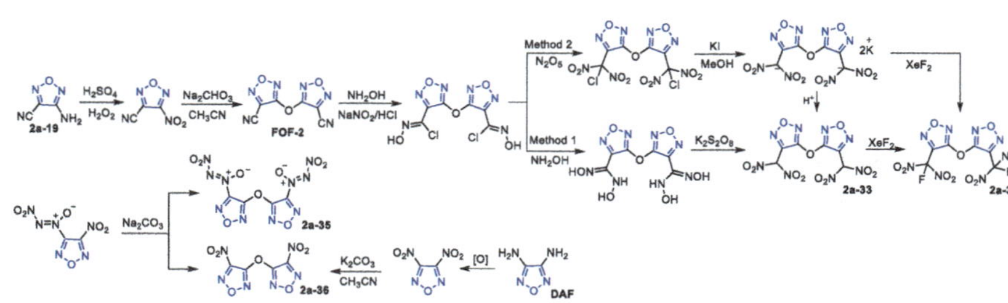

Scheme 2.4 Synthesis of oxy bridged bicyclic 1,2,5-oxadiazole derivatives

and 6%, respectively. It is noteworthy that **2a-36** can also be obtained from dinitrofurazan, which was prepared from the oxidation of **DAF** in a high yield (86%). However, the energetic properties of all those compounds have not been reported.

Compared with furanzan, 1,2,4-oxadiazole have been studied very rarely as energetic materials because its heat of formation (75 kJ mol^{-1}) is just followed by furazan of four oxadiazole isomers. The presence of fewer readily cleaved N–O bonds in the 1,2,4-oxadiazole ring leads to a higher thermal stability and lower sensitivity than furazan [29]. Based on these properties, it is expected to obtain an insensitive and thermally stable compound by combining a 5-methyl-1,2,4-oxadiazole structure with a furazan ring. As shown in Scheme 2.5, 3-amino-4-(5-methyl-1,2,4-oxadiazol-3-yl)furazan (**AMOF**) was synthesized by reacting 4-amino-N′-hydroxy-1,2,5-oxadiazole-3-carboximidamide (**AAOF**) and acetamide. In order to obtain a balance between the explosive energy and safety, the trinitroethyl was introduced to **AMOF** to give **2a-37** through the Mannich reaction of electron deficient trinitroethylethanol and **AMOF** catalyzed by Lewis acid. Moreover, the further nitrating reaction of **2a-37** to nitramino compound **2a-38** was performed using the mixture of pure nitric acid and acetic anhydride. When **AMOF** was treated with 100% nitric acid, the nitramino compound **2a-39** was also obtained [30]. The 1,2,4-oxadiazole-furazan compounds show good thermal stabilities (**2a-37**, 200 °C; **2a-38**, 172 °C), low sensitivities to impact and friction (**2a-37**, IS > 60 J; FS > 360 N) and acceptable detonation performances (**2a-37**, D = 8118 m s^{-1}; **2a-38**, D = 8602 m s^{-1}). It is noteworthy that the melting point of compound **2a-38** is even lower than the room temperature (T_m < 25 °C) while the melting point of **2a-37** is 125 °C. However, the detonation performance of compound **2a-38** is lower due to the existence of methyl group. Moreover, 3-(4-aminofurazan-3-yl)-4H-[1,2,4]oxadiazol-5-one (**2a-40**) exhibits herbicidal activity on broad leaf (dicotyledons) weeds. It can be synthesized through a three-step reaction including cyclization, ammoniation and oxidation starting from **AAOF** in a 67% yield [31]. In view of **AMOF**, the derivatives of **2a-40** may possess better performance than corresponding compound **2a-(37-39)**.

Furthermore, the dinitromethyl group was also introduced to 1,2,4-oxadiazole (Scheme 2.5). The first attempt to prepare 4-amino-3-(5-acetic acid methyl ester-1,2,4-oxadiazolyl)-furazan (**2a-41**) was performed through the substitution reaction of 3-amino-4-aminoximinofurazan and methyl malonyl chloride in the presence of potassium carbonate in a yield of 72.4%. In order to seek a broader prospective for practical applications, a more economical route to synthesize **2a-41** was developed by replacing the expensive methyl malonyl chloride with dimethyl malonate, which gave **2a-41** in a slightly lower yield of 67.8%. The nitration of **2a-41** using the mixture of nitric acid and sulfuric acid gives **2a-42** in a yield of 87.5%. Then, a new series of energetic salts based on **2a-42** were designed and synthesized [32].

In order to obtain a high level of detonation performance while fulfilling the requirement of insensitivity, the combination of furazan and 1,2,4-oxadiazole containing two nitramino groups was reported. As shown in Scheme 2.5, **AAOF** was treated with cyanogen bromide in aqueous ethanol to give the diamino compound **2a-43**. Compound **2a-43** was further nitrated with fuming nitric acid at − 5 °C to form the dinitramino compound, which was extracted from the reaction mixture

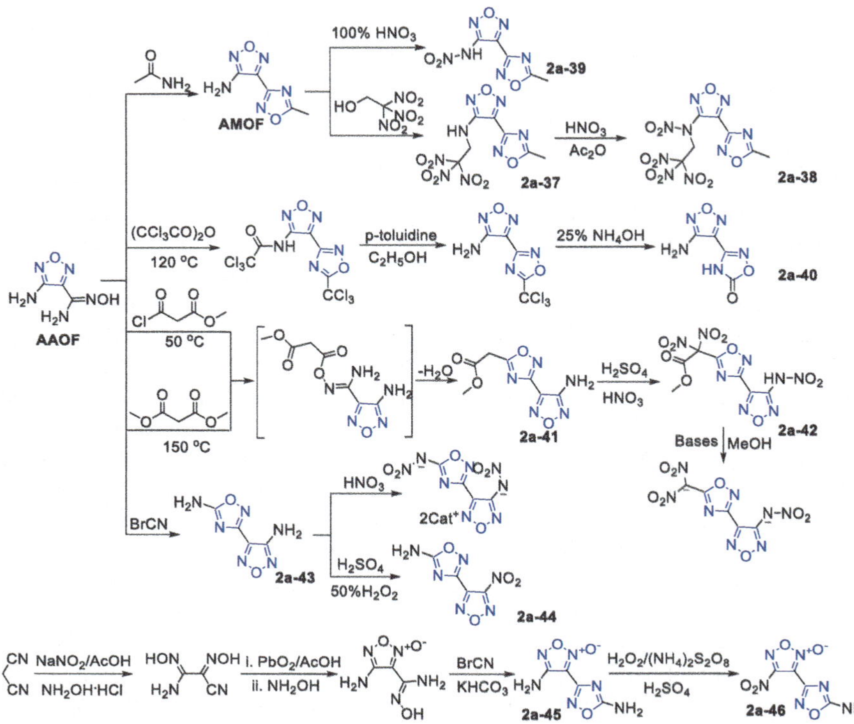

Scheme 2.5 Synthesis of C–C bonded bicyclic 1,2,5-oxadiazole-based 1,2,4-oxadiazole derivatives

with diethyl ether. The solution in diethyl ether was treated with a series of base directly to form corresponding energetic salts due to the unstabilitivity of neutral compound [1]. The selectively oxidation of **2a-43** could be proceeded in the mixture of 50% H_2O_2 and 98% H_2SO_4 using Na_2WO_4 as a catalyst to obtain nitro product 3-(4-nitro-1,2,5-oxadiazol-3-yl)-1,2,4-oxadiazol-5-amine (**2a-44**) [33]. Similar to the synthetic route to **2a-44**, the combination of furoxan and 1,2,4-oxadiazole was also prepared. The 4-amino-3-(5-amino-1,2,4-oxadiazol-3-yl)-1,2,5-oxadiazole 2-oxide (**2a-45**) was obtained through a three-step reaction. The first synthetic step of **2a-45** was involved of the addition of nitrile with hydroxylamine in the presence of $NaNO_2$ starting from malononitril to give intermediate dioxime. Then the amidoxime of furoxan was obtained by cyclization of dioxime with PbO_2 and hydroxylamine solution. In the final step, the **2a-45** was obtained by the cyclization of amidoxime using BrCN in base condition. Further oxidated product **2a-46** was obtained through the oxidation of **2a-45** in the mixture of H_2O_2, H_2SO_4 and $(NH_4)_2S_2O_8$ [34].

The synthesis of high energy–density melt-castable explosive is a particularly challenging area in HEDMs. Melt-cast explosives are sought because they must allow for scalable and efficient manufacturing processes. Recently, the bis(isoxazole) ring

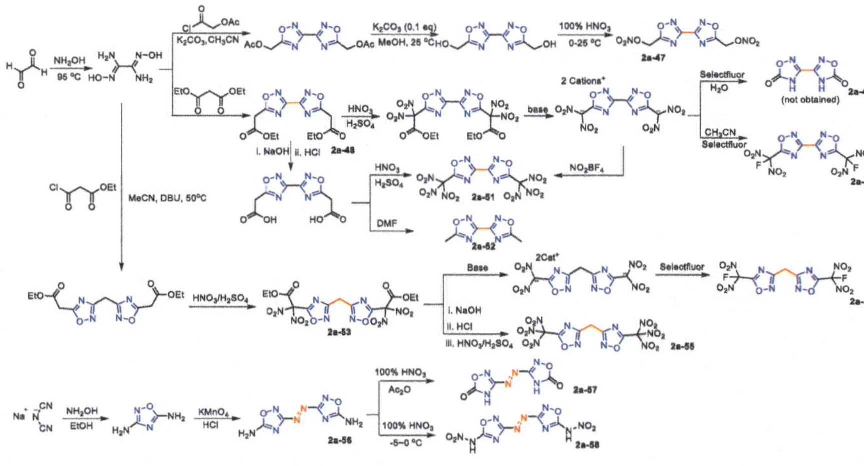

Scheme 2.6 Synthesis of bicyclic 1,2,4-oxadiazole derivatives containing different linkages

system was found to offer tandem melt-castability and energetic propellant plasticization properties potentially. A powerful stand-alone melt-castable explosive material as well as an energetic plasticizing ingredient with nitrocellulose-based propellant formulation dinitrate compound **2a-47** was synthesized through an efficient and scalable route by Johnson [35]. As shown in Scheme 2.6, a four-step reaction including oximation, cyclization, alcoholization and oxidation starting from glyoxal was proceeded to result in **2a-47**. Compound **2a-47** has a melting point of 84.5 °C and a thermal onset of decomposition at 183.4 °C. The density (1.83 g cm^{-3}) and theoretical detonation velocity (8180 m s^{-1}) of **2a-47** are significantly higher than those of **TNT**.

The introduction of oxadiazoles-formally derivative with high oxygen balance of five-membered ring systems is an efficient approach for the development of energetic oxidizers. A combination of 3,3′-bi(1,2,4-oxadiazoles) and other energetic groups containing high oxygen content such as fluorodinitromethyl and trinitromethyl moieties was synthesized as a candidate for oxidizer (Scheme 2.6). Similar to the synthetic route to **2a-41**, the salt 5,5′-bis(dinitromethanide)-3,3′-bi(1,2,4-oxadiazole) was prepared by the nitration of **2a-48** followed by alkalization. The **2a-48** was obtained through the oximation of glyoxal and then cyclization using diethyl malonate [36]. The synthesis of **2a-50** was performed by fluorination of this salt using Selectfluor in anhydrous acetonitrile with moderate yield (66%). When using water as solvent for the reaction, the known 3,3′-bi(1,2,4-oxadiazol-5-one) (**2a-49**) crystallised from the solution after 20 h due to the existence of alkaline hydrolysis. As for compound **2a-51**, there are two different methods to prepare it. One is the nitration of salt 5,5′-bis(dinitromethanide)-3,3′-bi(1,2,4-oxadiazole) using nitronium tetrafluoroborate in anhydrous acetonitrile in a low yield of 27%. The other is that the decarboxylative nitration of the bicarbonic acid using fuming nitric acid in concentrated sulfuric acid furnishes **2a-51** in much better yield (89%). It is interesting that the attempted crystallization of bicarbonic acid in DMF furnished crystals suitable for X-ray diffraction of the decarboxylated product 5,5′-dimethyl-3,3′-bi(1,2,4-oxadiazole) (**2a-52**) in quantitative yield [37]. Because of the high combined nitrogen and oxygen content and the multiple nitro groups of compounds **2a-50** and **2a-51**, the calculated detonation velocities of compounds **2a-50** (8814 m s^{-1}) and **2a-51** (8367 m s^{-1}) are comparable to **RDX** (8750 m s^{-1}). For a practical use as energetic oxidizers, the benchmark sensitivities are defined to be not higher than that of **PETN** (pentaerythrityl tetranitrate: IS = 4 J, FS = 80 N). Thus, compounds **2a-50** and **2a-51** have low sensitivities (**2a-50**, IS = 10 J, FS = 192 N; **2a-51**, IS = 10 J, FS = 80 N). Compound **2a-51** melts at 119 °C with subsequent decomposition at 124 °C. The **2a-50** melts at 67 °C and has a higher decomposition point at 151 °C than **2a-51**. In addition, the I_{sp} (specific impulse) for propellant compositions with aluminium and binder (**PBAN**, polybutadiene acrylonitrile) were calculated for comparison with the rocket-motor composition for solid-rocket boosters used by the NASA Space Shuttle program [38]. The composition consists of 70% ammonium perchlorate (**AP**) as oxidizer, 16% aluminium as high-performing fuel which additionally increases the combustion temperature, 6% polybutadiene acrylic acid, 6% polybutadiene acrylonitrile both as binder, and 2% bisphenol-A ether as curing agent, and achieves a I_{sp} value of 260 s. For **2a-51**, the 75/11 mixture achieves the highest I_{sp} value of 255 s. The I_{sp} values for compounds **2a-50** and **2a-51** can be increased by reducing the binder system in the compositions to its half (from 14 to 7%). The result may be meaningful for future prospects for the design of propellant. The results indicate that compounds **2a-50** and **2a-51** are possible candidates as oxidizer for replacing **AP**.

The incorporation of methylene linkage to energetic backbone was proved to be an effective method to improve thermal stability, reduce sensitivity and increase density [39]. A series of methylene-bridged 1,2,4-oxadiazole rings with polynitro groups were synthesized shown in Scheme 2.6. The key precursor, diethyl 2,2′-(methylenebis(1,2,4-oxadiazole-3,5-diyl))diacetate (**2a-53**) was obtained from the reaction of methyl malonyl chloride and N1,N3-dihydroxymalonimidamide in the presence of 2,3,4,6,7,8,9,10-octahydro-pyrimido[1,2-a]azepine (**DBU**). The routes of further modification to fluorodinitromethyl and trinitromethyl derivatives are similar to the synthesis of compound **2a-50** and **2a-51**, respectively [40]. Compared with bi(1,2,4-oxadiazole) derivatives, the fluorodinitromethyl and trinitromethyl functionalized with methylene-bridged 1,2,4-oxadiazoles exhibit higher heat of formation, lower sensitivity and better detonation performance, respectively (entry compd **2a-(54-55)**, Table 2.1).

Not only the planar structure is maintained, but also the heat of formation is enhanced when heterocyclic azoles are connected by azo bridges. Additionally, although both of the isomers of oxadiazole have a planar structure, they have different properties in both physical and energetic performance such as 5,5′-diamino-3,3′-azo-1,2,5-oxadiazole (**2a-9**) and 5,5′-diamino-3,3′-azo-1,2,4-oxadiazole (**2a-56**). The further nitration product of **2a-9** (**2a-10**, Scheme 2.2) exhibits a high density and excellent detonation properties but high sensitivity to impact and friction. Based on these attractive properties of **2a-10**, 5,5′-dinitramino-3,3-azo′-1,2,4-oxadiazole

(**2a-58**) was synthesized by the nitration of **2a-56** using 100% nitric acid and washing with trifluoroacetic acid. The carbonyl substituted product **2a-57** was formed using the mixture of 100% nitric acid and acetic anhydride as nitration reagent (Scheme 2.6) [41]. Compound **2a-58** possesses comparable detonation performances (D = 9190 m s^{-1}, P = 37.5 GPa) and sensitivities (IS = 2 J, FS = 10 N) to **2a-10**. Compound **2a-54** exhibits very low impact sensitivity (IS > 40 J) but not ideal detonation velocity (7683 m s^{-1}) [42].

Although 1,3,4-oxadiazole has a slightly lower energy than 1,2,5-oxadiazole, it could have a stabilizing effect on nitroamino furazan derivatives. Therefore, 1,3,4-oxadiazole was incorporated into a molecule with nitroamino furazan to pursue an optimum balance between energy and safety. As shown in Scheme 2.7, compound **2a-59** was prepared by treating 4-amino-3-carbohydrazide-furazan with cyanogen bromide under base condition. Further nitration of **2a-59** using 100% nitric acid led to the formation of 3-nitroamino-4-(5-amino-1,3,4-oxadiazol-2-yl)furazan monohydrate (**2a-60**), which exists as a zwitterionic tautomer. Moreover, the attempts to oxidize two amino groups of **2a-59** to nitroamino groups by various nitrating reagents were unsuccessful according to Tang [43], since the amino group bonded to the 1,3,4-oxadizazole ring is very unreactive. Nitro-based **2a-61** was obtained by treating **2a-59** with the mixture of 50% hydrogen peroxide, 98% H_2SO_4 and Na_2WO_4 [44]. The compound **2a-60** possesses low sensitivities to impact and friction (IS = 15 J, FS = 120 N) among the nitroamino compounds, which is agreed with the above discussion. In researchers unremitting efforts to pursue more powerful, less sensitive and eco-friendly energetic materials, a trinitromethyl or fluorodinitromethyl group was incorporated into the nitrogen-rich bi(1,3,4-oxadiazole) backbone. As shown in Scheme 2.7, the synthetic route to **2a-62** is same with other trinitromethyl compounds discussed above. Only a series of energetic salts of bi(1,3,4-oxadiazole) containing two nitro were synthesized while the synthetic route to **2a-63** have not been reported [45]. The performances of **2a-62** and **2a-63** were computed by abinitio calculations using the GAUSSIAN 09 program package. They all exihibit high densities above 1.90 g cm^{-3}, and high performances (**2a-62**, D = 8764 m s^{-1}, P = 33.2 GPa; **2a-63**, D = 8440 m s^{-1}, P = 29.7 Gpa.) Moreover, compound **2a-62** would be a suitable candidate as an oxidizer in solid rocket propellants due to its high specific impulse (I_{sp} (neat) = 252 s, I_{sp} (10% Al, 10% **PBAN**) = 267 s) [46].

2.2 Bicyclic Tetrazole-Based Energetic Compound

The tetrazole moiety has received special attention in HEDMs due to its several energetic features including large numbers of N–N bonds often combined with the stability of aromatic systems, and the formation of N_2 as the main detonation product. Depending on different substituents or both anionic and cationic for the formation of salts, the sensitivities of many energetic compounds can range from extremely

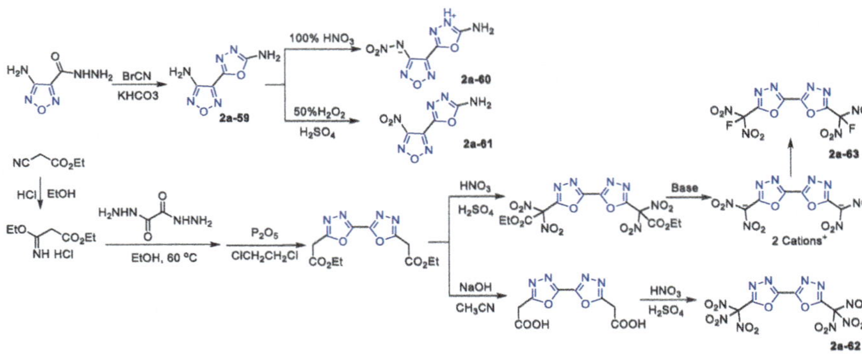

Scheme 2.7 Synthesis of C–C bonded bicyclic 1,3,4-oxadiazole derivatives

sensitive to completely insensitive, making these compounds owning widely potential applications such as primary and secondary explosives, gun propellant additives, pyrotechnic colorants.

Bistetrazoles are of special interest because of their potential to act as chelating ligands to form five- or six-membered rings with the coordinated metal ion. This coordination behaviour is similar to that observed for nitramino- and nitriminotetrazole derivatives. As shown in Scheme 2.8, 5-(tetrazol-1-yl)-2H-tetrazole (**2b-1**) and 2-methyl-5-(tetrazol-1-yl)-2H-tetrazole (**2b-2**) were easily synthesized from the cyclization of corresponding aminotetrazole and sodium azide using triethyl orthoformate as a catalyst [47]. As shown in Table 2.2, the energetic performance of **2b-1** belongs to the class of secondary explosive, whereas its sensitivities belong to the class of primary explosive. Compound **2b-2** is a promising combustible material, which could be used as a nitrogen-rich fuel in modern smokeless propellant compositions.

The synthesis of dihydroxylammonium 5,5′-bistetrazole-1,1′-diolate (**TKX-50**), a classic bistetrazoles compound, was reported by Klapötke [48]. There are two synthetic routes (A and B) to **TKX-50** through the acid–base reaction of the intermediate **1,1′-BTO** and hydroxylamine (Scheme 2.8). The first route A of which, the oxidation of the bistetrazole ring with aqueous potassium peroxymonosulfate can lead to **2,2′-BTO** accompanied by the formation of **1,1′-BTO** in a poor yield (11%). In order to enhance the yield, another of route B was presented. The **1,1′-BTO** can be also isolated through the cyclization reaction of commercially available glyoxal. This method provide a higher yield of **TKX-50** and an easier operation without refinement.

N-amino compounds possess high heats of formation due to the additional N–N single bond. In addition, the new donor sites and functionalities can be introduced by the amino group to yield new potential nitrogen ligands for high energy-capacity transition metal complexes. *O-p*-toluenesulfonyl-hydroxylamine (**THA**) is a well-known powerful amination agent to aminate electron-poor systems. Both 1,1′-diamino-5,5′-bistetrazole and 1,2′-diamino-5,5′-bistetrazole (**2b-3** and **2b-4**) were

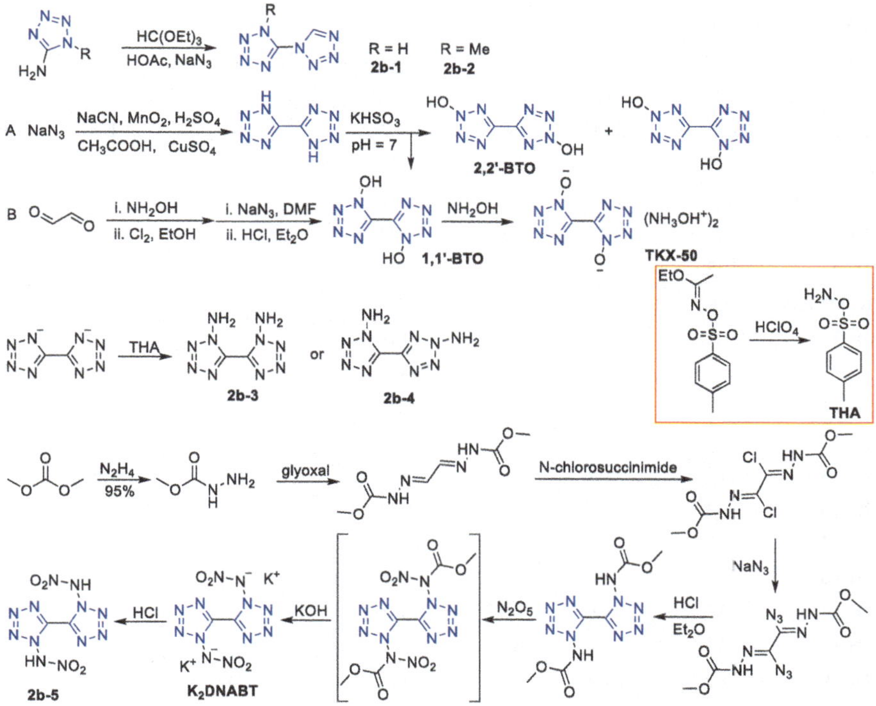

Scheme 2.8 Synthesis of C–C bonded bicyclic tetrazole derivatives

prepared through the amination of the bistetrazole anion with **THA** [49]. To enhance the energy of bistetrazole, a nitramino group was introduced to diaminobistetrazole backbone. In theory, unprotected 1,1′-diamino-5,5′-bistetrazole could be nitrated, but the amination of 5,5′-bistetrazole is a procedure with low yield and high effort. Therefore, an alternative route to 1,1′-dinitramino-5,5′-bistetrazole (**2b-5**) through introducing the removable protecting group was reported (Scheme 2.8). First, the carbonate was reacted with hydrazine hydrate to afford methyl carbazate. Then the condensation reaction of the carbazate and half an equivalent of glyoxal have been done to form ester, which is subsequently oxidized with NCS (N-chlorosuccinimide) to obtain the corresponding chloride. The substitution of the chlorine atom in chloride by sodium azide offers the diazide, which is then cyclized with hydrochloric acid in a diethyl ether suspension. The N-methoxycarbonyl-protected 1,1′-diamino-5,5′-bistetrazole can be gently nitrated with N_2O_5, then the intermediate was treated with KOH to precipitate dipotassium **K_2DNABT**. Last, target product **2b-5** can be formed by treating **K_2DNABT** with HCl [50].

The energetic properties of bistetrazole-based energetic compounds are shown in Table 2.2 compounds **2b-(1-5)**. These compounds possess positive heat of formations ($\Delta_f H > 326$ kJ mol^{-1}), high detonation velocities ($D > 7593$ m s^{-1}) and pressures ($P > 21.3$ GPa) especially for compound **2b-5** ($D = 10{,}142$ m s^{-1}; P

Table 2.2 Energetic properties of tetrazole derivatives

Compound	T_d (°C)	D (m s^{-1})	P (GPa)	Δ_fH (kJ mol^{-1})	IS (J)	d (g cm^{-3})
C–C bonded bicyclic-1,2,5-oxadiazole derivatives						
2b-1	145	8355	28.2	622	< 1	1.73
2b-2	135	7593	21.3	592	9	1.57
TKX-50	221	9698	42.4	447	20	1.88
2b-3	185	8898	32.6	798	3	1.76
2b-4	175	8934	32.9	809	3	1.76
K_2DNABT	200	8330	31.7	326	1	2.11
2b-5	107	10,142	45.6	932	< 1	1.93
Alkyl or alkoxy bridged bicyclic tetrazole derivatives						
2b-11	223	8255	24.1	639	25	1.65
2b-12	232	7767	21.2	498	25	1.65
2b-13	215	7886	21.5	523	25	1.63
2b-14	194	9329	38.2	1038	10	1.86
2b-15	173	8374	28.1	1032	10	1.66
2b-16	145	7963	24.7	989	15	1.58
2b-17	157	9897	46.7	1088	1	1.90
2b-18	134	9200	38.4	1103	1.5	1.81
2b-19	183	8460	32.1	420	23	1.79
2b-20	143	8256	29.3	414	36	1.73
2b-21	166	7996	26.3	382	> 40	1.66
2b-22	147	8734	35.7	498	11	1.84
2b-23	203	8892	34.6	487	4	1.85
2b-24	130	8909	35.1	435	< 2	1.83
Imine bridged bicyclic tetrazole derivatives						
H_2bta	250	9120	34.3	633	> 30	1.86
Azo bridged bicyclic tetrazole derivatives						
2b-25	80	9185	/	1030	< 1	1.74
2b-26	127	7320	21.0	986	/	1.48
2b-27	/	9184	39.0	1092	< < < 1	1.80
HBT	208	8523	27.7	2463	> 30	1.84
2b-28	170	9548	42.4	883	< 1	1.90
2b-29	120	8711	31.2	390	1	1.71
2b-30	179	8723	/	429	H_{50} < 4 cm	1.66
2b-31	/	9330	40.0	1058	/	/
Methylene, triazene and tetrazene bridged bicyclic tetrazole derivatives						
2b-33	182	7825	20.9	514	3	1.53
2b-34	201	7220	/	1499	10	1.56

(continued)

Table 2.2 (continued)

Compound	T_d (°C)	D (m s^{-1})	P (GPa)	Δ_fH (kJ mol^{-1})	IS (J)	d (g cm^{-3})
2b-36	288	8226		590	> 40	1.74
2b-37	247	8307		646	25	1.71

= 45.6 GPa). However, neutral bistetrazole compounds are extremely sensitive to impact and friction, which limits their applications.

In researchers' continuing efforts to seek more powerful, less sensitive, eco-friendly energetic materials, two tetrazole-derived compounds with a bridge were reported. Substitution of the heterocyclic tetrazole ring with amino groups is one of the simplest methods to enhance safety and stability, Shreeve described that the reaction of primary amine with cyanogen azide led to diaminotetrazole containing a series of bridges in a good yield, such as alkoxy bridge (**2b**-(**8-10**)) [51], alkyl bridge (**2b**-(**6-7**)) [52] and carbonyl bridge (**2b**-(**11-13**)) [53] (Scheme 2.9). The treatment of aminotetrazole compounds with excess 100% nitric acid led to nitroiminotetrazoles in good yields [54]. Moreover, the incorporation of one or more trinitroethyl fragments into the framework of aminotetrazoles through the Mannich reaction were described. As shown in Table 2.2, these diaminotetrazoles with carbonyl bridge (compounds **2b**-(**11-13**)) show relatively high decomposition temperature above 215 °C. All those diaminotetrazole derivatives exhibit low sensitivity to impact and friction but low detonation performance. Furthermore, the nitramino and trinitroethyl groups were introduced to diaminotetrazole backbone to enhance their energy. It can clearly see that the nitroiminotetrazoles (compounds **2b**-(**14-18**)) possess higher heat of formation than correspondingly trinitroethy tetrazoles (compounds **2b**-(**19-22**)), which results in higher detonation velocity and pressure of nitroiminotetrazole. Furthermore, the alkyl-bridged compounds also present higher heat of formation than correspondingly alkoxy-bridged compounds.

Apart from nitramino and trinitroethyl groups, nitro groups or polynitro groups were also combined with tetrazole skeletons. The presence of nitro groups will afford an increase in acidity of N–H in azole rings, which can result in storage problems. Linking the acidic nitrogen atoms on these azole rings through a variety of linkages is an efficient method to decrease the acid of azole ring. This structural modification not only increases the stability, but also improves detonation properties. The synthesis of a series of ether-bridged compounds (**2b-23** and **2b-24**) was described as shown in Scheme 2.9 [16]. 2,2′-(Oxybis(methylene))bis-(5-(trinitromethyl)-1H-tetrazole) (**2b-24**) was isolated from the condensation of 5-trinitromethyltetrazole with paraformaldehyde in concentrated sulfuric acid, followed by precipitation which resulted from pouring the solution into ice water. However, some ether-bridged compounds could not be prepared using this route because of either the low reactivity of the azoles or the difficulty in obtaining the neutral compound. Therefore, the reactions of bis(chloromethyl)ether with stoichiometric amounts of sodium 5-nitrotetrazolate dihydrate was utilized to obtain the ether-bridged compound **2b-23**. Moreover, these ether-bridged compounds have low melting points (**2b-23**, T_m =

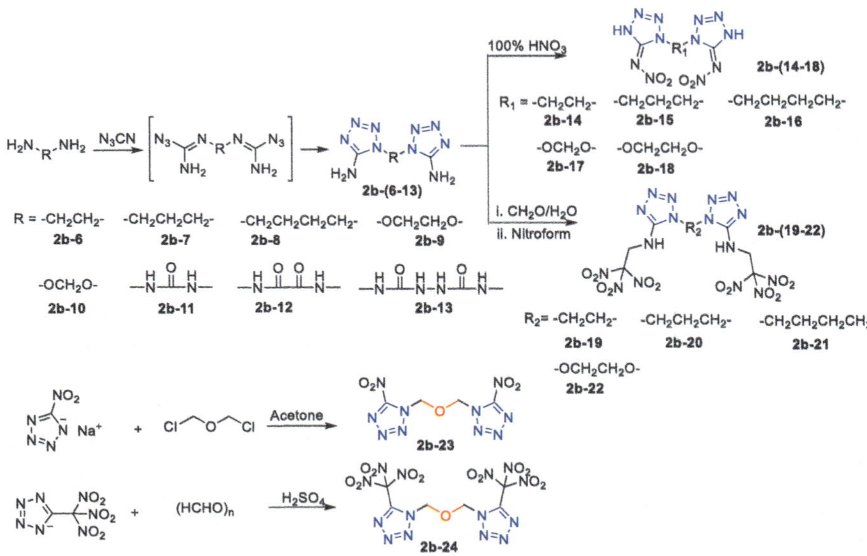

Scheme 2.9 Synthesis of alkyl and alkoxy bridged bicyclic polynitrotetrazole derivatives

148 °C; **2b-24**, $T_m = 75$ °C), moderate decomposition temperatures (**2b-23**, $T_d =$ 203 °C; **2b-24**, $T_d = 130$ °C) and high detonation properties (**2b-23**, $D = 8892$ m s^{-1}, $P = 34.6$ GPa; **2b-24**, $D = 8909$ m s^{-1}, $P = 35.1$ GPa), which suggests them as possible candidates as melt-cast explosives.

Promisingly, the introduction of a new energetic linkage to known frameworks and substituents will provide a strategy for structurally diverse energetic materials. In the case of high-nitrogen explosives, the nitrogen chain linkages have great advantages in high nitrogen contents and positive heats of formation. The heats of formation increase stepwise upon increasing the number of nitrogen atoms and N=N double bonds in linkages as shown in Fig. 2.2. In this regard, the synthesis of a series of energetic compounds containing nitrogen linkages such as –NH–, –N=N–, –N=N–NH–, –NH–N=N–NH– and –N=N–NH–N=N– have been reported.

Many tetrazole derivatives have constituted anionic moieties in energetic salts due to their weak acidity such as 5,5′-bis(1H-tetrazolyl)amine (**H$_2$bta**). There are three synthesis procedures for **H$_2$bta** (Scheme 2.10). (1) In situ reaction of hydrazoic acid (prepared from sodium azide and a weak acid like trimethylammonium chloride, boric acid, ammonium chloride) with sodium dicyanamide; (2) the reaction of sodium dicyanamide with sodium azide in the presence of a catalyst like zinc chloride, bromide or perchlorate, followed by an acidic work-up; (3) the reaction of 5-amino-tetrazole and cyanogen bromide under base-catalyzed conditions to form the 5-cyaniminotetrazolinediide anion, followed by a subsequent cycloaddition of hydrazoic acid under acidic conditions [55]. Compound **H$_2$bta** with a high nitrogen content of 82.3% exhibits very low sensitivities (IS > 30 J, FS > 360 N), high decomposition temperature ($T_{dec} = 250$ °C) and high detonation velocity and pressure (D =

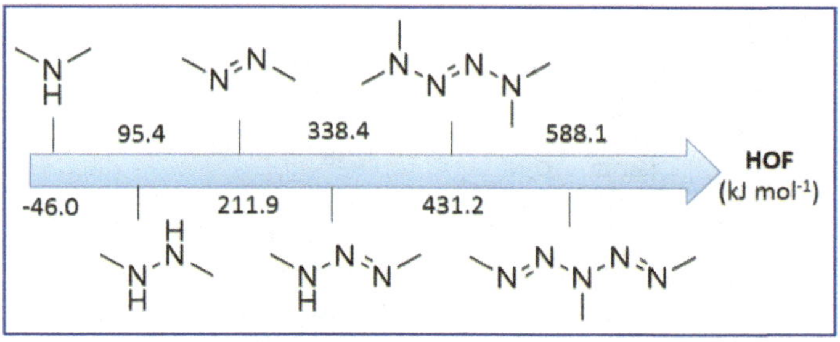

Fig. 2.2 Nitrogen linkages and their heats of formation

9120 m s^{-1}, P = 34.3 GPa). Furthermore, a series of energetic salts based on **H_2bta** are also synthesized [56].

As we known, energetic compounds containing N–N bonds in contiguous nitrogen atoms systems have even higher positive heats of formation than their counterparts with noncontiguous nitrogen atoms. The azo homocoupling of 1-aminotetrazole gave the first fully characterized compound containing ten linked nitrogen atoms. In Scheme 2.11, 1,1′-azobis(tetrazole) (**2b-25**) as a precipitate was obtained from the oxidative coupling reaction of 1-aminotetrazole in the presence of an aqueous acidic solution of sodium dichloroisocyanurate after dilution with a dilute bicarbonate solution and filtration [57]. Furthermore, the coupling reaction was also attempted using t-BuOCl as an oxidant but failed. Compound **2b-25** exhibits high detonation velocity (D = 9185 m s^{-1}) but sensitive to impact (IS < 1 J) and low decomposition temperature (T_d = 80 °C) due to the ten contiguous nitrogen atoms.

In a quest for the synthesis of a highly stable high-nitrogen compound, 1, 1′-azobis(5-methyltetrazole) (**2b-26**), a relatively stable ten-nitrogen catenated azo compound was synthesized by three different azo coupling reagents (sodium dichloroisocyanurate (SDIC), tert-butyl hypochlorite (t-BuOCl) and trichloroisocyanuric acid (TCICA)) from 5-methyl-1-aminotetrazole by G. Cheng (Scheme 2.11) [58]. The incorporation of methyl group enhanced the thermal stabilitity of **2b-26** (T_d = 127 °C) compared with compound **2b-25** (T_d = 80 °C).

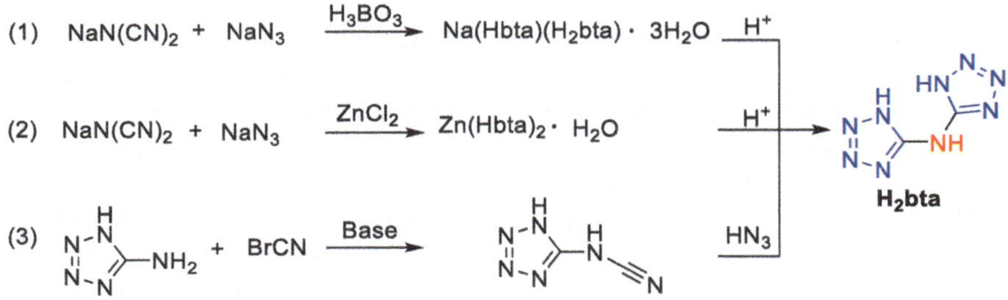

Scheme 2.10 Synthesis of imine bridged bicyclic tetrazole derivatives

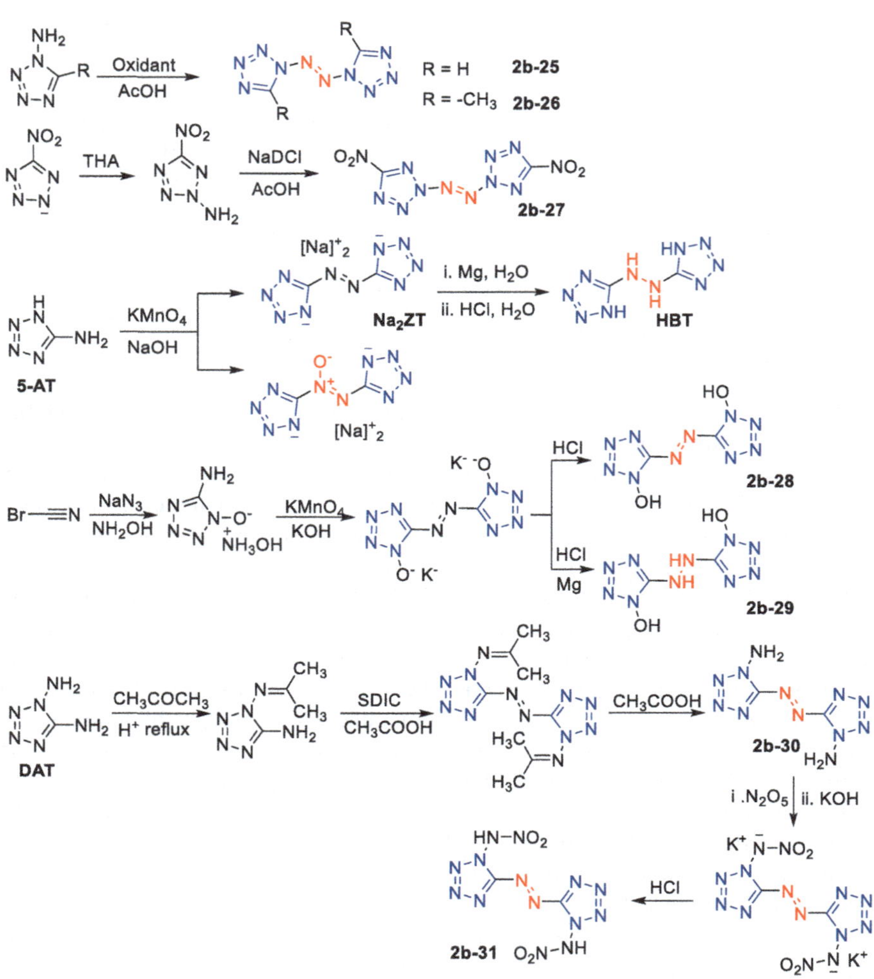

Scheme 2.11 Synthesis of azo bridged bicyclic tetrazole derivatives

Another azo compound containing eight linked nitrogen atoms (**2b-27**) have been prepared through the azo coupling of 2-amino-5-nitrotetrazole using sodium dichloroisocyanurate and acetic acid [49]. Like other compounds containing long catenated nitrogen atom chains, it is sensitive to impact and friction because continuous N–N bonds tend to be more easily broken (IS < 1 J).

Herein, a large-scale synthesis for the 5,5′-hydrazinebistetrazole (**HBT**), a high-performance insensitive energetic material, was described by Klapötke [59]. As shown in Scheme 2.11, **HBT** was formed from the reduction reaction of 5,5′-azotetrazolate pentahydrate in the presence of an excess of magnesium powder in aqueous solution and subsequently treated with hydrochloric acid. **HBT** exhibits low sensitivity to impact (IS > 30 J) and friction (FS > 360 N) due to extensive hydrogen-bondings in **HBT**. Also, the two tunable properties density and heat of formation of a material is dependent. The performance of an energetic material is heavily dependent on density, and the increases of density tend to outweigh the adverse effect on the heat of formation caused by strong hydrogen-bonding. This is the principal reason that the prospective insensitive materials with multiple hydrogen bonds possess high densities and stability.

Moreover, 5-amino-1H-tetrazole (**5-AT**) was oxidized in sodium hydroxide solution with potassium permanganate not only yielding sodium 5,5′-azotetrazolate pentahydrate (**Na_2ZT**) but also 5,5′-azoxytetrazolates [60]. The N-oxide formation at one of the azo bridge nitrogen atoms can change the properties of the **Na_2ZT** notably. On the one hand, the crystal density is dramatically affected by further possibilities to form hydrogen-bonds in the solid state, on the other hand, the increased oxygen content makes the materials suitable alternatives when residue-free combustion of the compound is desired. Other N-oxide 5,5′-azotetrazolate-1,1′-dioxide (**2b-28**) and bis(1-hydroxytetrazolyl)hydrazine (**2b-29**) were obtained from the reaction of aqueous hydroxylamine and cyanogen azide then azo-coupled by using permanganate in basic medium [61].

The oxidative coupling of **DAT** was studied by Z. Du using a series of oxidizers such as sodium dichloroisocyanurate (SDIC), $KMnO_4$ and NaClO, but failed. It might be because the tetrazole ring is broken when both amino groups in **DAT** are oxidized. Subsequently, another amino group of **DAT** was protected by acetone to form 1-isopropylideneamino-1-H-tertrazole-5-amine, and then coupled with SDIC result in the formation of 1,1′-diamino-5,5′-azotetrazole (**2b-30**) [62]. Further nitration of **2b-30** to 1,1′-dinitramino-5,5′-azotetrazole (**2b-31**) using N_2O_5 as a nitration reagent was reported by Klapötke [63]. While the properties of **2b-31** have not been tested by experiment, only some energetic performances were calculated by Zhang [64].

As shown in Fig. 2.2, the triazene bridge possesses an increase of 126.5 kJ mol^{-1} in heat of formation than azo bridge. The synthesis of a series of tetrazole derivatives with triazene bridge have been reported shown in Scheme 2.12. Initially, 5-amino-tetrazole derivatives were smoothly converted into corresponding 1,3-bis(2-tetrazol-5-yl)triazene derivatives (**2b-(32-34)**) without observation of the known diazonium salt intermediate in the presence of nitrite acid [65–68]. It is worth mentioning that the continuous N_{11} chain in compound **2b-34** reported by G. Cheng and H. Yang's group is the longest nitrogen sequence reported so far. However, the neutral compound containing continuous N_{11} chain is unstable in air.

Moreover, the tetrazene compound 1,4-bis-[1-methyltetrazol-5-yl]-1,4-dimethyl-2-tetrazene (**2b-35**) is readily prepared through a one pot reaction starting from 1-methyl-5-(methylamino)-1H-tetrazole. In the first step, 1-methyl-5-(methylamino)-1H-tetrazole is converted to the *N*-nitrosamino-*1H*-tetrazole, which is then subjected to reduction with Zn/CH_3COOH to give the corresponding 1,1-substituted hy-drazine derivative. In-situ oxidation with the help of bromine in concentrated acetic acid results in the formation of **2b-35**. Compound **2b-35** is surprisingly stable and melts without decomposition at 160 °C. It is neither sensitive to friction (FS > 360 N), impact (drop hammer test; energy > 30 Nm) nor heat [67].

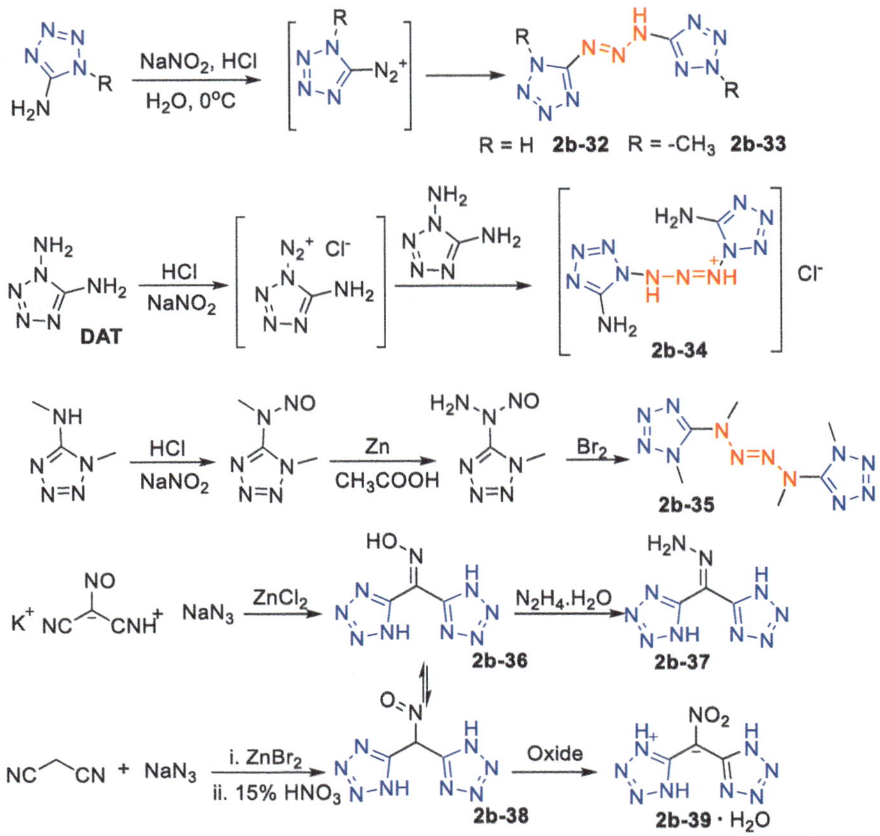

Scheme 2.12 Synthesis of methylene, triazene and tetrazene bridged bicyclic tetrazole derivatives

Compared with oxadiazole-based energetic compounds, tetrazole-based compounds above discussion tend to be more sensitive to impact and friction, which often precludes practical applications. As shown in Scheme 2.12, Chand's group reported the facile and inexpensive synthesis of di(1H-tetrazol-5-yl)methanone oxime (**2b-36**) and 5,5′-(hydrazonomethylene)bis(1H-tetrazole) (**2b-37**) precursors to a family of new nitrogen-rich energetic tetrazoles linked by a methylene bridge [69]. Oxime groups are introduced into methylene bridge to increase thermal stability and decrease sensitivity. As a result, target compounds contianing oxime groups appear to be a good balance between the stability and energy which may hold some promise for application as secondary explosives. Furthermore, nitroso group in methylene backbone (**2b-38**) could be oxidized to the nitro compound 5,5′-(nitromethylene)bis(1H-tetrazole) (**2b-39**). Another route to this nitro compound was also reported by Stierstorfer including the cyclization of malononitrile and sodium azide using zinc bromide as a catalyst and then nitration using 15% HNO_3 [70]. But the energetic properties of **2b-38** and **2b-39** have not been characterized.

2.3 Bicyclic Triazole-Based Energetic Compound

Triazole compounds have been studied over the last couple of years with growing interest due to their high positive heats of formation and the high level of environmental compatibility. Numerous 5,5-bistriazole compounds linked by C–C bond possess promising properties as energetic materials. As shown in Scheme 2.13, Klapötke's group reported a simple and straightforward method for N-oxide introduction in triazole compounds. The method is based on the reaction of oxalic acid and aminoguanidinium bicarbonate in the presence of concentrated hydrochloric acid and subsequent cyclization in basic media. The oxidation of **DABT** was accomplished by the well-known Sandmeyer reaction via diazotization in sulfuric acid and subsequent reaction with sodium nitrite at 40 °C. Then 3,3′-dinitro-5,5′-bis(1H-1,2,4-triazole) (**DNBT**) was successfully oxidized to 3,3′-dinitro-5,5′-bis-1,2,4-triazole-1,1′-diol (**2c-1**) in a buffered aqueous solution of oxone. Compared with other oxidation agents like organic peracids, perborates, hydrogen peroxide, or hyperfluoric acid, the use of oxone as oxidant has unique advantages like suitable oxidation potential, moderate costs, simple handling, and sufficient long-time stability [71]. Moreover, nitrate ester **2c-2** was synthesized through a simple and efficient N-functionalization approach based on **2c-1** using the mixture of pure nitric acid and acetic anhydride [72]. The novel energetic inner salt 5,5′-bis(3-diazo-1,2,4-trizole) (**BDTZ**) was also synthesized if the oxidation mixture cooled to 0–5 °C [73]. Compared with **DNBT** ($D = 8413\ m\ s^{-1}$, $P = 32.0$ GPa, IS = 10 J, FS = 360 N, $T_d = 251$ °C), the N-hydroxy compound **2c-1** ($D = 8729\ m\ s^{-1}$, $P = 36.2$ GPa, IS = 10 J, FS = 360 N, $T_d = 191$ °C) shows higher detonation performances, the same sensitivities toward impact and friction and a lower decomposition temperature. The nitration condition was optimized by using the mixture of concentrated sulfuric and nitric acid in a ratio of 3:1 resulting in **2c-3**. The diazotization reaction of **DABT** proceeded in sulfuric acid and subsequent reaction with an excess of sodium azide to afford compound **2c-4** [74]. In spite of the good calculated performance for **2c-5**, the high moisture sensitivity and low thermal stability result in being unstable when stored at room temperature. Therefore, another modification method is to construct nitramino compound to enhance energetic performance. The 5,5′-dinitro-2*H*,2′*H*-3,3′-bis(1,2,4-triazole)-2,2′-diamine (**2c-6**) was prepared through the amination of the ammonium salt of **DNBT** using O-tosylhydroxylamine. The *N,N′*-dinitroamino-functionalized bis(1,2,4-triazole) (**2c-7**) with good hydrolytic stability was obtained through the nitration of **2c-6** in mixed acid (sulfuric acid and fuming nitric acid) [75].

Similar to the above method to C–C bond connection of bis(1,2,4-triazole), 4,4′,5,5′-tetramino-3,3′-bis(1,2,4-triazole) **2c-8** was obtained by the condensation reaction of 1,3-diaminoguanidine hydrochloride and oxalic acid in polyphosphoric

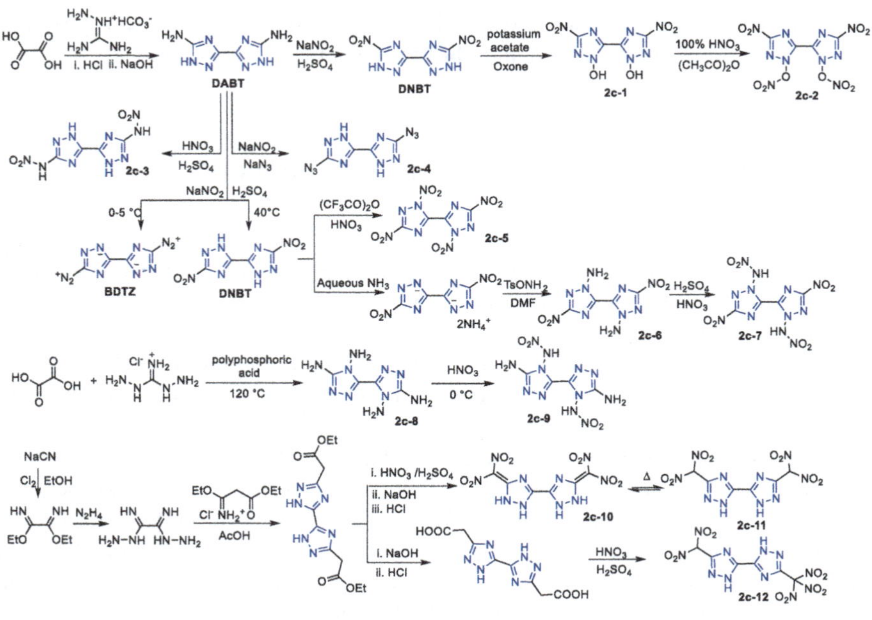

Scheme 2.13 Synthesis of C–C bonded bicyclic 1,2,4-triazole derivatives

acid. To increase the energetic performance, the further selective N-amino nitration of **2c-8** into 5,5′-diamino-4,4′-dinitramino-3,3′-bis(1,2,4-triazole) (**2c-9**) was achieved using 100% HNO_3. Meanwhile, the high energetic compound **2c-9** (D = 8846 m s^{-1}, P = 34.1 GPa) preserved high thermal stability (T_d = 259 °C) due to strong hydrogen-bonding network [76].

The synthesis of the dinitromethyl-bis(1,2,4-triazole) **2c-10** (**DNMBT**) was achieved through a completely different synthetic pathway. The chlorination of sodium cyanide in ethanol and subsequent reaction with hydrazine leads to the formation of the oxalimidohydrazide. This intermediate was reacted with the commercially available 3-ethoxy-3-iminopropionic acid ethyl ester hydrochloride to form the 3,3′-(diethyl acetate)-5,5′-bis-1,2,4-triazole. The introduction of nitro groups into this ester was accomplished by nitration with nitric acid in a mixture of sulfuric acid and oleum (1:1). The ethyl ester hydrolyses rapidly, whereas stirring the nitrated intermediate product in basic media and subsequent acidification with concentrated hydrochloric acid leads to the release of CO_2 and the formation of **DNMBT**. The existence of different isomers of compound **DNMBT** could be demonstrated by recording the NMR and IR spectra at elevated temperatures [74]. Furthermore, the introduction of trinitromethyl group to bis(1,2,4-triazole) to enhance oxygen balance was reported by Fan' group. The desired product 5,5′-bis(trinitromethyl)-3,3′-bis(1,2,4-triazole) (**2c-12**) was produced from the nitration of 5,5′-bis(acetic acid)-3,3′-bis-1,2,4-triazole in the mixture of concentrated sulphuric acid and 96% nitric acid [77]. All discussed bis(1,2,4-triazole) derivatives (**2c-(1-12)**) have a highly positive heat of formation in the range from 114 to 1256 kJ mol^{-1}. They all exhibit acceptable detonation performances (D, 7780–9234 m s^{-1}; P, 25.0–38.2 GPa) especially for **2c-7** (D = 9234 m s^{-1}; P = 38.2 GPa).

The two constitutional triazole isomers, 1,2,3-triazole and 1,2,4-triazole, have been fully investigated. In comparison, 1,2,3-triazole features higher heat of formation and better thermal stability, as well as enhanced detonation performance arising from tremendous energy storage in N–N and N=N backbones. A variety of functionalizations of 4,4′-bis(5-nitro-1,2,3-2H-triazole) such as methylation, amination, and N-oxidation strategies were employed to develop more powerful, less sensitive, and eco-friendly energetic materials. In Scheme 2.14, the expected 5,5′-dinitro-4,4′-bis(1,2,3-triazole) (**2c-13**) was obtained through the cyclization reaction of diacetates and sodium azide in methanol. Diammonium 4,4′-bis(5-nitro-1,2,3-triazolate), the precursor for methylation, amination, and N-oxidization reactions, was synthesized by bubbling gaseous ammonia to a diethyl ether solution of **2c-13**. Four isomers of **2c-14** were obtained through the methylation of diammonium salt using dimethyl sulfate in DMF. The reaction of freshly prepared *o*-*p*-toluenesulfonyl-hydroxylamine (THA) with this salt led to the formation of three isomers **2c-15**. When this salt was treated with Oxone in a buffer solution (pH 4–5) at 40 °C, **2c-16** was isolated by extracting with ethyl acetate [78]. As expected, all the nitrogen-rich compounds derived from **2c-13** show relatively high thermal stabilities ($T_d > 191$ °C), high densities, and heats of formation ($\Delta_f H > 373$ kJ mol^{-1}). The introduction of N-NH_2 and the N–O group were found to be effective methods to improve the density and detonation property of the parent ammonium salt. The compounds with substituent groups in the N_3 position of 4,4′-bis(5-nitro-1,2,3-2H-triazole) system possess slightly better detonation properties than the other isomers with the same substituent group at different positions.

Considering the advantages of 1,2,3-triazole ring, the combination of 1,2,3-triazole and 1,2,4-triazole will be helpful to enhance detonation properties. Yang's group described the synthesis of some compounds consisting of 1,2,3-triazole and 1,2,4-triazole rings carrying energetic moieties like amino, nitroimino, and nitro. As shown in Scheme 2.15, the intermediate methyl 4-nitro-2H-1,2,3-triazole-5-carboxylate was prepared by modifying three-step reactions including nitration, cyclization and oxidation starting from ethyl hydrogen malonate. In the presence of

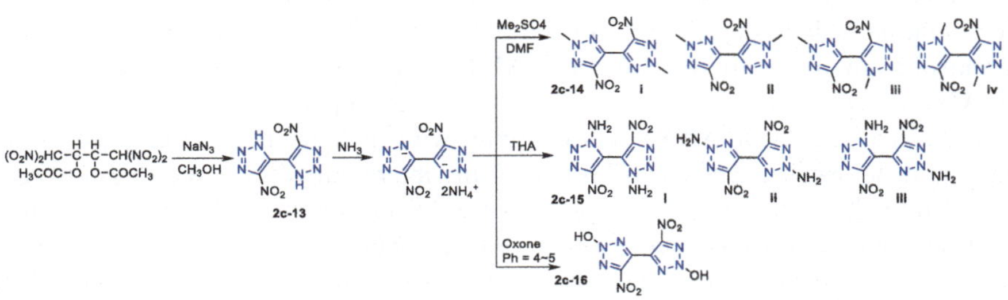

Scheme 2.14 Synthesis of C–C bonded bicyclic 1,2,3-triazole derivatives

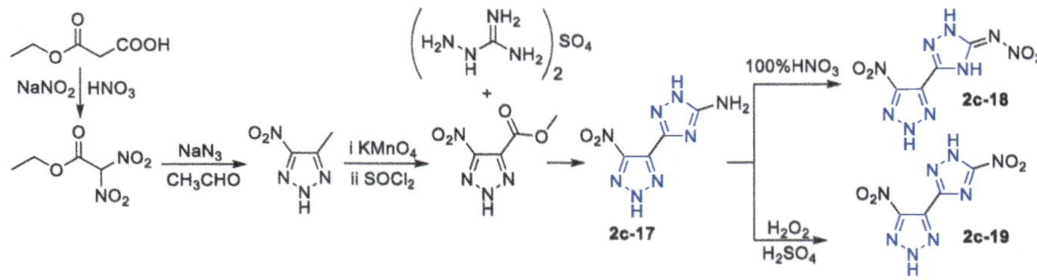

Scheme 2.15 Synthesis of C–C bonded bicyclic 1,2,3-triazole-based 1,2,4-triazole derivatives

two equivalents of aminoguanidinium sulphate in NaOMe solution, 4-nitro-2H-1,2,3-triazole-5-carboxylate was readily converted to a C–C bonded 1,2,3-triazole-(1,2,4-triazole) bicyclic compound **2c-17** via a one-pot cascade cyclization. Nitroimino-triazole 4-nitro-5-(5-nitroimino-1,2,4-triazol-3-yl)-2H-1,2,3-triazolate (**2c-18**) and 4-nitro-5-(5-nitro-1,2,4-triazol-3-yl)-2H-1,2,3-triazolate (**2c-19**) were obtained by the nitration of **2c-17** using 100% nitric acid and the mixture of 30% H_2O_2 and concentrated H_2SO_4, respectively [79]. With a large number of inherently energetic C–N and N–N bonds in these structures, compounds **2c-(17-19)** possess acceptable thermal stability ($T_d > 168$ °C), good impact and friction sensitivities (IS > 16 J, FS > 180 N), and acceptable detonation pressures (P > 27.4 GPa) and velocities (D > 8384 m s^{-1}).

As known, the heat of formation of a C–C single bond is lower than that of a C–N single bond. Therefore, the C–N bonded bisheterocyclic compounds are expected to be more energetic. The strategy for constructing C–N bonded bisheterocyclic compounds is usually through reaction of N-nitroazoles and nucleophilic reagents. As shown in Scheme 2.16, a series of C–N bonded bisheterocyclic compounds containing triazole have been synthesized and fully characterized. The 1,3-dinitro-1,2,4-triaozle was prepared by the treatment of 3-nitro-1,2,4-triazole with ammonium nitrate and trifluoroacetic anhydride (TFAA) in dichloromethane. The **2c-20** and **2c-21** were obtained by reacting dinitrotriaozle with corresponding sodium salts in the presence of Na_2CO_3 in methanol. Further treatment of **2c-21** with *O*-*p*-toluenesulfonyl-hydroxylamine (THA) resulted in two isomers, **2c-22** and **2c-23**, where amino groups is bonded at either N-1 or N-2 of the nitro triazole ring [80].

Constructing azole based on C–NH_2 in heterocyclic ring is also an efficient method to synthesis bisheterocyclic compounds linked by C–N bond. The primary amino group of 5-amino-3-nitro-1,2,4-triazole (**ANTA**) is reactive and three synthetic routes to **ANTA** were summarized as shown in Scheme 2.16 [81]. Licht's group reported that the diazotation of **ANTA** and coupling with methazonic acid resulted in an intermediate. Subsequently cyclization of this intermediate gave the dinitrobitriazole 4-nitro-2-(5-nitro-1,2,4-triazole-3-yl)-1,2,3-triazole (**2c-24**). Moreover, another dinitrobitriazole **2c-25**, isomer of **2c-24**, could be obtained by the same method [82]. The detonation velocities and pressures of all bistriazole compounds linked by C-N bond lie in the range 7867–8980 m s^{-1} and 16.8–33.1 GPa, respectively. They also

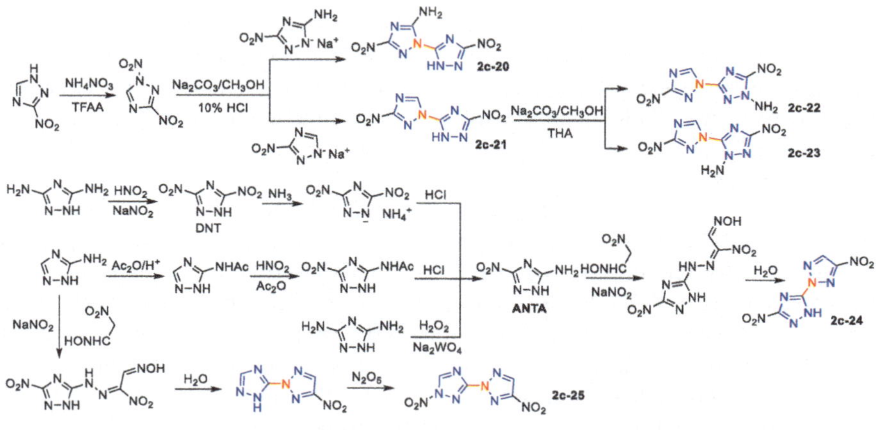

Scheme 2.16 Synthesis of C–N bonded bicyclic triazole derivatives

exhibit acceptable decomposition temperature above 164 °C. Except **2c-24** and **2c-25**, other bistriazole compounds possess low sensitivities to impact and friction (IS > 22 J, FS > 160 N).

The oxidative coupling reaction of N–NH_2 on heterocycle was employed to create a rather long chain of catenated nitrogen atoms. Such a structure could dramatically increase the heats of formation of high-nitrogen compounds. S. Pang's group reported the effective synthesis and characterization of novel tetrazene-linked bistriazoles. As shown in Scheme 2.17, the tetrazene-linked bistriazoles (**2c-26**) was obtained through the oxidation coupling reaction of 4-amino-1,2,4-triazole using sodium dichloroisocyanurate (SDIC) as an oxidative reagent in a high yield. The different product **2c-27** could be formed through different the molar ratios of SDIC to 4-amino-1,2,4-triazole. Moreover, other oxidative reagents such as $KMnO_4$ in HCl or NaOH, MnO_2 and H_2O_2 were also attempted to oxidize 4-amino-1,2,4-triazole, but failed to afford the desired oxidation coupling product [83]. In order to enhance the performance, four carbon atoms in triazole ring could be modified by judicious selection of the substituent such as azido and amino. To introduce azido or amino groups into heterocycle **2c-26**, the chloriation reaction using SDIC as a chlorating agent is necessary achieved. Then azides **2c-(28-31)** with enhanced detonation velocity and pressure were obtained through azidation reaction using sodium azide in DMF. Amino derivatives **2c-(32-35)** with higher stability were prepared through a two-step reaction involved substitution of these azides using triphenylphosphine (PPh_3) as a reducing agent and then hydrolyzation in 1 M HCl. Meanwhile, a one-pot method to synthesize iminophosphorane intermediates from **2c-26** without isolating azides was also developed [84]. Subsequently, in order to obtain triazole salt **2c-36**, the protonation of 4,4′-azo-1,2,4-triazole was attempted through traditional methylation and azo coupling reaction based on 4-amino-1,2,4-triazole, but failed. Instead, 3,3′-dione-4,4′-azo-1,2,4-triazole (**2c-37**) was obtained [85]. Moreover, the high-nitrogen compound **2c-38** containing eight directly linked nitrogen atoms was synthesized through the oxidation coupling reaction of 1-amino-1,2,3-triazole using SDIC at a low temperature in 78% yield. Surprisingly, it is a photochromic molecule that undergoes a reversible color change when subjected to irradiation, so it could be a prospective "green" photochromic material. This unique photochromic feature of **2c-38** arises from its trans–cis photoisomerization, which was confirmed by Raman spectroscopy [86].

According to the method of oxidative coupling of N–NH_2 groups, Shreeve's group reported the synthesis of a new nitrate ester and azide energetic compounds containing tetrazene-linked bistriazole. 3,3′,5,5′-Tetra(hydroxymethyl)-4,4′-azo-1,2,4-triazole or 3,3′,5,5′-tetra(azidomethyl)-4,4′-azo-1,2,4-triazole (**2c-39**) were formed by oxidation coupling reaction of N–NH_2 compounds using one equivalent of SDIC in acetic acid. In addition, the hydroxy groups can be nitrated with fuming nitric acid in acetic anhydride as mentioned above. By using an analogous method, the corresponding tetranitrate, 3,3′,5,5′-tetra(nitroxymethyl)-4,4′-azo-1,2,4-triazole (**2c-40**) was generated [87]. As shown in Table 2.3 compounds **2c-(28-40)**, all those tetrazene-linked (N-N′-azo) bistriazoles possess relatively high heats of formation due to the catenated nitrogen atoms. Although the densities of these compounds fall into the range from 1.53 to 1.76 g cm^{-3}, they exhibit good detonation performances (D, 7470–9370 m s^{-1}; P, 19.2–38 GPa).

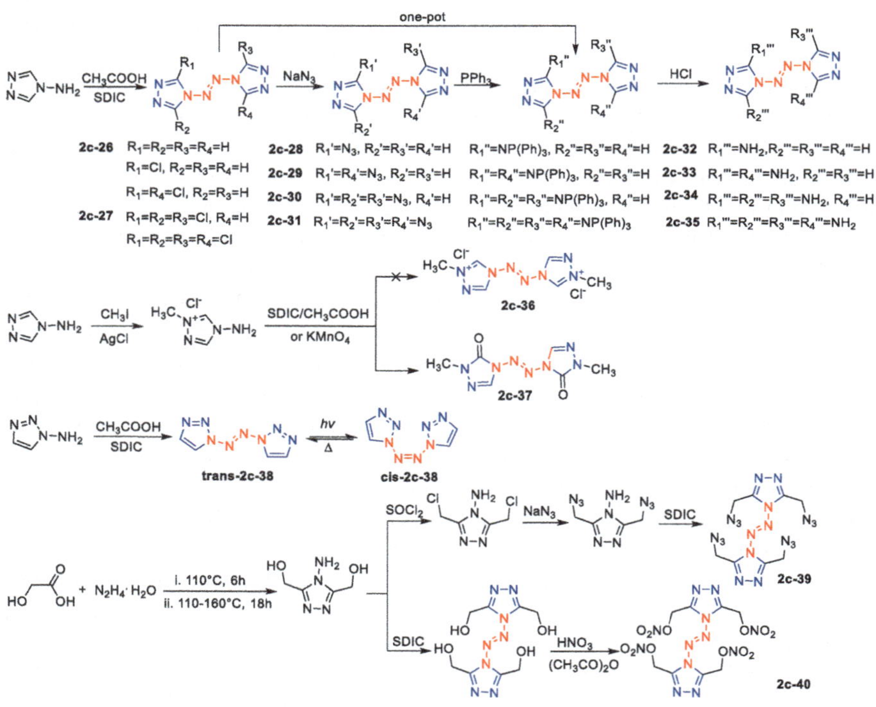

Scheme 2.17 Synthesis of N,N′-azo bridged bicyclic triazole derivatives

Table 2.3 Energetic properties of bicyclic triazole-based derivatives

Compound	T_d (°C)	D (m s^{-1})	P (GPa)	$\Delta_f H$ (kJ mol^{-1})	IS (J)	FS (N)	d (g cm^{-3})
C–C bonded bicyclic 1,2,4-triazole derivatives							
DNBT	251	8413	32.0	285	10	360	1.90
2c-1	191	8729	36.2	290	10	360	1.91
2c-2	176	8639	32.7	114	6	60	1.83
2c-3	194	8355	30.0	405	3	108	1.80
2c-4	201	7944	25.0	971	3	48	1.70
BDTZ	120	7780	26.7	1256	/	/	1.73
2c-6	271	8677	31.8	440	40	360	1.83
2c-7	121	9234	38.2	592	3	40	1.88
2c-9	259	8846	31.2	692	9	120	1.76
2c-10	121	8499	34.1	298	20	360	1.80
2c-12	148	9073	36.2	332	23	252	1.89
C–C bonded bicyclic 1,2,3-triazole derivatives							
2c-13	155	/	/	439	/	/	/
2c-14(i)	227	7563	21.3	373	> 40	/	1.62
2c-14(ii)	275	7642	22.0	403	> 40	/	1.63
2c-14(iii)	229	/	/	397	/	/	/
2c-14(iv)	237	7626	21.9	391	> 40	/	1.63
2c-15(i)	198	8930	34.5	581	9	/	1.85
2c-15(ii)	235	8884	33.9	576	10	/	1.85
2c-15(iii)	191	8887	34.0	579	9	/	1.84
C–C bonded bicyclic 1,2,3-triazole-based 1,2,4-triazole derivatives							
2c-17	263	8384	27.4	521	> 40	> 360	1.76
2c-18	168	9067	36.2	593	16	180	1.88
2c-19	211	8961	35.2	560	24	360	1.86
C-N bonded bicyclic 1,2,3-triazole-based 1,2,4-triazole derivatives							
2c-20	252	7867	24.3	86.2	> 40	> 360	1.68
2c-21	250	8364	29.0	361.8	22	160	1.78
2c-22	167	8630	31.5	476	25	240	1.81
2c-23	242	8577	30.9	465	28	240	1.80
2c-24	250	8980	28.9	190	4	> 36 (kp)	1.89
2c-25	164	8980	30.8	516	2	8.1 (kp)	1.82
N,N′-azo bridged bicyclic triazole derivatives							
2c-28	190	8280	27.0	1205	12	/	1.63
2c-29	186	8540	30.0	1549	8	/	1.66
2c-30	136	8950	34.0	1889	< 3	/	1.74

(continued)

Table 2.3 (continued)

Compound	T_d (°C)	D (m s^{-1})	P (GPa)	$\Delta_f H$ (kJ mol^{-1})	IS (J)	FS (N)	d (g cm^{-3})
2c-31	136	9370	38.0	2245	< 3	/	1.76
2c-32	248	7640	22.0	830	> 40	/	1.53
2c-33	245	7840	23.0	799	> 40	/	1.59
2c-34	246	7730	22.0	774	> 40	/	1.57
2c-35	290	7700	22.0	779	> 40	/	1.57
2c-37	258	7470	19.2	344	> 40	> 360	1.62
2c-39	136	7693	20.9	1940	< 2	< 5	1.53
2c-40	146	8194	28.5	247	6	80	1.70
C,C′-azo bridged bicyclic triazole derivatives							
DNAT	/	8500	/	407	12.5	250–360	1.88
2c-41	209	8784	33.7	647.7	2	20	1.85
2c-42	150	8964	36.7	555	1.5	/	1.83
2c-43	165	8742	33.8	506	5.5	/	1.78
Ether bridged bicyclic triazole derivatives							
2c-46	280	7401	22.9	119	28	240	1.86
2c-47	320	7543	21.5	167	20	240	1.64
Ethylene bridged bicyclic triazole derivatives							
2c-48	307	8069	24.5	315	> 40	360	1.72
2c-49	107	8816	34.3	516	10	160	1.84
N–N bonded bicyclic triazole derivatives							
2c-50	202	6850	15.2	628	18	260	1.41
2c-51	185	7208	17.5	664	17	290	1.51

Besides N,N′-azo linkage, a series of bistriazoles containing C,C′-azo linkages were also synthesized. As shown in Scheme 2.18, 5,5′-dinitro-3,3′-azo-1H-1,2,4-triazole was synthesized through oxidizing **ANTA** with permanganic acid accompanied by the formation of minor product 5-chloro-5′-nitro-3,3′-azo-1H-1,2,4-triazole by Naud's group [88]. Subsequently, a series of corresponding nitrimino compound with enhanced performance were preparaed. First, the treatment of 3,5-diamino-1H-1,2,4-triazole with acetic anhydride provided 1-acetyl-diaminotriazole. The desired 5-acetylaminotriazole was obtained in a nearly quantitative yield with thermal isomerization by heating a suspension of 1-acetyl-diaminotriazole in decaline. Then, the synthesis of 5,5′-diamino-3,3′-azo-(1H-5-amino-1,2,4-triazole) (**DAAT**) was performed with a stoichiometric amount of potassium permanganate at 100 °C. After the removal of the generated manganese oxide by filtration, acidifying the solution leads to the precipitation of compound **DAAT** as an orange solid. The synthesis of the novel 5,5′-dinitrimino-3,3′-azo-1H-1,2,4-triazole (**2c-41**) was accomplished in good yields using a volume ratio H_2SO_4/HNO_3 of 6:1 and two equivalents of nitric acid per amine group [89]. In the same year, the corresponding trinitromethyl compounds of azo-bridged triazole were prepared by Shreeve's group. In Scheme 2.18, the intermediate aminotriazolylacetic acid was synthesized by intramolecular condensation of the malonyl derivative and aminoguanidine bicarbonate in alkaline medium. Then the trinitromethyl compound 5,5′-bis(trinitromethyl)-3,3′-azo-1,2,4-triazole (**2c-42**) was obtained by a two-step reaction involved oxidative coupling reaction of this intermediate using potassium permanganate and then nitration in mixed acids of nitric acid and sulfuric acid. Moreover, the methylation of hexanitro **2c-42** into 1,1′-dimethyl-5,5′-bis-(trinitromethyl)-3,3′-azo-1,2,4-triazole (**2c-43**) was carried out using various methylation reagents, but only using trimethylsilyl diazomethane ($TMSCH_2N_2$) as a methylation reagent can reaction successed [90]. As shown in Table 2.3, 3,3′-azo-(1H-5-nitro-1,2,4-triazole) (**DNAT**) and **2c-(41-43)** exhibit high heats of formation and excellent energetic properties due to the introduction of nitro groups. The detonation pressures and detonation velocities of these energetic triazoles were calculated based on their heats of formation and the densities. The calculated detonation pressures and velocities of triazoles lie in the range between 33.7–36.7 GPa and 8500–8964 m s^{-1}, respectively. The impact sensitivities range from those of the relatively less sensitive **2c-43** (IS = 5.5 J) and **DNAT** (IS = 12.5 J) to sensitive compounds **2c-41** (IS = 2 J) and **2c-42** (IS = 1.5 J).

As above described, the presence of nitro groups or polynitro groups will lead to an increase in acidity of N–H in azoles which can result in storage problems and thus decrease their practical value. In addition to salt formation and eliminating the acidic proton by replacement with an amino group or a methyl group, using a variety of chains to link the acidic nitrogen atoms in these azoles is another efficient deprotonation method. This structural modification not only increases the stabilities, but also improves detonation properties. A series of triazoles with high thermal stability and low sensitivities containing an ether bridge or alkyl bridge were described shown in Scheme 2.19. A series of polynitrotriazoles (**2c-(44-47)**) were also obtained following the similar synthetic route to

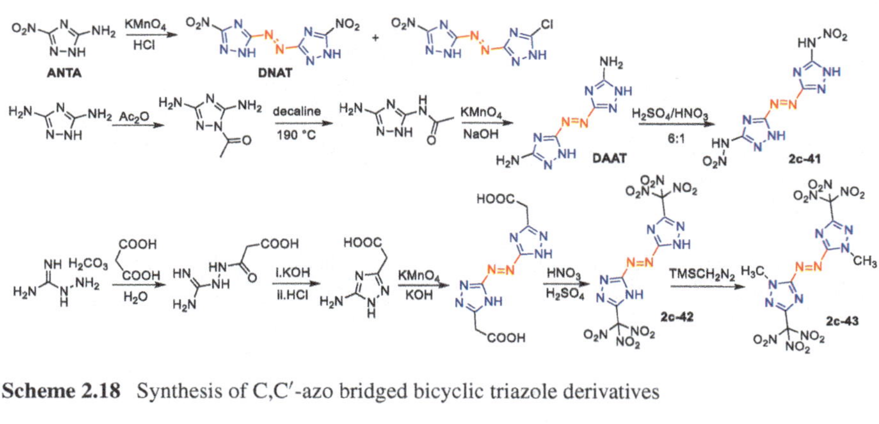

Scheme 2.18 Synthesis of C,C′-azo bridged bicyclic triazole derivatives

2b-23 and **2b-24**. Both 1,1′-(oxybis-(methylene))bis(5-nitro-3-(trinitromethyl)-1H-1,2,4-triazole) (**2c-44**) and 1,1′-(oxybis(methylene))bis(5-nitro-3-(trinitromethyl)-2H-1,2,4-triazole) (**2c-45**) can be synthesized through the substitution reaction of paraformaldehyde and 5-nitro-3-(trinitromethyl)-1H-1,2,4-triazole, but it is difficult to separate the two compounds due to the similar polarity [16]. Apart from ether-bridged compounds, a new family of ethylene-bridged bis(nitramino-1,2,4-triazolates) were also formed from **ANTA**. Using tetraethylammonium bromide (TEAB) as a phase-transfer catalyst and KOH as the base, the reaction of **ANTA** and dibromoethane formed 1,1′-(ethane-1,2-diyl)-bis(3-nitro-1H-1,2,4-triazol-5-amine) (**2c-48**) in good yield and high purity. The corresponding nitramine product **2c-49** was obtained through the nitration reaction of **2c-48** using 100% nitric acid [91].

Another approach to extend the nitrogen chain for high-nitrogen systems from both academic and practical considerations have been reported by G. Cheng's group. They employed the bis(1,2,3-triazole) formation reaction from tosylhydrazones and N-amino (N–NH_2) azole instead of the primary amine (C–NH_2)-based Sakai reaction to synthesize some novel nitrogen chain compounds with a N_6 or N_7 structure. The intermediate was obtained from the reaction of α,α-dichlorotosylhydrazones and N-NH_2 of the amino azole in the presence of Et_3N. Then, the second step reaction of the formation of target compound by utilizing DiPEA (N,N-diisopropylethylamine) was further investigated [92].

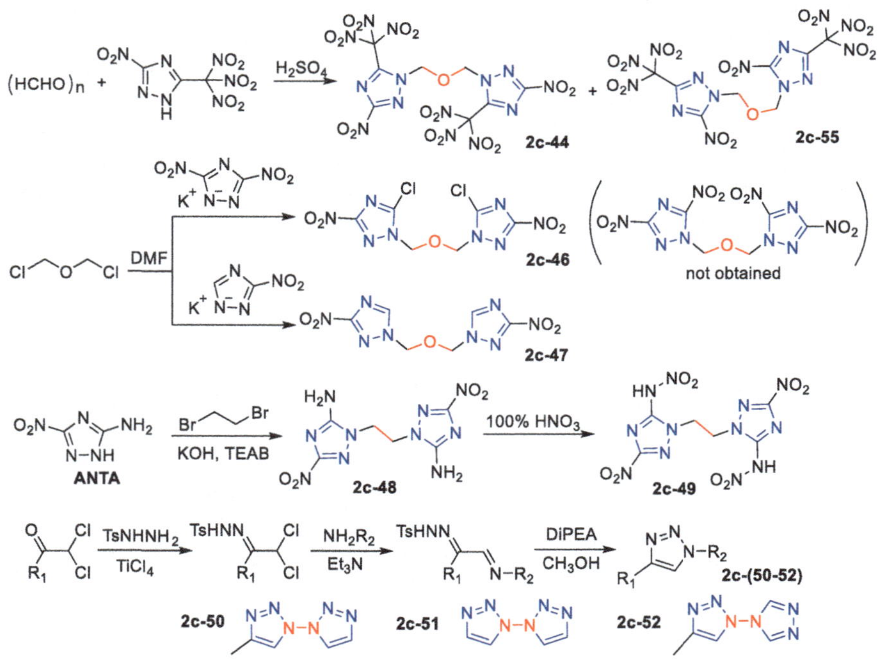

Scheme 2.19 Synthesis of bicyclic triazoles derivatives with other linkages

2.4 Bicyclic Azine-Based Energetic Compounds

Recently, the azine-based energetic compounds have been investigated due to the large fraction of nitrogen. The energy of azine-based compounds is derived from their very high positive heat of formation, not from the combustion of the carbon backbone or the ring/cage strain like traditional explosive. The high heat of formation is directly attributable to the large number of inherently energetic N–N and C–N bonds. As mentioned above, high-nitrogen compounds containing polyazides possess even higher heats of formation because their energy content rapidly increases with the number of energetic azido groups in the molecule. The synthesis and properties of novel 4,4′,6,6′-tetra(azido)hydrazo-1,3,5-triazine (**2d-1**) and 4,4′,6,6′-tetra(azido)azo-1,3,5-triazine (**2d-2**) were reported by Huynh's group. As shown in Scheme 2.20, the rapid reaction occurs between 4,4′,6,6′-tetra(chloro)hydrazo-1,3,5-triazine and an excess of hydrazine monohydrate in CH_3CN to give 4,4′,6,6′-tetra(hydrazino)hydrazo-1,3,5-triazine which underwent diazotization to yield **2d-1**. A suspension of **2d-1** in $CHCl_3$ solution was oxidized by chlorine gas at room temperature to **2d-2** [93]. The compound **2d-2** has a very high measured heat of formation of 2171 kJ mol^{-1}, but the detonation pressure and velocity did not be calculated. Shreeve's group also reported a series of high-nitrogen triazine-derived compounds through incorporating an energetic nitroguanidyl group into the backbone of the 1,3,5-triazine ring. As shown in Scheme 2.20, N-methyl-N-nitroso-N′-nitroguanidine was firstly synthesized to provide a nitroguanidyl group through the substitution reaction of commercially available nitroguanidine and methylamine then nitrosating by sodium nitrite. Subsequently, the aqueous solutions of hydrazine derivatives of 1,3,5-triazine-base compounds were reacted with N-methyl-N-nitroso-N′-nitroguanidine to afford the nitroguanidyl-functionalized 1,3,5-triazine derivative (**2d-3**) in good yields [94]. This nitroguanidyl-functionalized triazine compound is insensitive to impact and friction (IS > 40 J, FS = 360 N), further highlighting its application potentials as new-generation insensitive energetic materials.

As known, energetic compounds containing a reactive NH group enable the straightforward formation of high-energy salts with nitrogen-rich bases and ions of various metals. As shown in Scheme 2.20, 4,6-diazido-N-(4,6-diazido-1,3,5-triazin-2-yl)-1,3,5-triazin-2-amine (**2d-4**), a triazine compound containing a NH group, was synthesized through the substitution reaction of aminotriazine and cyanuric chloride in the presence of NaOH starting from 4,6-dichloro-N-(4,6-dichloro-1,3,5-triazin-2-yl)-1,3,5-triazin-2-amine and then azidation with NaN_3 in acetone in nearly quantitative yield [95].

The tetrazine ring is also an extremely useful building block containing 68.3% of nitrogen to design high-energy and low-sensitivity explosives. Strong oxidizers (hypofluorous acid and peroxytrifluoroacetic acid) were employed to obtain azo or azoxy group from 3-amino-1,2,4,5-tetrazines, but no circumstance for the formation of an azo or azoxy linkage was observed. Thus, the synthesis of azo tetrazine using a different oxidizing reagent and further N-oxidation were achieved. As shown in Scheme 2.21, the hydrazo compound was prepared by treatment of readily available

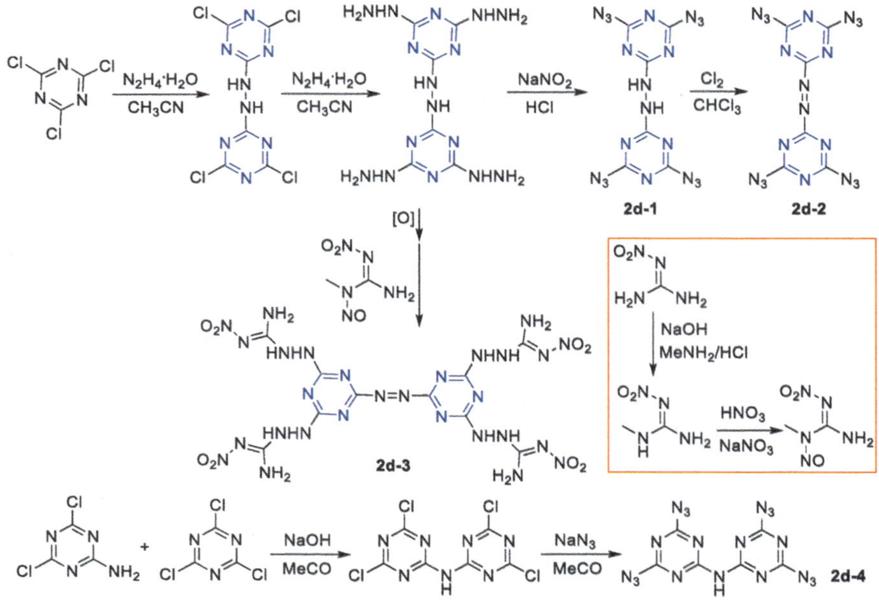

Scheme 2.20 Synthesis of bicyclic triazine derivatives

3,6-bis(3,5-dimethylpyrazol-1-yl)-1,2,4,5-tetrazine with 0.5 equivalent of hydrazine. A variety of oxidizing reagents typically such as Br_2, NO_2, MnO_2, HgO, HONO were used to oxidize this hydrazo compound into an azo compound, but failed. The azo compound can be obtained only using N-bromosuccinimide (NBS) as the oxidizing reagent. And the compound **2d-5** was isolated through the ammoniation reaction of azo compound and ammonia in acetonitrile then conducted in dimethyl sulfoxide (DMSO) [96]. The N-oxide reaction of **2d-5** using different oxidants were attempted to improve its density and explosive performance. The result demonstrated that the peroxytrifluoroacetic acid is an efficient oxidant to oxidize the nitrogen atom in compound **2d-5**. Unfortunately, a complex mixture of various N-oxides of **2d-5** were obtained such as **2d-6**, and the pure compound containing N-oxides have not been isolated (Table 2.4) [97].

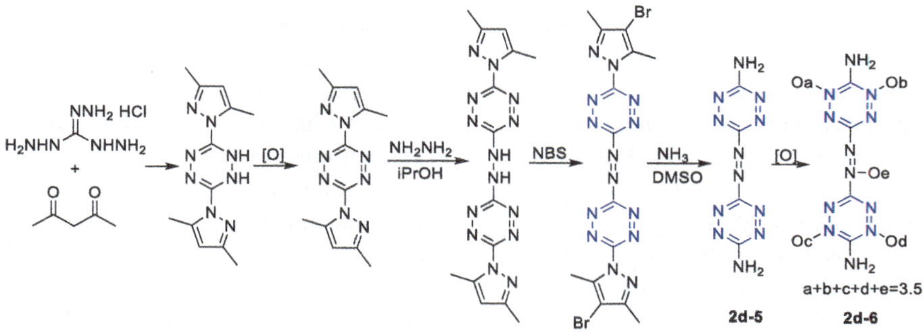

Scheme 2.21 Synthesis of bicyclic tetrazine derivatives

Table 2.4 Energetic properties of triazine and tetrazine derivatives

Compound	T_d (°C)	D (m s^{-1})	P (GPa)	Δ_fH (kJ mol^{-1})	IS (J)	FS (N)	D (g cm^{-3})
Bicyclic triazine derivatives							
2d-1	202	/	/	1753	18.3 cm (H_{50})	29	1.65
2d-2	200	/	/	2171	6.2 cm (H_{50})	24	1.72
2d-3	201	7812	23.9	598	> 40	360	1.71
2d-4	/	7190	22.2	1941	/	/	1.71
Bicyclic tetrazine derivatives							
2d-5	252	/	/	862	70 cm (H_{50})	360	1.84

2.5 Bicyclic Imidazole and Pyrazole Based Energetic Compounds

As two of five membered azole-based energetic backbones, imidazole and pyrazole have received significant interest in the exploration of energetic materials. Various energetic bisimidazole-based and bispyrazole-based energetic materials were synthesized and further study associated with physicochemical properties. Recently, a multi-functionalized strategy to improve the energetic performance of various heterocyclic imidazole and pyrazole was demonstrated. As shown in Scheme 2.22, the nitration of precursor 1H,1′H-2,2′-bisimidazole into polynitro compound **TNBI** was achieved by employing sodium nitrate and concentrated sulfuric acid. And the N-amination of the bisimidazole **TNBI** was also accomplished using THA as an amination reagent. Subsequently nitration of the amino compound **2e-1** into N-nitramino functionalized bisimidazole was successed in the mixture of sulfuric acid and nitric acid. Because of the highly reactive nature of the N-amino groups, the compound *N,N′*-dinitramino-4,4′,5,5′-tetranitro-bisimidazole (**2e-2**) was obtained by the nitration reaction of **2e-1** in mixed acid at − 15 to − 10 °C [98]. According to experimental results and theoretical analysis, the diazonium compound could be easily initiated by shock and impact, and therefore may be promising environmentally friendly primary explosive candidates. The synthesis of two new ternary CNO-inner bis(diazonium) salts, 4,4′-dinitro-5,5′-diazo-2,2′-bisimidazole zwitterion (**2e-4**) containing two stages was described by Shreeve. The first stage of **2e-4** synthesis consisted of ammoniation of **TNBI** to 4,4′-dinitro-5,5′-diamino-1H,1′H-2,2′-bisimidazole (**2e-3**) in aqueous ammonia. In the final stage, **2e-4** was obtained through the nitration of **2e-3** in a mixture of trifluoroacetic anhydride (TFAA) and 100% nitric acid at 0 °C. Moreover, concentrated sulfuric acid was used as a nitration reagent to enhance the yield of **2e-4** up to 80% [99]. Apart from the oxidation, amination, and nitration of bisimidazoles, the trinitroethylation of bisimidazole was also attempted. In Scheme 2.22, the amination of bisimidazoles **2e-5** and **2e-6** could be processed using O-tosylhydroxylamine and hydroxylamine-O-sulfonic acid as amination reagent, respectively. However, the attempted trinitroethylation of these bisimidazoles (**2e-1**, **2e-7**, **2e-8**) functionalizated with trinitroethanol did not yield the desired trinitroethyl product **2e-9** [100].

Chavez's group reasoned that N-methylation would disrupt any hydrogen bonding in the parent imidazole compound, which could potentially lead to a lower melting point. They investigated two synthesis routes to 1,1′-dimethyl-4,4′,5,5′-tetranitro-2,2-bisimidazole (**2e-12**). The first route began with the nitration of methylated bisimidazole **2e-11** to **2e-12** in the mixture of sulfuric acid and nitric acid accompanied by the formation of the trinitrobisimidazole intermediate. Another two-step route with more pure product starting from **TNBI** was studied [101]. The desired product **2e-12** was prepared from the methylation of the ammonium salt of **TNBI**, using neat dimethylsulfate as methylation reagent at 70 °C [102].

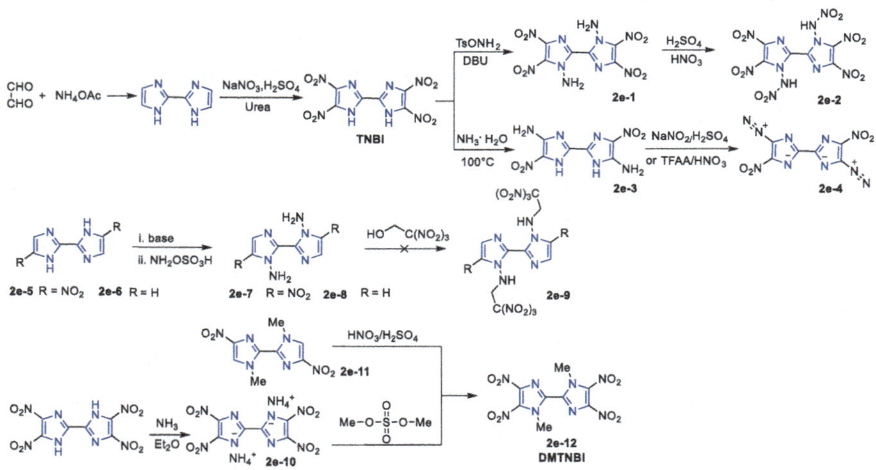

Scheme 2.22 Synthesis of C–C bonded bicyclic imidazole derivatives

Apart from polynitro bisimidazoles, polynitro bispyrazoles have also attracted considerable attention due to their remarkable comprehensive properties including low hygroscopicity and low toxicity, insensitivity to external stimuli, outstanding thermal and chemical stability, and high energy performance. The synthesis of a series of polynitro bispyrazoles linked by C–N bond was reported by Zhou's group (Scheme 2.23). Bispyrazole (**2e-13**) was firstly prepared via direct substitution reaction of trinitropyrazole (**TNP**) and *NH*-pyrazole [103]. Then the polynitro bispyrazole **2e-14** was synthesized by the nitration of **2e-13** in the mixture of 100% nitric acid and concentrated sulphuric acid. Given the strong electron withdrawing ability of the fragment 3,5-dinitropyrazole, the substitution of chlorine in the molecule of **2e-14** was easier than that of 4-chloro-3,5-dinitropyrazole, which was demonstrated by the success in preparing corresponding ammonium salt with 25% aqueous ammonia at a much lower reaction temperature. Compound **2e-15** was obtained by the acidification of this ammonium salt. Furthermore, this ammonium salt can be successfully aminated to compound **2e-17** by freshly prepared **THA** under mild conditions in moderate yields. As shown in Table 2.5 compounds **2e-(14-17)**, compound **2e-14** has the highest thermal stability ($T_d = 308$ °C), followed by **2e-16** ($T_d = 297$ °C), which are even better than that of traditional explosive **TNT** (295 °C) [104].

Apart from polynitro bispyrazoles rings linked by a C–N bond, continuing efforts toward the synthesis and reactivity studies of nitropyrazoles linked by a C–C bond was paid. According to previous research, various methods were deemed to be efficient for the introduction of amino group into a pyrazole ring such as vicarious nucleophilic substitution of hydrogen atom, cinesubstitution of an N-nitro group, Hofmann rearrangement of amide group, reduction of a nitro group, and nucleophilic ipso-substitution of a nitro group. Dalinger's group summarized a series of synthetic routes of nitro-substituted bispyrazoles from unsubstituted 3,3′-bispyrazole through vicarious nucleophilic substitution of hydrogen atom [105]. As shown in Scheme 2.24, unsubstituted 3,3′-bispyrazole was synthesized in two steps from oxalyl chloride and ethyl vinyl ether. Direct acidic nitration of unsubstituted 3,3′-bispyrazole by heating in a mixture of concentrated sulfuric and nitric acids resulted in the introduction of two nitro groups at positions 4 of both pyrazole ring (**2e-18**). Then the introduction of amino group to **2e-18** was attempted by treating **2e-18**

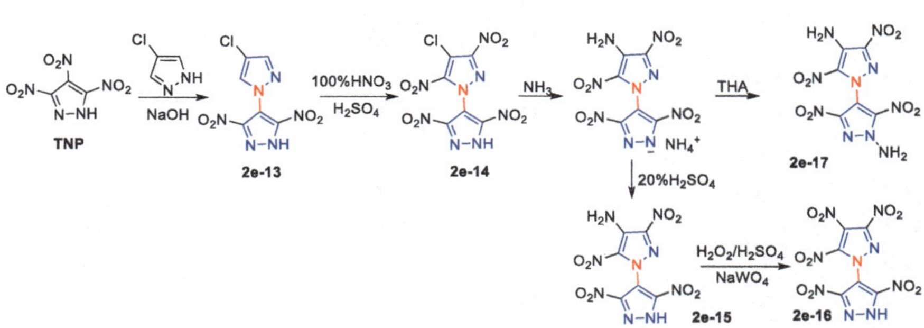

Scheme 2.23 Synthesis of C–N bonded bicyclic pyrazole derivatives

Table 2.5 Energetic properties of bicyclic pyrazole and imidazole derivatives

Compound	T_d (°C)	D (m s^{-1})	P (GPa)	$\Delta_f H$ (kJ mol^{-1})	IS (J)	FS (N)	d (g cm^{-3})
C–C bonded bicyclic imidazole derivatives							
2e-1	217	9012	36.6	309	15	160	1.93
2e-2	116	9350	40.1	476	3	20	1.94
2e-4	199	8653	31.1	820	1	20–40	1.83
2e-12	258	7604	24.2	145	> 40	> 292	1.70
C-N bonded bicyclic pyrazole derivatives							
2e-14	308	8724	36.0	185	> 40	/	1.96
2e-15	242	8600, 8789, 8626[a]	35.0, 35.9, 36.0[a]	388	> 40	/	1.89
2e-16	297	8814, 8760, 8981[a]	37.0, 35.0, 36.0[a]	824	28	/	1.82
2e-17	284	8648, 8760, 8653[a]	35.1, 35.4, 33.9[a]	478	> 40	/	1.87
C–C bonded bicyclic pyrazole derivatives							
2e-18	365	8120	26.9	221	> 40	> 360	1.83
2e-19	206	8869	34.5	407	2	40	1.85
2e-22	183	8301	28.8	414	20	240	1.81
2e-23	376	8026	26.2	229	> 40	> 360	1.81
2e-24·H_2O	243	8556	32.3	42.7	30	360	1.84
2e-25	152	8926	35.4	444	2	60	1.86
2e-26	228	7779	23.0	212	> 60	> 360	1.72
2e-28	150	8943	34.0	987	1	20	1.85
N–N′-azo bridged bicyclic pyrazole or imidazole derivatives							
2e-29	240	8928	34.8	687	5	60	1.81
2e-30	223	8964	35.1	687	5	80	1.82
2e-31	195	8834	34.6	644	2	40	1.80
2e-32	258	8979	35.1	592	15	160	1.85
2e-33	232	8932	35.1	612	10	120	1.83
2e-34	271	7632	24.9	578	25	360	1.86
2e-35	251	8255	30.6	666	4	40	1.92
2e-36	154	8702	31.4	1259	2	40	1.77
N,N′-ethylene bridged bicyclic pyrazole derivatives							
2e-37	311	8189	27.9	219	> 40	> 360	1.77
2e-38	80	8753	34.3	381	7	80	1.84
2e-39	247	7803	24.2	442	20	80	1.72
2e-40	250	8759	34.1	307	25	160	1.84
2e-41	112	8797	33.4	1234	4	60	1.78

(continued)

Table 2.5 (continued)

Compound	T_d (°C)	D (m s^{-1})	P (GPa)	Δ_fH (kJ mol^{-1})	IS (J)	FS (N)	d (g cm^{-3})
2e-42	319	7877	27.0	230	> 40	> 360	1.88
2e-43	135	8558	31.0	1014	3	60	1.76
2e-44	256	8127	27.3	238	> 40	> 360	1.75
2e-45	81	8710	33.7	368	6	60	1.83
N,N′-methylene bridged bicyclic pyrazole derivatives							
2e-46	310	8332	29.6	205	11	> 360	1.80
2e-47	205	9304	39.1	379	4	144	1.93
2e-48	226	8016	26.0	497	1.5	40	1.73
Ether bridged bicyclic pyrazole derivatives							
2e-49	284	7813	26.2	129	30	> 360	1.84
2e-50	298	7904	26.0	163	> 40	> 360	1.69
2e-51	299	7874	25.7	141	10	240	1.69
2e-52	203	8068	27.1	127	25	> 360	1.74
2e-53	270	8067	27.8	75	> 40	> 360	1.76

[a]Detonation velocity calculated by the EXPLO5 program, modified nitrogen equivalent method, and Kamlet-Jacobs equations

with trimethylhydrazinium iodide (TMHI) in the presence of t-BuOK, but did not result in the formation of diamino derivative 4,4′-dinitro-1H,1′H-[3,3′-bipyrazole]-5,5′-diamine **2e-26**. N-nitrobispyrazole **2e-19** was prepared by the nitration of **2e-18** under two different nitration conditions. Using acetyl nitrate in trifluoroacetic acid as nitration reagent exhibits a higher yield than the use of potassium nitrate and $(CF_3CO)_2O$ in 1,2-dichloroethane. And N-dinitrobispyrazole **2e-19** in the presence of anionic C- and N-nucleophiles under mild conditions underwent a cinesubstitution reaction to provide 5,5′-disubstituted derivatives **2e-20** and **2e-21**. Acidic hydrolysis of the cyano groups in bispyrazole **2e-20** using concentrated sulfuric acid formed dicarboxamide intermediate then rearranged through the Hoffman reaction to give target amine **2e-26**. Another method to this dicarboxamide intermediate comprising cyclization, nitration, esterification and ammoniation using diethyl oxalate as a raw material (shown in red frame of Scheme 2.24) was reported by Shreeve [106]. In addition, the desired diamine **2e-26** can also be obtained by the ammoniation of the phosphazine intermediate of diazide **2e-21** in aqueous ammonia solution with a moderate yield.

Furthermore, the reduction of a nitro group is also a generally reliable method for the introduction of amino groups in a pyrazole ring. Suitable starting compound tetranitrobispyrazole **2e-24** was synthesized from 3,3′-bispyrazole through three steps. And the symmetrical isomer **2e-25** could also be prepared in the different nitration condition of dinitrobipyrazole **2e-23**. Selective reduction of nitro groups of **2e-24** was achieved by using hydrazine and activated carbon in the presence of $FeCl_3$

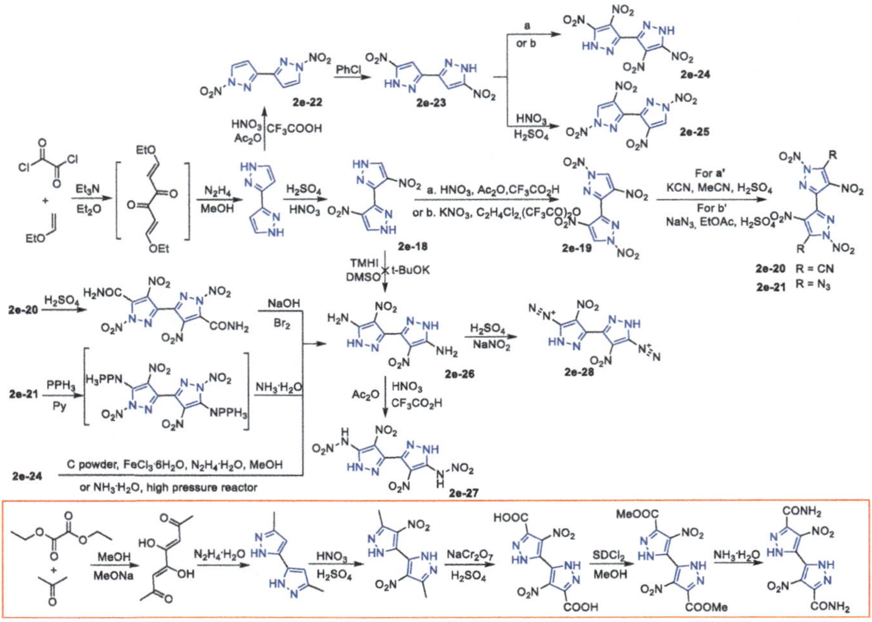

Scheme 2.24 Synthesis of C–C bonded bicyclic pyrazole derivatives

as a catalyst. It is noteworthy that the substitution of nitro group with an amino group in compound **2e-24** can be proceeded under high reaction pressure and resulted in selective substitution of both nitro groups in positions 5 to form the desired diamine **2e-26**. The aminobispyrazole **2e-26** can be smoothly nitrated at the C-amino groups with acetyl nitrate in trifluoroacetic acid to give a good yield of tetranitrodiamine **2e-27**. In the same year, Shreeve's group also reported the synthesis of these polynitrated bispyrazoles. The bispyrazole zwitterion (**2e-28**) was prepared through the similar synthetic route of **2e-4** starting from **2e-26** [107]. The detonation performance of polynitro bispyrazoles 2e-(**18-28**) was given in Table 2.5. These polynitro compounds possess high detonation velocities (D > 7779 m s^{-1}) and pressures (P > 23.0 GPa). They all exhibit low sensitivities to impact and friction (IS > 20 J, FS > 240 N) except **2e-19** and **2e-25**. Considering to the high sensitivities of diazonium **2e-4** and **2e-28**, they may be candidates for the next generation of green lead-free primary explosives due to good thermal stabilities and excellent detonation performances.

A study of N-diazo-bridged nitroazoles with enhanced oxygen balance and detonation performance was also reported. The new family of N-diazo-bridged energetic materials show excellent thermal stabilities and relatively low impact sensitivities, which makes them more conducive to storage. A series of nitroazoles including nitropyrazoles, nitroimidazoles and nitrotriazole have been synthesized. And further reactions involving neutralization with ammonia and amination using O-tosylhydroxylamine ($TsONH_2$) gave rise to N-amino nitroazoles. Moreover,

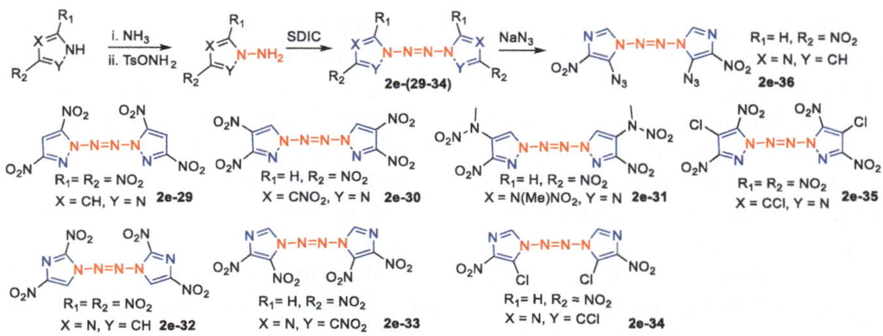

Scheme 2.25 Synthesis of N-N′-azo bridged bicyclic pyrazole or imidazole derivatives

employing the same sodium dichloroisocyanurate (SDIC) system above mentioned, oxidative coupling of nitroazoles could result in N–N′-azo bridged compounds **2e-(29–35)** (Scheme 2.25). Further reaction of **2e-34** and sodium azide led to **2e-36** while similar attempts to introduce azide into **2e-35** produced an extremely unstable azido-substituted product which could not be isolated from the reaction [108]. With a large number of N=N and N≡N bonds, the heats of formation of N-diazo bridged nitroazoles are higher than those of unsubstituted nitroazoles, ranging from 578 to 1259 kJ mol^{-1}. The impact sensitivities of N_6 derivatives (**2e-(29-31)** and **2e-35**) are relatively high (2–10 J) while some N_4 compounds (**2e-(32-34)**) have moderate sensitivities (10–25 J). The friction sensitivity of these compounds ranges from very sensitive to insensitive (40–360 N). In Table 2.5 compounds **2e-(29-36)**, all these compounds show good detonation pressures (P) and velocities (D) range from 24.9 to 35.6 GPa and from 7632 to 8979 m s^{-1}, respectively.

Apart from the synthesis of N,N′-ethyl-bridged bis(1,2,4-triazoles) and bis(tetrazoles), an effective synthesis of N,N′-alkyl-bridged nitropyrazoles with good oxygen content and diversified functionalities were also demonstrated. Shreeve's group attempted to prepare *N,N′*-ethylene-bridged nitropyrazoles through the ethylation of several nitropyrazoles using dibromoethane as ethylation reagent in the presence of potassium hydroxide [109]. The result implied that the electron-donating group in nitro-substituted aminopyrazole could enhance the activity of coupling reaction of amino group. As shown in Scheme 2.26, the precursor ammonium 4-amino-3,5-dintropyrazolate (**N**) was prepared through a three-step reaction included chlorination, nitration and ammoniation starting from pyrazole. 1,1′-(Ethane-1,2-diyl)bis(3,5-dinitro-1H-pyrazol-4-amine) (**2e-37**), N,N′-ethylenebridged nitropyrazole, was synthesized by the ethylation reaction of **N** using dibromoethane as a reagent without the presence of a base. As a phase transfer catalyst, the addition of tetraethylammonium bromide (**TEAB**) to ethylation reaction does improve the reaction yield and shorten the reaction time. *N,N′*-[1,1′-(ethane-1,2-diyl)bis(3,5-dinitro-1H-pyrazole-4,1-diyl)]dinitramide (**2e-38**) was prepared through the nitration of **2e-37** in 100% nitric acid. However, reaction of **2e-37** using sodium nitrite/sulfuric acid and sodium azide resulted in the formation of diazonium salt **2e-39**, rather than the desired diazido product. In the presence of sulfuric acid and 50% hydrogen peroxide, **2e-37** was oxidized through the green nitroso intermediate to give the bis(pyrazole) **2e-40**. The 5,5′-diazido product **2e-41** was obtained through the substitution of nitropyrazole **2e-40** using sodium azide as nucleophile. Encouraged by this result, *N,N′*-ethylenebis(4-chloropyrazole) was successfully prepared from 4-chloro-pyrazole by employing the same synthesis conditions. Further nitration of the bis(4-chloropyrazole) in the mixture of nitric acid and sulphuric acid resulted in 1,2-bis(4-chloro-3,5-dinitro-1H-pyrazol-1-yl)ethane (**2e-42**) and then azidation of **2e-42** with sodium azide gave 1,2-bis(4-azido-3,5-dinitro-1H-pyrazol-1-yl)ethane (**2e-43**). Furthermore, the reaction of 5-amino-3,4-dinitropyrazole and dibromoethane yielded *N,N′*-ethylene-bridged 5,5′-diaminobis(nitropyrazole) (**2e-44**). Its nitramine derivative *N,N′*-[1,1′-(ethane-1,2-diyl)bis(3,4-dinitro-1H-pyrazole-5,1-diyl)]dinitramide (**2e-45**) was obtained by treatment of **2e-44** with 100% nitric acid.

A series of methylene-bridged pyrazoles with promising performance and sensitivity characteristics were synthesized based on this ethylation route [39]. The sodium salt **M** was treated with diiodomethane to give methylene-bridged pyrazole, bis(4-amino-3,5-dinitropyrazolyl)methane (**2e-46**), which could be further oxidized to bis(3,4,5-trinitropyrazolyl)methane (**2e-47**) by stirring in a H_2O_2/H_2SO_4 mixture. The final synthetic step to bis(4-diazo-5-nitro-3-oxopyrazolyl)methane (**2e-48**) was carried out by the diazotation of **2e-46** with the mixture of $NaNO_2$ and H_2SO_4. Moreover, a series of alkoxy-bridged compounds **2e-(49-53)** containing pyrazole and imidazole were also synthesized with the same synthetic route to **2b-23** and **2b-24** above mentioned [16].

These energetic compounds (**2e-(40-53)**) display various properties because of the diversified functions. Diaminobispyrazoles exhibit good thermal stabilities, highly insensitivities, and favorable energetic performances, which were represented as **2e-37** (T_{dec} = 311 °C, IS > 40 J, D = 8189 m s^{-1}) and **2e-46** (T_d = 310 °C, IS = 11 J, D = 8332 m s^{-1}). By contrast, N,N′-alkylbridged dinitraminobis(pyrazoles) and diazidobis(pyrazoles) have better detonation performance, but with higher impact and friction sensitivity especially for **2e-47** (T_d = 205 °C, IS = 4 J, D = 9304 m s^{-1}).

2.6 Bicyclic Mixed Heterocyclic-Based Energetic Compounds

Hereto, all these N-heterocyclic backbones including oxadiazole, tetrazole, triazole, pyrazole, imidazole, triazine and tetrazine have been discussed individually in the aspect of structure and property features. The combination of these rings has been employed to satisfy the demand of modern HEDMs.

As for the combination of oxadiazole and tetrazole ring, a series of nitrogen-rich energetic salts based on 3-amino-4-tetrazolefurazan (**HATF**) have been synthesized. As shown in Scheme 2.27, 4-amino-3-(5-tetrazolyl)furazan (**HATF**) was easily

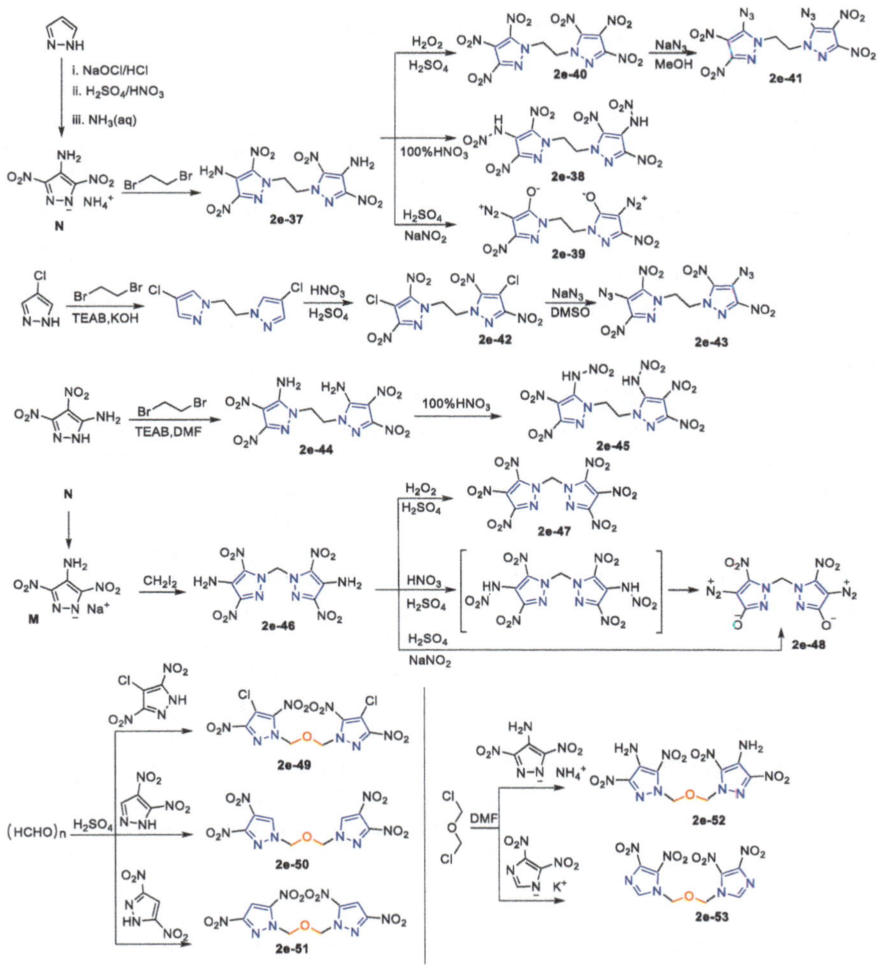

Scheme 2.26 Synthesis of N,N′-alkyl bridged and ether bridged bicyclic pyrazole derivatives

synthesized starting from **AAOF** through the oxidation using PbO_2 as an oxidant and followed cycloaddition using NaN_3 [110]. Only a series of salts of 3-nitramino-4-tetrazolefurazan were obtained by the nitration of **HATF** with mixed fuming nitric acid and concentrated sulfuric acid and followed by alkalization. The neutral compound 3-nitramino-4-tetrazolefurazan have not been reported probably due to its instability [111]. A nitroamine group and a hydroxyl moiety were introduced into an energetic backbone to boost the energetic performance. 5-(4-Amino-furzan-3-yl)-1-hydroxytetrazole (**2f-1**) was prepared from **AAOF** through substitution by azide then cycloaddition with NaN_3. Further nitration reaction of **2f-1** was proceeded with an excess of fuming nitric acid to afford **2f-2** which is stable in diethyl ether solution but decomposed in air at room temperature [112]. Apart from furazan, furoxan ring was also connected to tetrazole to adjust the balance between energy and safety. 4-Nitro-3-(5-tetrazole)-furoxan (**HTNF**, **2f-4**) has been designed and synthesized by Zhou's group. 4-Amino-3-cyanofuroxan was treated with sodium azide and $ZnBr_2$ in water to form 4-amino-3-(5-tetrazole)furoxan (**2f-3**). And the desired product **2f-4** was finally prepared by the oxidation of **2f-3** in the mixture of 50% H_2O_2 and H_2SO_4 using Na_2WO_4 as a catalyst [113]. The target compound **2f-4** possesses high detonation performance ($D = 8681$ m s^{-1}, $P = 34.3$ GPa) but high sensitivity to impact with an IS value below 2 J.

Many studies on the combination of oxadiazole and triazole were also reported. Recently, Xu et al. reported the straightforward synthesis and characterization of a series of energetic compounds consisting of the triazole moiety and oxadiazole rings functionalized with methyl, amino, azo, nitro, and nitramino groups [114]. As shown in Scheme 2.28, hydrazide compound **a** was reacted with 1-methyl-1-nitroso-2-nitroguanidine (MNNG) in aqueous solution to give the nitroguanidyl-functionalized intermediate **b**, subsequently intramolecular cyclization of **b** in aqueous alkaline solution results in the formation of target compounds **2f**-(**5-7**) [115]. The further modified compounds **2f-8** and **2f-9** were prepared by the nitration using 100% nitric acid and oxidation using 30% hydrogen peroxide starting from **2f-7**, respectively. Notably, the one-pot reaction of **c** from **a** was demonstrated to increases the efficiency of this procedure. And a series of polynitro-triazolylfurazanate compounds were synthesized based on this cyclization reaction shown in Scheme 2.28. The intermediates acid and ethyl ester were obtained through the substitution reaction of 3-carbohydrazide-4-aminofurazan and ethyl 3-ethoxy-3-iminopropionate (EEIP) followed by the cyclization reaction using glacial acetic acid. Besides, the ester can also be obtained from the acid through esterification reaction in the presence of thionyl chloride. These esters were all nitrated using mixed nitric acid and sulfuric

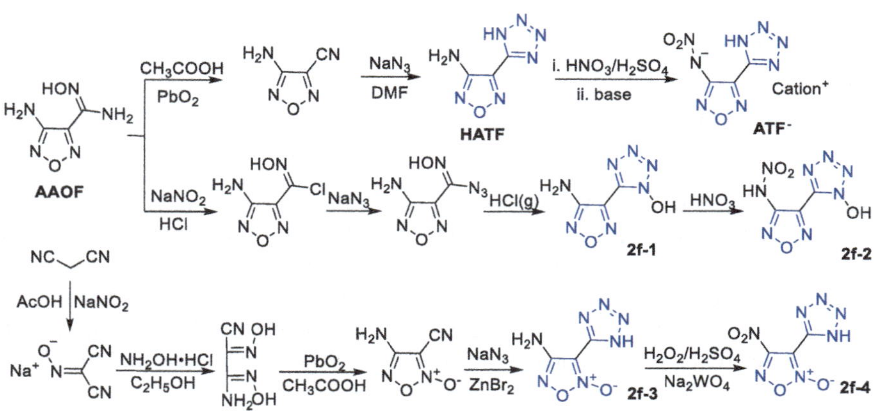

Scheme 2.27 Synthesis of C–C bonded bicyclic oxadiazole-tetrazole derivatives

acid as the reaction medium to give 3-nitramino-4-(3-(trinitromethyl)-1,2,4-triazol-5-yl)furazan (**2f-10**) and **2f-11**, respectively. The gem-dinitro salts were given by the treatment of compounds **2f-10** and **2f-11** with base. Then the neutral compound **2f-12** was obtained by acidifying gem-dinitro salt with hydrochloric acid [116]. As shown in Table 2.6, bicyclic oxadiazole-triazole compounds **2f-(5-12)** exhibit high detonation performances (D, 8079–9278 m s^{-1}; P, 24.9–39.0 GPa) and acceptable sensitivities to impact and friction (IS > 9 J, FS > 100 N) except **2f-10** (IS = 4 J, FS = 60 N).

The combination of a triazole ring and a tetrazole ring linked by the C–C bond can lead to energetic materials with tunable properties. The synthesis of 5-amino-1H-1,2,4-triazole-3-carbonitrile (**ATCN**) comprises three-step reaction including

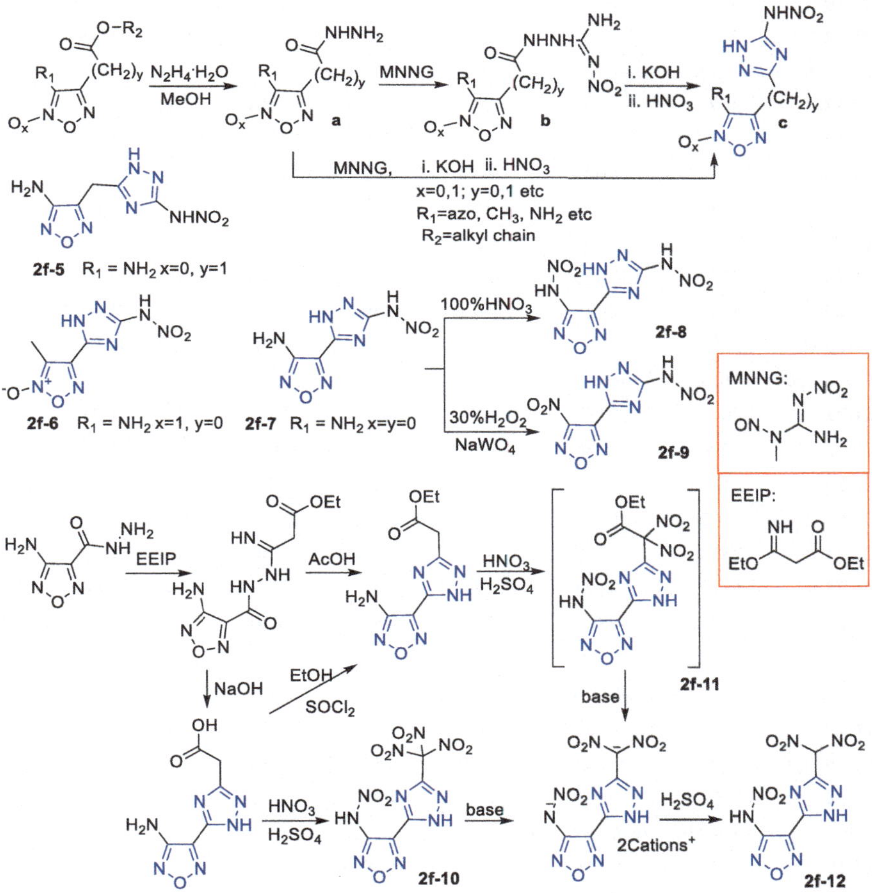

Scheme 2.28 Synthesis of C–C bonded bicyclic oxadiazole-triazole derivatives

Table 2.6 Energetic properties of mixed heterocyclic-based energetic compounds

Compound	T_d (°C)	D (m s^{-1})	P (GPa)	$\Delta_f H$ (kJ mol^{-1})	IS (J)	FS (N)	d (g cm^{-3})
C–C bonded bicyclic oxadiazole-tetrazole derivative							
2f-4	169	8681	34.3	500	< 2	/	1.81
C–C bonded bicyclic oxadiazole-triazole derivatives							
2f-5	197	8079	24.9	449	> 40	> 360	1.70
2f-6	181	8236	28.2	444	> 40	> 360	1.74
2f-7	196	8332	27.7	468	35	360	1.74
2f-8	128	9258	39.1	549	18	250	1.92
2f-9	163	9025	36.0	525	20	280	1.85
2f-10	67	9278	39.0	549	4	60	1.89
2f-12	91	9005	36.3	537	9	100	1.85
C–C bonded bicyclic tetrazole-triazole derivatives							
2f-13	347	7193	18.7	511	> 40	> 360	1.61
2f-14	211	8020	26.5	519	25	288	1.73
2f-15	215	8097	26.6	577	< 1	18	1.71
2f-16	164	7874	24.1	859	< 1	20	1.66
C-N bonded bicyclic tetrazole-triazole derivatives							
2f-18	246	8316	28.8	487	> 40	> 360	1.77
2f-20	324	8423	24.2	440	> 60	/	1.69
2f-28·H_2O	239	7852	23.0	223	> 40	> 360	1.65
N-methylene-C bridged bicyclic tetrazole-triazole derivatives							
2f-21	204	8187	24.9	476	15	240	1.71
2f-22	159	8750	31.8	285	10	160	1.81
2f-23	218	8213	25.2	462	18	360	1.72
N,N′-ethylene bridged bicyclic tetrazole-triazole derivatives							
2f-24	255	7395	18.8	383	> 40	> 360	1.58
2f-25	181	8113	25.6	477	38	> 360	1.69
2f-26	277	7807	21.6	359	> 40	> 360	1.66
2f-27	96	8690	32.2	512	12	240	1.80
C-N bonded bicyclic pyrazole-triazole derivatives							
2f-29	200 (melt)	6662	16.8	328	> 40	> 360	1.66
2f-30·H_2O	330	8262	29.0	140	> 40	> 360	1.75
2f-32	270	9167	37.8	833	9	240	1.84
Bicyclic pyrazole-tetrazole derivatives							
2f-19	222	9261	38.3	848	25	168	1.85
2f-33	279	9289	38.6	856	35	240	1.86
2f-34	136	7824	23.1	463	39	360	1.65

(continued)

Table 2.6 (continued)

Compound	T_d (°C)	D (m s^{-1})	P (GPa)	$\Delta_f H$ (kJ mol^{-1})	IS (J)	FS (N)	d (g cm^{-3})
2f-35·H_2O	130	7890	25.9	242	24	360	1.75
2f-36	257	7889	23.2	334	> 40	360	1.67
2f-37	265	7768	21.8	356	> 40	360	1.65
2f-38	135	8172	27.1	431	30	240	1.71
2f-39	134	7959	24.7	437	> 40	360	1.67
2f-40	221	7448	21.7	393	> 40	> 360	1.71
2f-41	159	8185	26.1	785	10	120	1.66
2f-42	254	7072	16.8	356	> 40	> 360	1.53
2f-43	232	7516	20.6	395	> 40	> 360	1.59
2f-44	157	8114	26.7	486	34	240	1.69
2f-46	127	7917	23.1	1058	3	80	1.63
2f-47	239	7883	23.0	369	> 40	> 360	1.68
Bicyclic azole-tetrazine derivatives							
2f-49	192	8429	27.3	352	35	360	1.76
2f-50	252	8180	23.6	348	35	360	1.76
2f-51	237	8304	26.4	427	27	360	1.79
2f-52	221	8438	27.7	449	24	360	1.80
2f-53	168	8707	32.0	431	17	240	1.84
2f-54	211	8413	27.5	328	19	360	1.82
2f-55	161	8731	30.9	781	8	60	1.82
2f-56	181	8884	32.4	539	14	120	1.85

esterification, ammoniation and cyanation starting from 5-amino-1H-1,2,4-triazole-3-carboxylic acid. Then [3+2] cycloaddition of **ATCN** using sodium azide proceeded readily in water with zinc chloride as catalyst to afford **2f-13**. Compound **2f-13** was then converted into energetic derivatives (**2f**-(**14-16**)) by the introduction of nitro-, nitramino-, and azido moieties (Scheme 2.29) [117].

Apart from C–C bonded tetrazole-triazole compounds, C-N bonded tetrazole-triazole compounds were also synthesized. Now a variety of secondary amines as versatile intermediates for the synthesis of substituted tetrazoles have been prepared (Scheme 2.29). A series of amines were cleanly converted into the tetrazoles **2f-(17-20)** in the presence of cyanogen azide, which was synthesized from the reaction of cyanogen bromide and dried sodium azide in acetonitrile [118–121]. As shown in Scheme 2.29, the synthesis of N-methylene-C linked triazole and tetrazole (**2f-21** and **2f-23**) could be achieved from aminotetrazole according to the same synthetic route to **2f-5**. Further nitration product **2f-22** was prepared by the nitration of **2f-21** in 100% nitric acid. And several attempts to nitrate **2f-23** using various nitrating reagents

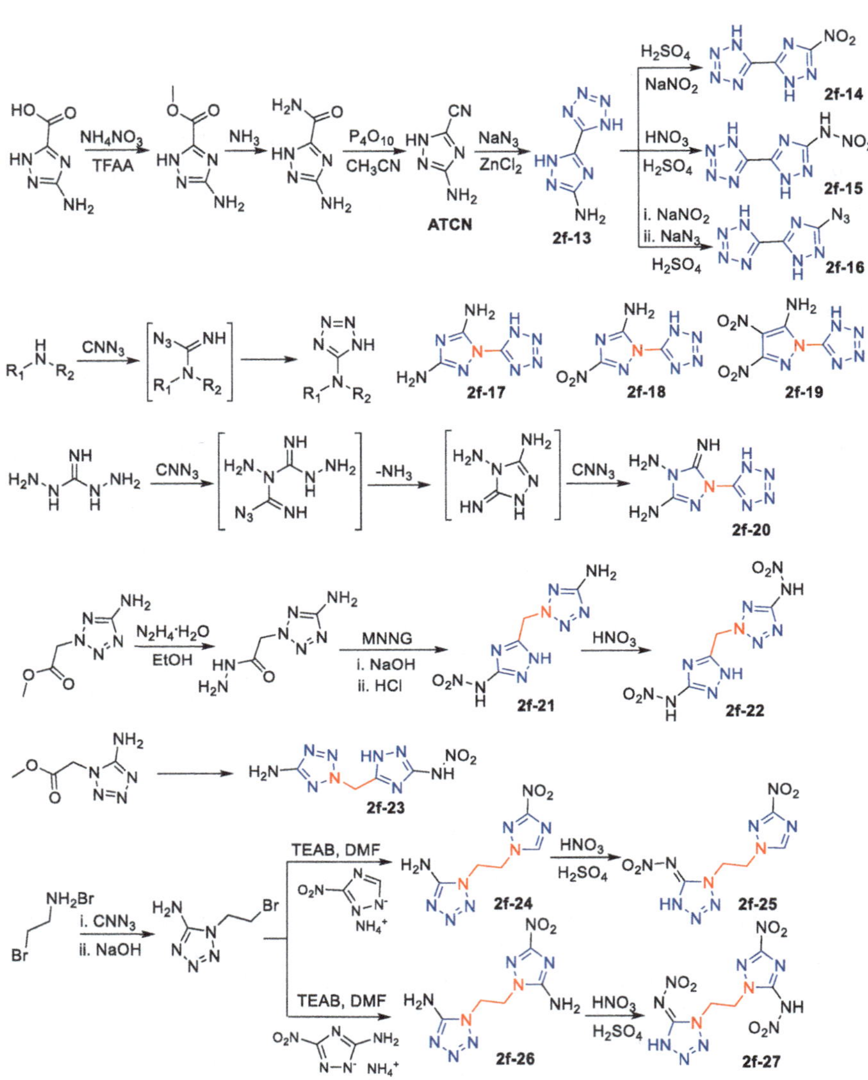

Scheme 2.29 Synthesis of bicyclic tetrazole-triazole derivatives

were unsuccessful [122]. Also, the synthesis of asymmetric *N,N′*-ethylene bridged 5-aminotetrazole and 1,2,4-triazole moieties have been examined. First, the treatment of haloethylamines with cyanogen azide gave aminotetrazole compound containing a haloethyl group. Then, the nucleophilic reaction of the triazole species and 2-haloethyltetrazole formed bicyclic-based compounds containing an *N,N′*-ethylene bridge (**2f-24** and **2f-26**). The nitroamino compounds **2f-25** and **2f-27** were obtained through the nitration reaction of **2f-24** and **2f-26** in the mixture of nitric acid and sulfuric acid, respectively [123]. Most of those compounds **2f**-(**13-27**) possess high thermal stabilities with decomposition temperature above 200 °C and low sensitivities except **2f-15** (IS > 10 J, FS > 160 N) and **2f-16** (IS < 1 J, FS < 20 N).

As mentioned above, the substitution reaction of N-nitroazoles and nucleophilic reagents is an efficient strategy for constructing C–N linked bisheterocyclic compounds. Similar to the synthetic route to **2c-20**, a series of nitrosubstituted 1,2,4-triazole (**2f**-(**28-30**)) have been synthesized from 1-nitro-1,2,4-triaozle (Scheme 2.30). Compounds **2f-30** and its isomer **2f-31** were formed through the reaction of 1,3,4-trinitropyrazole and sodium 3-nitrotriazolate in the presence of NH_4NO_3 and TFAA [80]. Continuing effort to synthesize heat-resistant materials with good energetic performances and low sensitivities, **2f-32** was synthesized through the substitution reaction of **ANTA** and polynitrated pyrazole [124]. As expected, compounds **2f-30** and **2f-32** possess high detonation performance (D, 8262 and 9167 m s^{-1}; P, 29.0 and 37.8 GPa), high thermal stability (T_d, 330 and 270 °C) and acceptable sensitivity to impact (IS, > 40 and 9 J).

Apart from the C–N bonded bicyclic pyrazole-tetrazole compound **2f-19**, the synthesis of a series of N-bridged and alkyl-bridged bicyclic pyrazole-tetrazole compounds have also been reported. As shown in Scheme 2.31, 3,5-dinitro-4-aminopyrazole (**LLM-116**) was prepared by the nitration of 4-chloropyrazole in the mixture of concentrated sulfuric and nitric acids and then amination in 25% aqueous ammonia. The subsequent oxidation of **LLM-116** using 30% H_2O_2 and sulfuric acid furnished the desired **TNP**. The synthesis of a N-bridged pyrazole-tetrazole (**2f-33**)

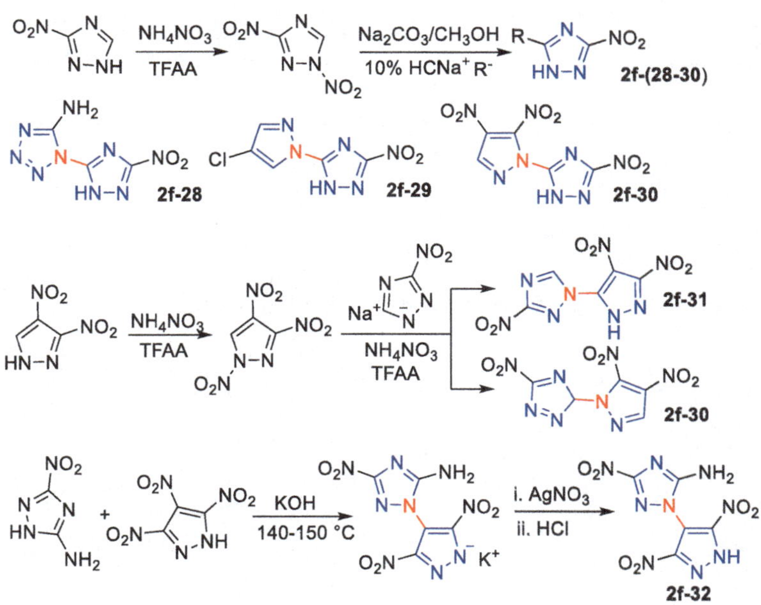

Scheme 2.30 Synthesis of C–N bonded bicyclic pyrazole-triazole derivatives

was synthesized through the reaction of **TNP** and 5-amino-1H-tetrazole (**ATZ**) in the presence of KOH at 140–160 °C [125]. Notably, this target compound **2f-33** features not only insensitivities to impact and friction (IS = 35 J, FS = 240 N) and super heat-resistance (T_d = 279 °C) but also high energy properties (D = 9289 m s^{-1}, P = 38.6 GPa).

As mentioned above, Shreeve's group has been interested in developing energetic compounds based on the symmetric *N,N′*-ethylene bridged bis(aminotetrazoles) and bis(nitropyrazoles). Moreover, they have also reported the synthesis of asymmetric *N,N′*-ethylene bridged 5-aminotetrazole and pyrazole moieties through a two-step reaction including the substitution of pyrazolate and then cyclization using cyanogen azide. The synthesis of a series of *N,N′*-ethylene bridged bicyclic pyrazole-tetrazole compounds **2f-(34-48)** are shown in Scheme 2.31. Further modification has been done to enhance their performance by the introduction of energetic groups such as nitro, nitroimino, nitramino and azide [126]. As shown in Table 2.6, the detonation velocities and pressures of *N,N′*-ethylene bridged bicyclic pyrazole-tetrazole compounds **2f-(34-47)** lie in range from 7072 m s^{-1} and 16.8 GPa to 8185 m s^{-1} and 27.1 GPa, respectively. They also exhibit low sensitivities to impact and friction

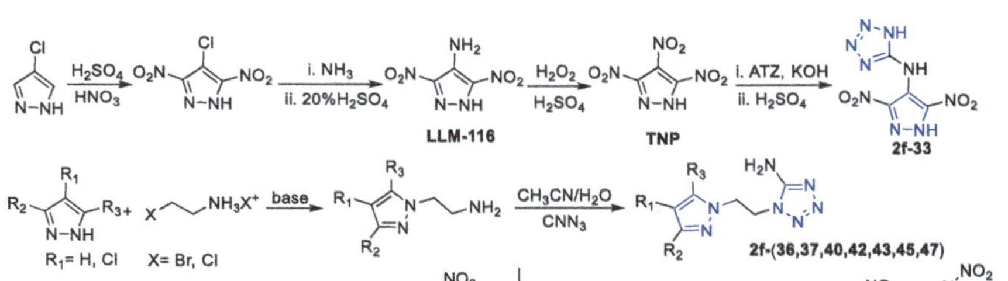

Scheme 2.31 Synthesis of bicyclic pyrazole-tetrazole derivatives

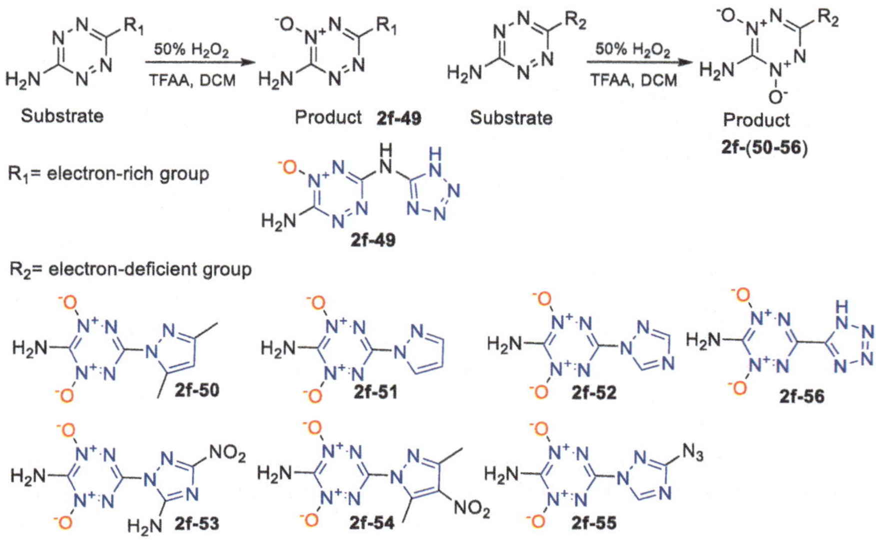

Scheme 2.32 Synthesis of bicyclic azole-tetrazine derivatives

(IS > 24 J, FS > 240 N) except **2f-41** (IS = 10 J, FS = 120 N) and **2f-46** (IS = 3 J, FS = 80 N) due to azide group in molecule.

The N–O bond is a relatively strong bond possessing significant double bond character owing to π-back-bonding. In order to understand the influence of a substituent group on the oxidation of 1,2,4,5-tetrazine. Shreeve's group selected a few tetrazines as theoretical subjects shown in Scheme 2.32. All substrates gave N-oxide products using 50% hydrogen peroxide and trifluoroacetic acid anhydride in dichloromethane (DCM) at room temperature [127]. It is noteworthy that the N-oxidation of tetrazine derivative with electron-rich groups can give the mono-N-oxide product while the substrate containing electron-deficient groups can be oxidized to the di-N-oxide products (**2f-(50-55)**). All these N-oxide compounds exhibit good detonation velocity range from 8180 to 8884 m s^{-1}, acceptable decomposition temperature from 161 to 252 °C and impact sensitivity from 8 to 35 J.

2.7 Experimental Section

4,4′-Diamino-3,3′-bifurazanyl (**2a-1**): Dichloroglyoxime (4.20 g, 27 mmol) was added to a stirred solution of sodium cyanide (3.00 g, 54 mmol) and sodium hydroxide (1.86 g, 46 mmol) in water (50 ml) in small portions below 10 °C. The total amount of sodium hydroxide used was 54 mmol, with 8 mmol derived from the sodium cyanide. Once complete dissolution of all solid components was achieved, the solution underwent a gradual warming process to reach a temperature of 20 °C. At this point, ice cooling was employed to acidify the solution with concentrated sulfuric acid until a pH of 2 was achieved. The resulting mixture was then cooled to 0 °C, at which point the solid precipitate was collected via filtration. Subsequently, the solid was thoroughly washed using 10 mL of ice water and finally dried. This procedure yielded dicyano-glyoxime in the amount of 1.90 g (51% yield), with a melting point (T_m) of 141 °C and decomposition temperature (T_d) of 145 °C.

A solution of hydroxylamine hydrochloride (2.78 g, 40 mmol) in water (10 ml) was combined with a separate solution of sodium hydroxide (1.60 g, 40 mmol) in water (10 ml). Subsequently, dicyano-glyoxime (1.40 g, 10 mmol) was introduced into the resulting mixture, which was then heated at 50 °C for a duration of 1.5 h. Upon addition of acetic acid (2.4 ml), the mixture was cooled to 0 °C. The resulting product was collected via filtration, washed with water, and subsequently dried. This process successfully yielded 1,4-diamino-1,2,3,4-tetraoximinohutane in the amount of 1.47 g (T_m, 180 °C).

A solution of sodium hydroxide (0.53 g, 13 mmol) in water (10 mL) was prepared. To this solution, 1,4-diamino-1,2,3,4-tetraoximinobutane was added. The resulting solution was then heated at a temperature range of 165–170 °C for a duration of one hour using a stainless-steel autoclave. After completion of the reaction, the vessel was rapidly cooled to 0 °C in an ice bath. The pressure was released, and the resulting product was collected by filtration. It was subsequently washed with water and dried to obtain the final product **2a-1** (0.31 g, T_d, 305 °C).

4,4′-Dinitro-3,3′-bifurazanyl (**2a-2**): To a slurry of hydrogen peroxide (0.9 mL, 33 mmol) in dichloromethane (20 mL) was added trifluoro-acetic anhydride (5.0 mL, 35 mmol) with stirring and ice cooling. After the exotherm had subsided, **2a-1** (0.31 g, 1.8 mmol) was added and the mixture was heated under reflux for 20 h. Following the completion of the reaction, the solution was carefully extracted with saturated sodium carbonate solution until the aqueous extracts exhibited basic properties when tested with litmus paper. The organic layer was subsequently dried using magnesium sulfate. The solvent was removed via evaporation using a rotary evaporator, leaving behind a residue. This residue was subjected to chromatography over silica gel, utilizing a mixture of chloroform and hexane (50/50) as the eluent. This process yielded 0.10 g of a pure product, as confirmed by thin-layer chromatography. The compound displayed a melting point of 85 °C. This compound is sensitive explosive that should be handled with care. elemental analysis calcd (%) for $C_4N_6O_6$ (228): C 21.06, N 36.85; found: C 21.01, N 36.78.

3,3′-Dinitramino-4,4′-bifurazane (**2a-3**): 1.01 g (6 mmol) **2a-1** were added in small portions to 100% nitric acid (8 mL) at -5–0 °C. After the addition, the resulting solution was stirred for 45 min at 0–5 °C. The suspension was then poured onto a bed of ice (50 g), and the resulting precipitate was collected by filtration. The precipitate was subsequently washed with ice-water (10 mL) and dried under ambient conditions. This process yielded 1.61 g (80%) of **2a-3** as a colorless solid. Additionally, by extracting the filtrate with ethyl acetate, followed by drying the organic phase with magnesium sulfate and evaporating the solvent, an additional 366 mg (16%) of **2a-3** could be isolated. DSC (5 °C min^{-1}): T_d, 80 °C; IR (ATR): n 3207 (w), 1739 (m), 1622 (m), 1551 (w), 1420 (w), 1366 (w), 1304 (w), 1280 (s), 1226 (w), 1217 (w),

1088 (w), 1000 (m), 976 (w), 914 (w), 897 (w), 860 (m), 777 (vs), 745 (w), 731 (w), 705 cm^{-1} (w); Raman (1064 nm, 200 mW, 25 °C): n 3218 (20), 1634 (12), 1619 (100), 1552 (98), 1454 (30), 1382 (49), 1336 (19), 1285 (32), 1033 (26), 988 (13), 913 (23), 775 (14), 749 (19), 610 (10), 491 (36), 453 cm^{-1} (12); ^{1}H NMR (DMSO-d_6, 25 °C): d = 11.49 ppm; ^{13}C NMR (DMSO-d_6, 25 °C): d = 153.5, 141.7 ppm; elemental analysis calcd (%) for $C_4H_2N_8O_6$ (258.11): C 18.61, H 0.78, N 43.41; found: C 18.62, H 0.78, N 43.92.

3,3-Diamino-4,4-bisfuroxane (**2a-4**): Diaminotetraoximobutane (787 mg) was dissolved in the minimum amount of 1 M hydrochloric acid possible. Dropwise addition of Br_2 (1.233 g) was carried out at a temperature of 0 °C over a period of 45 min. The resulting mixture was then stirred for an additional 45 min at 0 °C. The precipitated product **2a-4** was collected by filtration and subsequently dried under ambient conditions, yield 602 mg (78%). DSC (5 °C min^{-1}): T_d, 160 °C. IR (ATR): n = 3388 (w), 3273 (w), 3175 (w), 1653 (vs), 1599 (s), 1491 (s), 1432 (s), 1322 (w), 1049 (s), 957 (s), 892 (m), 717 (m) cm^{-1}. Raman (1064 nm, 300 mW, 25 °C, cm^{-1}): n = 3261 (17), 3154 (8), 1681 (13), 1618 (2), 1595 (100), 1469 (97), 1382 (10), 1260 (14), 1025 (10), 895 (11), 761 (8), 709 (18), 521 (4), 356 (34), 344 (18), 249 (3) cm^{-1}. ^{1}H NMR (DMSO-d_6, 25 °C): 3.5–6.8 ppm. ^{13}C NMR (DMSO-d_6, 25 °C): 141.8, 122.5 ppm. MS (DEI$^+$): m/z = 200.1 $[M]^+$. $C_4H_4N_6O_4$ (200.11): calcd C 24.01, H 2.01, N 42.00; found C 23.78, H 2.06, N 40.34.

3,3-Dinitramino-4,4-bisfuroxane (**2a-5**): **2a-4** (1 g, 5 mmol) was added to 100% HNO_3 (4.2 mL, 100 mmol) in portions while carefully maintaining the temperature within the range of – 10 to 5 °C. After stirring for 30 min, the resulting precipitate was collected by filtration and subsequently washed with trifluoroacetic acid (3 mL). The solid product was then dried under vacuum for a duration of 1 h and stored in the refrigerator. Pale yellow solid, yield: 79.3%. T_d: 74.5 °C. ^{1}H NMR (CD_3CN, ppm): 7.55 (2H, s, NH). ^{13}C NMR (CD_3CN, ppm): 125.9, 155.9 ppm. IR (KBr): n = 3240, 1612, 1493, 1445, 1393, 1317, 1071, 978, 885, 801, 752, 737, 700, 500, 447 cm^{-1}. EA ($C_4H_2N_8O_8$, 290.11): calcd C 16.56, H 0.69, N 38.62; Found C 16.11, H 0.70, N 37.51.

4,4-Diamino-3,3-bisfuroxane (**2a-6**): A suspension was prepared by combining compound **2a-4** (1000 mg) with dioxane (30 mL). The resulting mixture was then heated at reflux for a duration of 4 h. Afterward, the solvent was evaporated, and cold water (30 mL) was added. Compound **2a-6** (410 mg, 41% yield) was subsequently isolated by filtration. DSC (5 °C min^{-1}): 170 °C (T_d). IR (ATR): n = 3441 (m), 3347 (m), 1617 (s), 1607 (s), 1571 (vs), 1546 (w), 1527 (s), 1438 (s), 1200 (m), 1192 (m), 923 (s), 845 (m), 737 (m) cm^{-1}. Raman (1064 nm, 300 mW, 25 °C, cm^{-1}): n = 3349 (40), 1628 (73), 1554 (17), 1520 (60), 1458 (7), 1443 (3), 1387 (77), 1291 (4), 1221 (4), 1189 (22), 1113 (3), 1004 (11), 862 (3), 788 (11), 713 (3), 692 (2), 664 (49), 637 (3), 586 (19), 562 (10), 456 (6), 377 (2) cm^{-1}. ^{1}H NMR (DMSO-d_6, 25 °C): 6.65 ppm. ^{13}C NMR (DMSO-d_6, 25 °C): 157.1, 98.8 ppm. MS (DEI$^+$): m/z = 200.1 $[M]^+$. $C_4H_4N_6O_4$ (200.11): calcd C 24.01, H 2.01, N 42.00; found C 24.95, H 2.06, N 40.63.

4,4-Dinitro-3,3-bisfuroxane (**2a-7**): A solution of compound **2a-6** (1 g) was prepared by dissolving it in concentrated sulfuric acid (5 g). This solution was then added dropwise to a suspension of $(NH_4)_2S_2O_8$ (14 g) in 50% H_2O_2 (28 g), while maintaining the temperature below 30 °C. The resulting suspension was stirred at 30 °C for 4 h, followed by an additional hour of stirring at 50 °C. The solution was subsequently poured onto crushed ice (50 g) and extracted with dichloromethane (4 times with 25 mL each). After drying the organic phase with magnesium sulfate, the organic solvent was left to evaporate, allowing the formation of crystals. Within the resulting yellow liquid, a few individual crystals of **2a-7** suitable for X-ray diffraction analysis were observed.

4,4-Dinitramino-3,3-bisfuroxane (**2a-8**): **2a-6** (1 g, 5 mmol) was added to 100% HNO_3 (4.2 mL, 100 mmol) in portions while carefully maintaining the temperature within the range of − 10 to − 5 °C. Following a 30-min stirring period, the resulting precipitate was collected by filtration and subsequently washed with trifluoroacetic acid (3 mL). The solid product was then dried under vacuum conditions for 1 h and stored in a refrigerator. White solid, yield: 82.1%. T_d: 91.4 °C. ^{1}H NMR (CD_3CN, ppm): 9.15 (2H, s, NH). ^{13}C NMR (CD_3CN, ppm): δ = 103.3, 148.3 ppm. IR (KBr): ν = 3257, 3131, 1629, 1604, 1549, 1524, 1461, S4 1373, 1298, 1215, 1173, 1090, 982, 888, 816, 778, 709, 683, 527, 439 cm^{-1}. EA ($C_4H_2N_8O_8$, 290.11): calcd C 16.56, H 0.69, N 38.62; Found C 16.43, H 0.63, N 37.16.

3,3′-Diamino-4,4′-azofurazan (**2a-9**): The methanol solution of hydrazo-furazan (**b**), prepared as described previously, was subjected to continuous stirring at room temperature while air was bubbled through it for a duration of 20 h. Following this process, an orange precipitate formed and was subsequently collected by filtration. The precipitate was then washed with methanol and left to dry in ambient air. This resulted in the formation of 9.0 g (92%) of pure **2a-9**, which exhibited complete similarity to the compound reported previously. ^{1}NMR (deuteriomethyl sulfoxides) 6.89 (s, 4H); ^{13}C NMR (deuteriomethyl sulfoxide) 150.4, 155.6.

N,N′-bis(2,2,2-trinitroethyl)-3,3′-diamino-4,4′-azofurazan (**2a-11**): In a solution of hydrochloric acid (1 M, 50 mL), 3,3′-diamino-4,4′-azofurazan (0.49 g, 2.5 mmol) and 2,2,2-trinitroethanol (0.91 g, 5.0 mmol) were dissolved. The resulting reaction mixture was stirred at a temperature of 90 °C for a duration of 24 h. Upon cooling, a precipitate formed in the mixture. The precipitate was subsequently subjected to washing with water (20 mL × 3) and then air-dried to eliminate hydrochloric acid by volatilization. This process resulted in the formation of 1.25 g of pure product **2a-11**, which appeared as a light-yellow solid. (yield 96%). T_d, 230 °C; ^{1}H NMR (300 MHz, DMSO-d_6): δ = 7.27 ppm (t, J = 7.5 Hz, 2H, –NH–), 5.40 ppm (d, J = 6.0 Hz, 4H, –CH_2– $C(NO_2)_3$); ^{13}C NMR (300 MHz, DMSO-d_6): δ = 155.6, 150.4, 126.1, 48.0 ppm; IR: n = 3390, 2958, 2913, 1596, 1530, 1419, 1276, 1132, 1032, 855, 799, 711, 644, 579 cm^{-1}; elemental analysis calcd (%) for $C_8H_6N_{14}O_{14}$ (522.22): C 18.38, H 1.15, N 37.53; found: C 18.45, H 1.22, N 37.43.

Synthesis of N,N′-bis(2,2,2-trinitroethyl)-3,3′-dinitramino-4,4′-azofurazan (**2a-12**): To a vigorously stirred nitration reagent with 100% nitric acid and acetic anhydride (30 mL: 20 mL) was added **2a-11** (1.04 g, 2.0 mmol) at 0 °C, and temperature was strictly controlled below 5 °C during the feeding course. Then the mixture was stirred for 1 h at room temperature. The reaction was quenched by lowering the temperature of the reaction system with crushed ice to precipitate a yellow solid.

The precipitate was washed with water (30 mL × 3) and dired in the air. 1.12 g of the pure product **2a-12** was obtained as a yellow solid (yield 90%). T_d, 159 °C; ^{1}H NMR (300 MHz, Acetone-d_6): δ = 6.62 (s, 4H, $-CH_2-C(NO_2)_3$); ^{13}C NMR (300 MHz, Acetone-d_6): δ = 160.5, 145.8, 124.1, 54.5 ppm; IR: n = 3004, 2964, 1584, 1414, 1271, 1128, 1024, 933, 855, 803, 738, 699, 594 cm^{-1}; elemental analysis calcd (%) for $C_8H_4N_{16}O_{18}$ (612.21): C 15.68, H 0.65, N 36.59; found: C 15.61, H 0.62, N 36.52.

3,3′-Diamino-4,4′-azoxyfurazan (**DAAF**) (**2a-13**): Aqueous solution of DAF (500.0 g, 5.0 mol) and sodium bicarbonate (1260 g, 15 mol) in approximately 30 L of water was prepared at room temperature (around 22 °C) using a 50 L non-jacketed flask. Oxone (3075 g, 5 mol) was added to the solution in three equal portions of around 1000 g each. After stirring for about 2 h, the pH was measured and found to be 6.32. To adjust the pH, additional sodium bicarbonate (1260.0 g, 15 mol) was added, resulting in a pH of 7.36. Further Oxone (3075 g, 5 mol) was added, leading to a pH of 6.98. Stirring was continued for another approximately 2 h, at which point the pH was measured to be 6.78. The solid product that precipitated from the solution was filtered and washed with water. This process yielded 447 g (84%) of essentially pure **DAAF**, as confirmed by analysis using DSC, NMR, and elemental analysis. The mean particle size of the DAAF was determined to be 41 mm.

3,3′-Dinitramino-3,3′-azoxybifurazane (**2a-14**): In a stepwise manner, compound **2a-13** (1.06 g, 5 mmol) was dissolved in 100% nitric acid (6 mL) at a temperature range of − 5 to 0 °C. After the addition, the resulting solution was stirred for an additional 45 min, maintaining a temperature of 0–5 °C. The suspension was then poured onto a bed of ice weighing 40 g. The solid product was collected by filtration and subsequently dried in ambient air, resulting in the formation of **2a-14** (1.13 g, 75%) as a yellowish solid. An additional 183 mg (12%) of **2a-14** could be isolated by extracting the filtrate with ethyl acetate, followed by drying the organic phase with magnesium sulfate and evaporating the solvent DSC (5 °C min^{-1}): T_d, 70 °C; IR (ATR): n 3280 (w), 1738 (m), 1617 (s), 1521 (w), 1499 (w), 1437 (w), 1387 (m), 1362 (w), 1312 (vs), 1229 (m), 1046 (w), 1000 (w), 949 (m), 901 (w), 833 (m), 750 (m), 711 cm^{-1} (m); Raman (1064 nm, 200 mW, 25 °C): n 1592 (10), 1580 (18), 1521 (26), 1503 (100), 1468 (10), 1444 (85), 1382 (18), 1367 (65), 1324 (9), 1303 (20), 1047 (8), 871 cm^{-1} (8); ^{1}H NMR (DMSO-d_6, 25 °C): d = 11.57 ppm; ^{13}C NMR (DMSO-d_6, 25 °C): d = 155.6, 153.9, 151.2, 149.4 ppm; elemental analysis calcd (%) for $C_4H_2N_{10}O_7$ (302.12): C 15.90, H 0.67, N 46.36; found: C 15.93, H 0.75, N 45.68.

N,N′-bis(2,2,2-trinitroethyl)-3,3′-diamino-4,4′-azoxyfurazan (**2a-15**): In a solution of hydrochloric acid (1 M, 50 mL), 3,3′-diamino-4,4′-azoxyfurazan (0.212 g, 1.0 mmol) and 2,2,2-trinitroethanol (0.362 g, 2.0 mmol) were dissolved. The resulting mixture was stirred for a duration of 5 days at 100 °C. Afterward, the mixture was cooled to 20 °C, leading to the formation of a precipitate. The precipitate was washed with water (30 mL × 3) and subsequently dried. This process yielded 0.482 g of pure product **2a-15** as a light-yellow solid, with a yield of 89.6%. Alternatively, when the reaction time was extended to 7 days, 0.42 g of **2a-15** (yield 78.0%) were obtained through column chromatography on silica gel. (CH_2Cl_2/petroleum ether, 1:2). T_d,

192 °C; ^{1}H NMR (300 MHz, $(CD_3)CO$): δ = 7.40 (t, J = 6.0 Hz, 1H, –NH–), 7.66 (t, J = 6.0 Hz, 1H, –NH–), 5.41 ppm (t, J = 6.0 Hz, 4H, $–CH_2–C(NO_2)_3$); ^{13}C NMR (300 MHz, $(CD_3)CO$): δ = 154.21, 152.98, 152.29, 148.56, 127.09, 49.31, 49.11 ppm; IR: n = 3407, 3003, 2924, 2846, 1596, 1532, 1479, 1375, 1285, 1024, 854, 815, 762, 645 cm^{-1}; elemental analysis calcd (%) for $C_8H_6N_{14}O_{14}$ (538.22): C 17.85, H 1.12, N 36.43; found: C 17.90, H 1.08, N 36.40.

N,N′-bis(2,2,2-trinitroethyl)-3,3′-dinitramino-4,4′-azoxyfurazan (**2a-16**): In a vigorously stirred nitration reagent composed of 100% nitric acid and acetic anhydride (20 mL: 16 mL), 2a-15 (0.538 g, 1.0 mmol) was added at 0 °C. The temperature was strictly maintained below 5 °C throughout the addition process. After the completion of adding acetic anhydride, the reaction was kept at room temperature, and timely cooling was necessary if the reaction temperature sharply rose during the nitrification process. It was important to maintain a stable temperature for two hours. Upon completion of the reaction, the reaction system was quenched by lowering the temperature using crushed ice, resulting in the precipitation of a yellow solid. The precipitate was then washed with water (30 mL × 3) and dried in ambient air. This yielded 0.551 g of pure product **2a-16** as a yellow solid (yield 87.7%). T_d, 135 °C; ^{1}H NMR (300 MHz, $(CD_3)CO$): δ = 6.16 (s, 2H, $–N(O)=N–R–CH_2–C(NO_2)_3$), δ = 6.23 (s, 2H, $–N = N(O)–R–CH_2–C(NO_2)_3$); ^{13}C NMR (300 MHz, $(CD_3)CO$): δ = 162.67, 156.32, 153.12, 150.00, 148.06, 124.30, 54.18, 53.93 ppm; IR: n = 3005, 2963, 1589, 1541, 1505, 1406, 1348, 1273, 1210, 1150, 1122, 1032, 936, 854, 805, 766, 752, 636, 541 cm^{-1}; elemental analysis calcd (%) for $C_8H_4N_{16}O_{18}$ (628.21): C 15.30, H 0.64, N 35.64; found: C 15.34, H 0.62, N 36.57.

Dinitroazoxyfurazan (**2a-17**): Diaminoazoxyfurazan **2a-9** (2.12 g, 10 mmol) was added to a mixture of concentrated sulfuric acid (25 mL), hydrogen peroxide (30%, 30 mL), and ammonium persulfate (25.0 g). The addition was carried out at a temperature range of 10–20 °C with stirring. The resulting mixture was stirred at 50 °C for a period of 5 h. Afterward, the mixture was added to ice-cold water (200 mL). This led to the isolation of a yellow precipitate, which was subsequently washed with ice-cold water and dried. As a result, light-yellow needle-shaped crystals of **2a-17** were obtained (1.65 g, 60%), T_m, 110–112 °C (hexane/dichloromethane). IR (KBr): n 1572 (C=N–O), 1518 and 1351 (NO_2), 1036 (furazan ring). ^{13}C NMR (acetone-d_6,): 156.96, 154.68, 153.17, 149.29. El-MS m/z (%): 272 (M^+). Anal. calcd for $C_4N_8O_7$: C, 17.66; N, 41.18; 0, 41.16. Found: C, 17.78; N, 40.49; 0, 41.09.

Dinitroazofurazan (**2a-18**): A solution of diaminofurazan (4.0 g, 0.040 mol) in concentrated sulfuric acid (90 g) was added dropwise over a period of 30 min at a temperature range of 5–10 °C to a stirred suspension of ammonium persulfate (50 g) in hydrogen peroxide (30%, 60 g). The resulting mixture was allowed to warm up to 30 °C, and stirring was continued for 1.5 h at that temperature. Afterward, the reaction mixture was cooled using an ice bath, and the precipitate formed was filtered off. The filtrate was then poured into cold water (250 mL). The resulting water solution was subjected to extraction with dichloromethane (5 × 200 mL), followed by washing with water (100 mL) and drying using sodium sulfate. Evaporation of the solvent under vacuum yielded an orange oil, which was further purified through crystallization from water. This purification process resulted in the formation of light-yellow crystals weighing 1.8 g (35%), T_m 120°.

A stirred suspension of 3-amino-4-nitrofurazan (200 mg, 1.50 mmol) in concentrated hydrochloric acid (8 mL) was prepared. Over a period of 1 h at 40 °C, a solution of potassium permanganate (474 mg, 3.00 mmol) in water (53 mL) was added dropwise to the suspension. The resulting mixture was then stirred for 2 h at 55 °C. After cooling the solution, it was extracted with dichloromethane (3 × 50 mL). The combined organic fractions were subsequently washed with water (50 mL) and dried.

3-Amino-4-cyano-1,2,5-oxadiazole (**2a-19**): In a 100 mL round-bottom flask, the following components were added: 20 mL acetic acid, 20 mL acetonitrile, 20 mL water, and a magnetic stir bar. To this mixture, **AAOF** (8 g, 56 mmol) was added. The resulting mixture was heated to 35 °C with stirring. After reaching the desired temperature, manganese dioxide (5.36 g, 62 mmol) was added portion-wise over a period of 5 min. The contents of the flask were then heated to 55 °C for a duration of 20 min. The resulting mixture, which appeared black, was vacuum-filtered while still hot using celite as the filter medium. Next, the brown solution obtained from the filtration was covered with 30 mL of ether. Sodium carbonate hydrate (19 g, 150 mmol) was added to neutralize the solution, followed by filtration through celite to remove any precipitate. The filtrate was subjected to extraction using three portions of 40 mL ether. The organic layer was washed sequentially with 1 M $NaHCO_3$, water, and brine. It was then dried using magnesium sulfate and finally evaporated to dryness. The final product obtained weighed 5.5 g (85% yield) and exhibited a melting point of 85–87 °C. Its identity was confirmed by comparing its spectra with reported data. IR: 3414, 3338, 3208, 2266, 1684, 1644, 1589, 1513, 1407, 1260, 1021, 917, 866 cm^{-1}; ^1H NMR (acetonitrile-d_3 at 1.95): d (ppm) = 5.52 (br s); ^{13}C NMR (acetonitrile-d_3 at 1.3): d (ppm) = 108.6, 127.2, 157.8.

4,4′-Bischlorohydroximoyl-3,3′-azofurazan (**2a-21**): **2a-20** (2.16 g, 10 mmol) was suspended in 30 mL of ethanol, and a solution of 50% aqueous hydroxylamine (1.65 g, 25 mmol) was added. Stirring was continued for 30 min at room temperature. The solvent was then removed, and the resulting solid was washed with water and dried in ambient air. Next, the obtained solid (2.8 g) was dissolved in a mixture of concentrated hydrochloric acid (30 mL) and water (40 mL). The solution was cooled to 0 °C, and while maintaining a temperature below 0 °C, a solution of sodium nitrite (1.75 g, 25 mmol) in 10 mL of water was added. The reaction mixture was stirred for 2 h at the same temperature. The solid product was collected by filtration, washed with ice-water, and subsequently dried in ambient air. This yielded **2a-21** as a yellow solid. (1.96 g, 61%). T_m = 168 °C. T_d = 180 °C. ^1H NMR (300 MHz, DMSO-d_6): d = 13.73 ppm (s, 2H). ^{13}C NMR (75 MHz, DMSO-d_6): d = 160.7, 146.9, 123.3 ppm. IR (KBr): n 3344, 3029, 1697, 1614, 1546, 1479, 1384, 1267, 1237, 1032, 942, 887, 677, 610, 567 cm^{-1}. Elemental analysis calcd (%) for $C_6H_2Cl_2N_8O_4$ (321.04): C 22.45, H 0.63, N 34.90; found: C 22.68, H 0.56, N 35.48%.

Potassium 4,4′-bis(dinitromethyl)-3,3′-azofurazanate: A suspension of **2a-21** (1.96 g, 6.1 mmol) in $CHCl_2$ (20 mL) was added dropwise to a stirred mixture of 100% HNO_3 (5 mL) and trifluoroacetic acid anhydride (8.8 mL) at 0 °C. The resulting mixture was allowed to slowly warm up to room temperature and stirred for 2 h. The reaction mixture was then poured into ice water (60 mL) and extracted with $CHCl_2$ (3 × 30 mL). The organic phases were combined, washed with water and brine, dried over magnesium sulfate, and concentrated under vacuum. This provided the intermediate compound as a yellow oil. Next, the oil was dissolved in methanol (20 mL), and a solution of potassium iodide (2.02 g, 12.2 mmol) in methanol (30 mL) was added dropwise. The reaction mixture was stirred overnight at room temperature. The precipitate that formed was collected by filtration and washed with cold water (5 mL), methanol (5 mL) and ethyl ether (10 mL). This yielded potassium 4,4′-bis(dinitromethyl)-3,3′-azofurazanate (0.92 g, 33.5%) as an orange solid. T_d = 229 °C. ^{13}C NMR (75 MHz, DMSO-d_6): d = 162.1, 139.9, 121.9 ppm. IR (KBr): n 1746, 1630, 1557, 1500, 1463, 1398, 1225, 1145, 1067, 1001, 922, 820, 748, 698 cm^{-1}. Elemental analysis calcd (%) for $C_6K_2N_{10}O_{10}$ (450.32): C 16.00, H 0.00, N 31.10; found: C16.36, H 0.17, N 30.36%.

4,4′-Bis(fluorodinitromethyl)-3,3′-azofurazan (**2a-22**): Selectfluor™ (1.42 g, 3.8 mmol) was added to a suspension of 4,4′-bis(dinitromethyl)-3,3′-azofurazanate (0.45 g, 1.0 mmol) in anhydrous acetonitrile at a temperature of 25 °C. The resulting reaction mixture was stirred for a duration of 5 h, and then the solvent was evaporated by blowing air. The remaining solids were extracted with chloroform (15 mL × 3). After removing the solid by filtration, the filtrate was dried over anhydrous sodium sulfate and recrystallized from acetonitrile and chloroform to give **2a-22** (0.24 g, 58%). Yellow solid. T_m: 155 °C. T_d: 166 °C. ^{13}C NMR ($CDCl_3$, 126 MHz): d 161.4 (C–N=N–C), 137.7 (d, 30.2 Hz, C–C=N), 114.4 (d, 293.6 Hz, $C(NO_2)_2F$) ppm. ^{19}F NMR ($CDCl_3$, 470.5 MHz): d 94.0 ppm. IR (KBr): n 1607, 1482, 1409, 1355, 1306, 1212, 1051, 981, 914, 833, 796, 672, 628, 571 cm^{-1}. Elemental analysis for $C_6F_2N_{10}O_{10}$ (410.12): calcd C 17.57, H 0.00, N 34.15%. Found: C 17.59, H 0.22, N 32.51%.

4,4′-Bis(dinitromethyl)-3,3′-azofurazan (**2a-23**): a suspension of 4,4′-bis(dinitromethyl)-3,3′-azofurazanate (0.45 g, 1.0 mmol) in acetone (20 mL) was treated with hydrochloric acid (37%, 0.5 mL). The resulting reaction mixture was stirred for 1 h at a temperature of 25 °C. After the completion of the reaction, the mixture was filtered to remove any solid impurities. The filtrate was then concentrated, and the remaining solid material was subjected to recrystallization using a mixture of acetonitrile and chloroform as solvents. This process resulted in the formation of yellow crystals identified as compound **2a-23**. (0.31 g, 83%). Yellow solid. T_m: 108 °C. T_d: 140 °C. ^1H NMR (acetone-d_6): d 8.94 (s, 2H) ppm. ^{13}C NMR (acetone-d_6, 126 MHz): d 163.4 (C–N=N–C), 139.3 (C–C=N), 103.6 ($CH(NO_2)_2$) ppm. IR (KBr): n 3000, 2906, 1590, 1418, 1366, 1320, 1240, 1193, 1045, 904, 866, 809, 770, 700, 644, 590, 569 cm^{-1}. Elemental analysis for $C_6H_2N_{10}O_{10}$ (374.14): calcd C 19.26, H 0.54, N 37.44%. Found: C 18.73, H 0.50, N 35.41%.

3,3′-Diamino-4,4′-azofuroxan (**2a-24**): Into three bottles, mono-potassium malonyl hydrazine 31.2 g (0.20 mol), 250 mL dichloromethane, and 13.8 g (0.20 mol) $NaNO_2$ were added. The reaction was carried out at a temperature between 4 and 7 °C. Afterward, 50 mL HNO_3 and 25 mL H_2SO_4 were introduced in that order. This temperature was held for half an hour before being increased to the range of 25–30 °C for one hour. The temperature was then dropped to (− 22 ± 2) °C and a solution of 30 g $NaNO_2$ in 50 mL deionized water was added. This condition was held for an hour. The organic layer was isolated and extracted using 40 mL CCl_4. Following this, the volume of the CCl_4 solution was reduced to 40 mL through evaporation. At this point, 40 mL of 4-dioxane hexacyclod and 1.5 mL deionized water were introduced into a water bath maintained at 80 °C for 10 min. Upon cooling to ambient temperature, 80 mL of deionized water was incorporated, leading to the precipitation of 3.9 g of light-yellow crystal. The yield was 35%. IR (KBr): n 3340, 3321, 2207, 2169, 1663, 1609, 1519, 1367, 1158, cm^{-1}. Elemental analysis for $C_3H_2N_6O_3$: calcd C 21.60, H 1.39, N 49.20%. Found: C 21.18, H 1.19, N 49.41%.

Into 200 mL of acetone, 8.5 g (0.05 mol) of 4-amino-3-azide carbonyl azofuroxan was dissolved. Subsequently, 310 mL of concentrated HCl (37%) were poured in. The process was undertaken at a steady temperature of 20 °C. An aqueous solution of $KMnO_4$ (0.45 mol, 0.76 mol/L) was introduced while stirring. After the addition was complete, the system was stirred for an additional 2.5 h at the same temperature before oxalic acid was added until all MnO_2 had dissolved. The solution was then filtered and rinsed twice with 100 mL of water. The washed solids, along with 100 mL of dioxane and 3 mL of water, were placed into a 500 mL four-neck flask equipped with an agitator, thermometer, and dripping funnel. The mixture was heated to 80 °C and stirred for 15 min before allowing it to cool to ambient temperature. Following this, 120 mL of water was gradually introduced to precipitate the product. Upon completion of the precipitation process, a brick-red solid was obtained via filtration, which was found to be compound **2a-24** upon analysis. IR (KBr): n 3465, 3312, 1621, 1579 cm^{-1}. Elemental analysis for $C_4H_4N_8O_4$: calcd C 21.06, H 1.77, N 49.12%. Found: C 21.54, H 1.77, N 49.12%.

3,3′-Dinitro-4,4′-azofuroxan (**2a-25**): Pour 100 mL of high-concentration sulfuric acid and 360 mL of 90% hydrogen peroxide into the flask submerged in an ice bath. Next, incorporate 3 g of compound **2a-24**. Gradually raise the temperature to 30 °C while maintaining stirring for a duration of 3 h. Proceed with filtration, followed by rinsing and drying to yield a reddish-brown sediment. IR (KBr): n 1623, 1575, 1385 cm^{-1}. Elemental analysis for $C_4N_8O_8$: calcd C 16.68, N 38.90%. Found: C 16.67, N 38.58%.

N,N′-dinitro-N,N′-bis[3-chlorodinitromethyl-furazan-4-yl] methylenediamine (**2a-26**): To a solution of 1 M HCl (30 mL), add 3-amino-4-chloroximinofurazan **ACOF** (1.63 g, 10 mmol). Proceed to drip in 0.41 g (5 mmol) of 37% formaldehyde. Next, allow the blend to stir for a duration of 3 h. Collect the formed sediment by filtration and rinse with water. Air-drying results in the production of (3Z, 3′Z)-4,4′-(Methylenebis(azanediyl))bis(N-hydroxy-1,2,5-oxadiazole-3-carbimidoyl chloride) (3.07 g) appearing as a white solid substance, yield 91.1%. $T_d = 156$ °C. ^1H NMR: $\delta = 4.86$ (t, 2H), 6.64 (t, 2H), 13.47 (s, 2H) ppm; ^{13}C NMR: $\delta = 153.5$, 141.8, 126.5, 53.7 ppm; IR (KBr pellet): n = 3390, 3255, 3044, 2942, 2867, 2148, 1830, 1686, 1617, 1596, 1556, 1475, 1438, 1410, 1379, 1259, 1123, 1108, 1076, 1036, 1000, 947, 913, 868, 764, 712, 645, 564, 519, 485, 408 cm^{-1}; elemental analysis

(%) calcd for $C_7H_6Cl_2N_8O_4$ (337.08): C 24.94, H 1.79, N 33.24; found: C 25.06, H 1.84, N 33.63.

Start by dissolving the compound (3Z,3′Z)-4,4′-(Methylenebis(azanediyl))bis(N-hydroxy-1,2,5-oxadiazole-3-carbimidoyl chloride) (3.37 g, 10 mmol) in CH_2Cl_2 (30 mL). This solution is then incrementally introduced into a stirring blend of trifluoroacetic acid anhydride (14 mL) and 100% HNO_3 (8 mL), ensuring the reaction temperature remains below 0 °C. Once all the solution has been added, remove the ice bath and allow the blend to gradually warm to ambient temperature. Continue stirring for another 3 h. Subsequently, transfer the mixture into ice water (100 mL) and perform extraction with CH_2Cl_2 (3 × 20 mL). After combining the organic layers, wash them with water and brine. Allow drying over sodium sulfate before concentrating under reduced pressure to obtain the intermediate **2a-26** as a colorless oil.

N,N′-dinitro-N,N′-bis[3-dinitromethyl-furazan-4-yl] methylenediamine (**2a-27**): Compound **2a-26** was dissolved in 20 mL of methanol. Then, a solution of potassium iodide (6 g, 36 mmol) in 30 mL of methanol was added dropwise. The resulting mixture was stirred overnight at room temperature. The formed precipitate was collected by filtration and washed sequentially with 10 mL of cold water, 10 mL of methanol, and 10 mL of ethyl ether. This process yielded a yellow solid salt (4.06 g) with a 73% yield. Next, a mixture of the above salt (0.56 g, 1 mmol) and 0.5 mL of H_2O was stirred, and concentrated sulfuric acid (1 mL) was added dropwise. After the complete addition of sulfuric acid, the mixture was further stirred for 30 min. Then, it was extracted six times with 5 mL portions of CH_2Cl_2. The combined organic extracts were concentrated under vacuum, resulting in the formation of a white solid, **2a-27** (0.33 g), with a yield of 68.8%. Finally, it is recommended to store the obtained solid **2a-27** under vacuum.

N,N′-methylenebis(N-(4-(fluorodinitromethyl)-1,2,5-oxadiazol-3-yl)nitramide) (**2a-28**): To prepare compound **2a-27**, 0.48 g (1 mmol) of the compound was dissolved in 5 mL of acetonitrile. Ammonia gas was then bubbled into the solution. The resulting precipitate was filtered and washed with acetonitrile. It was then dried in air, yielding diammonium as a yellow solid with a 76% yield. Next, the salt (0.51 g, 1 mmol) was dissolved in 30 mL of anhydrous acetonitrile. Selectfluor (1.34 g, 3.79 mmol) was added to the solution at ambient temperature. The mixture, which turned yellow, was stirred until it became decolorized, which took approximately 2 h. The solvent was then evaporated under reduced pressure. To the yellowish residue, 50 mL of anhydrous diethyl ether was added. The resulting mixture was stirred in a closed vessel for 30 min. The solid was filtered, and the solvent was removed under reduced pressure. This process yielded compound **2a-28** as a white solid with a 72% yield. T_m = 83.4 °C. T_d = 127.6 °C. ^1H NMR: δ = 6.61 (s, 2H) ppm; ^{13}C NMR: δ = 148.7, 141.2, 140.9, 115.0, 112.7, 64.6 ppm; IR (KBr pellet): 3030, 2930, 2909, 1740, 1608, 1561, 1500, 1425, 1401, 1360, 1282, 1219, 1131, 1115, 1085, 1031, 991, 960, 924, 90, 836, 818, 795, 751, 727, 696, 660, 614, 559, 465, 437 cm^{-1}; elemental analysis (%) calcd for $C_7H_2F_2N_{12}O_{14}$ (516.16): C 16.29, H 0.39, N 32.56; found: C 16.57, H 0.41, N 31.93.

Methylene-bis(3-amino-4-nitrofurazan) (**2a-29**): A 50 mL round bottom flask, equipped with a magnetic stirring bar and reflux condenser, was prepared. Into the flask, 1.50 g of 3-amino-4-nitrofurazan (11.5 mmol), 0.60 g of 37% formaldehyde, 20 mL of water, and 8 drops of concentrated hydrochloric acid were added. The resulting mixture was refluxed for 10 min and then cooled to 0 °C. The product obtained was collected, washed with water, and dried. The yield of the product was 1.50 g (5.50 mmol), corresponding to a 96% yield. For further analysis, an analytical sample of the product was recrystallized from a mixture of acetone and water, T_m, 141 °C; IR (potassium bromide): 3432 (s), 1625 (sh), 1614 (vs), 1525 (s), 1449 (w), 1395 (m), 1385 (sh), 1340 (m), 1290 (w), 1197 (m), 1113 (m), 1061 (w), 1014 (m), 832 (m) cm^{-1}; ^1H NMR (acetone-d_6): 5.07 (t, J = 6 Hz,2 H), 7.2 (bs, 2 H, NH) ppm. Anal. calcd for $C_5H_4N_8O_6$: C, 22.07; H, 1.48; N, 41.17. Found: C, 22.10; H, 1.36; N, 41.14.

Methylene-bis(3-nitramino-4-nitrofurazan) (**2a-30**): In a 1 M solution of nitrogen pentoxide in methylene chloride (3.47 g, 3.52 mmol) maintained at – 25 °C, 119 mg of methylene-bis-(3-armino-4-nitrofurazan) (0.44 mmol) was added over a period of 15 min. The temperature was then raised to 0 °C, and the mixture was allowed to stir for 1.5 h. The resulting solution was stripped using a rotary evaporator without applying heat. This process yielded the product (150 mg, 0.42 mmol) with a 95% yield. For further analysis, an analytical sample of the product was recrystallized from a mixture of methanol and water, T_m, 99 °C; IR (potassium bromide): 3044 (w), 1616 (s), 1586 (s), 1547 (s), 1289 (s), 1229 (w), 1115 (m), 1081 (m), 915 (m), 836 (m); ^1H NMR (acetone-d_6): 6.85 (s) ppm. Anal. calcd for $C_5H_2N_{10}O_{10}$: C, 16.59; H, 0.56; N, 38.67. Found: C, 16.64; H, 0.53; N, 37.75.

Methylene-bis3-amino-4-methylfurazan (**2a-31**): In a 100 ml round bottom flask equipped with a magnetic stirring bar and a reflux condenser, 3-amino-4-methylfurazan (4.00 g, 40.4 mmol), 37% formaldehyde (1.60 g, 19.7 mmol), 1 mL of concentrated hydrochloric acid, and 50 ml of water were combined. Upon mixing, a white precipitate formed almost immediately. The mixture was then refluxed for 5 min and subsequently cooled to 0 °C using an ice bath. The resulting product was collected, washed with water, and dried. The yield obtained was 4.03 g (19.2 mmol), corresponding to a 98% yield. For further analysis, an analytical sample of the product was recrystallized from a mixture of dimethylformamide (DMF) and water, T_m, 198 °C; IR: 3350 (s), 1611 (s), 1563 (s), 1394 (m), 1241 (m), 1121 (w), 1063 (w) cm^{-1}; ^1H NMR (acetone-d_6): 2.20 (s, 6 H, CH_3), 4.65(t, J = 5 Hz, 2 H, CH_2), 7.21 (t, J = 5 Hz, 2H, NH) ppm. Anal. calcd for $C_7H_{10}N_6O_2$: C, 40.00; H, 4.80; N, 39.97. Found: C, 39.82; H, 4.88; N, 40.36.

Methylene-bis-(3-nitramino-4-methylfurazan) (**2a-32**): A 25 mL round bottom flask was prepared, into which 5.0 g of 100% nitric acid was placed. The flask was then cooled to – 5 °C, and 3.0 g of acetic anhydride was added over a period of 15 min. Methylene-bis-(3-amino-4-methylfurazan) (2.10 g, 10 mmol) was added in small portions over another 15 min. The resulting mixture was stirred for 30 min at – 5 °C and an additional 30 min at 15 °C. The mixture was then poured onto ice (50 g), causing the product to oil out. Over several days, the product gradually crystallized. It was collected and washed with cold methanol. The yield obtained was

2.85 g (9.5 mmol), corresponding to a 95% yield. For further analysis, an analytical sample of the product was recrystallized from methanol, T_m, 90 °C; IR (potassium bromide): 1595 (vs), 1507 (w), 1442 (w), 1421 (w), 1390 (w), 1093 (m), 898 (w), 768 (w), 752 (s), 734 (w), 672 (w), 632 (w) cm^{-1}; 1H NMR (acetone-d_6): 2.42 (s, 6 H), 6.70 (s, 2 H) ppm. Anal. calcd for $C_7H_8N_8O_6$: C, 28.01; H, 2.69; N, 37.32. Found: C, 28.04; H, 2.81; N, 37.14.

Oxy-bridged bis(dinitromethyl)furazan (**2a-33**): In a mixture of 50 mL of water and 25 mL of isopropanol, 3,3′-dicyanodifurazanyl ether (**FOF-2**) (4.16 g, 20.0 mmol) and hydroxylamine hydrochloride (2.85 g, 41 mmol) were added. Sodium carbonate anhydrous was then added in batches to the reaction mixture. The resulting mixture was stirred at room temperature for 1 h. The formed precipitate was filtered, washed with ice water, and dried under vacuum over P_2O_5. The obtained solid was recrystallized from methanol. This process yielded the product as white platelets, with a weight of 5.02 g (91.2% yield). T_m, 203–204 °C. ^{1}H NMR (DMSO-d_6, 500 MHz): 10.67 (s, 2H, OH), 6.28 (s, 4H, NH_2) ppm. ^{13}C NMR (DMSO-d_6, 125 MHz): 160.31 (C–O), 142.23 (C–C=N), 141.33 (C–NH_2) ppm. IR (KBr, cm^{-1}): n = 3495, 3454, 3349 (NH_2), 3172, 2919 (OH), 1680, 1656, 1525, 1101, 1021 (ether bond), 969 (furazan ring). Calc. for $C_6H_6N_8O_5$: C 26.67, N 41.18, H 2.24%; Found: C 26.62, N 41.07, H 2.18%.

To prepare compound 3,3′-bis(carboxamidoxymino)difurazanyl Ether, 5.40 g (20.0 mmol) of the compound was dissolved in a mixture of 55 mL of concentrated hydrochloric acid and 30 mL of water. Saturated sodium nitrite (2.38 g, 41.0 mmol) in water was added dropwise to the stirred solution of Compound 2 at 0 °C. The reaction mixture was then stirred for 2 h. After the initial stirring, the reaction mixture was heated to 20 °C for an additional 1.5 h until N_2 evolution ceased. A white precipitate formed during this process. The precipitate was filtered, washed with water, and dried under vacuum. The obtained solid was then recrystallized from a mixture of methanol and water (1:1). This recrystallization step yielded the product as white crystals, with a weight of 5.40 g (87.5% yield). T_m, 60–61 °C. ^{1}H NMR (DMSO-d_6, 500 MHz): 13.71 (s, 2H, OH). ^{13}C NMR (DMSO-d_6, 125 MHz): 158.99 (C–O), 143.03 (C–C=N), 123.40 (C–Cl). IR (KBr, cm^{-1}): n = 3533, 3167, 3020 (OH), 1570, 1518, 1126, 1024 (ether bond), 942 (furazan ring), 657 (C–Cl). Calc. for $C_6H_2N_6O_5Cl_2$: C 23.32, N 27.20, H 0.65%; Found: C 23.30, N 27.15, H 0.68%.

A suspension of compound 3,3′-bis(chlorohydroxyminomethyl)difurazanyl ether (0.55 g, 1.8 mmol) in 50 mL of acetonitrile was prepared at 20 °C. N_2O_5 (2.2 g, 20 mmol) was added to the suspension, and the mixture was heated to 35 °C and maintained at this temperature for 40 min. The solvent was then evaporated, leaving behind a residue. The residue was subjected to column chromatography on silica gel using ethyl acetate-pentane (1:6) as the eluent. This process allowed for the isolation of pure colorless crystals, specifically the synthesis of 3,3′-bis(chlorodinitromethyl)difurazanyl ether (0.23 g, 30.0% yield). T_m, 68–69 °C. ^{13}C NMR (DMSO-d_6, 125 MHz): 157.67 (C–O), 140.37 (C–C=N), 112.75 ($CCl(NO_2)_2$). IR (KBr, cm^{-1}): n = 1613, 1291 ($CCl(NO_2)_2$), 1049 (ether bond), 1582, 1515, 971 (furazan ring). Calc. for $C_6N_8O_{11}Cl_2$: C 16.72, N 26.00%; Found: C 16.82, N 25.94%.

3,3′-Bis(chlorodinitromethyl)difurazanyl ether (1.0 g, 2.3 mmol) was dissolved in 8 mL of methanol (MeOH) and treated with a solution of potassium iodide (KI) (1.5 g, 9.0 mmol) in 15 mL of MeOH at room temperature. The resulting mixture was then stirred at room temperature for 1 h. After the stirring period, the mixture was triturated with 20 mL of diethyl ether (Et_2O). The formed precipitate was collected and washed sequentially with ice-cold water, MeOH, and Et_2O. This process yielded a yellow solid with a weight of 0.81 g (85.7% yield). The obtained product can be directly used in the subsequent step without the need for further purification. T_d 245 °C (DSC measurement, 10 K min^{-1}). ^{13}C NMR (DMSO-d_6, 125 MHz): 160.77 (C–O), 142.31 (C–C=N), 118.67 (C–$(NO_2)_2$). IR (KBr, cm^{-1}): n = 1589, 1526, 1479, 1239 (C–$(NO_2)_2$), 1070 (ether bond) and 997 (furazan ring). Calc. for $C_6N_8O_{11}K_2$: C 16.44, N 25.57%; Found: C 16.15, N 25.36%.

In the experiment, a solution of the potassium salt of 3,3′-bis(dinitromethyl)difurazanyl ether (0.438 g, 1 mmol) in water (3 mL) was prepared. To this solution, 50% sulfuric acid (1 mL) was added. The resulting mixture was stirred for 10 min at room temperature. After stirring, the reaction mixture was subjected to extraction with ethyl ether (3 × 10 mL). The ethyl ether extracts were then washed with water and dried using magnesium sulfate. Subsequently, the ethyl ether solution was evaporated, resulting in the formation of a white solid. The weight of the obtained solid was 0.295 g (81.5% yield). ^1H NMR (DMSO-d_6, 500 MHz): d 10.49 (s, 2H, CH); ^{13}C NMR (DMSO-d_6, 125 MHz): 161.2, 142.8, 119.1; IR (KBr, cm^{-1}): n 3004, 2983, 1481, 1366, 1623, 1322, 1239, 1148, 1037, 1585, 1524, 999; calc. for $C_6H_2N_8O_{11}$: C 19.90, H 0.56, N 30.94%; found: C 19.84, H 0.52, N 30.78%.

3,3′-Bis(fluorodinitromethyl)difurazanyl Ether (**2a-34**): In this procedure, a suspension of the potassium salt of 3,3′-bis(dinitromethyl)difurazanyl ether (0.6 g, 1.37 mmol) in anhydrous acetonitrile was prepared at 20 °C. XeF_2 (0.93 g, 5.49 mmol) was added to the suspension and stirred for 48 h. The acetonitrile solvent was then removed under vacuum, leaving behind a residue. The residue was washed, resulting in the formation of a colorless solid. This solid was subjected to recrystallization from chloroform, yielding 0.23 g (42.6% yield) of **2a-34** as colorless needle crystals. The purity of the obtained product was determined to be above 98.5% by high-performance liquid chromatography (HPLC); ^{13}C NMR (DMSO-d_6, 125 MHz): 158.41 (s, C–O), 137.00 (d, $J^{13}C–^{19}F = 25$, C–C=N), 113.88 (d, $J^{13}C–^{19}F = 294.0$ Hz, CF$(NO_2)_2$). ^{19}F NMR (DMSO-d_6, 470.5 MHz): – 106.29. IR (KBr, cm^{-1}): n = 1616, 1576, 1515, 1310 (CF$(NO_2)_2$), 1196 (C–F), 1138, 1048 (ether bond) and 981 (furazan ring). Calc. for $C_6N_8O_{11}F_2$: C 18.10, N 28.15%; Found: C 18.26, N 27.92%.

Bis-3,3′-(nitro-NNO-azoxy)difurazanyl ether (**2a-35**) and 3,3-dinitrodifurazanyl ether (**2a-36**): In this experimental procedure, a solution of compound **2a-34** (0.41 g, 2 mmol) in 5 mL of CH_3CN was prepared under a dry atmosphere. To this solution, Na_2CO_3 (0.1 g, 1 mmol) was added. The resulting suspension was then stirred for approximately 0.5–1 h at a temperature of 50 °C. During the reaction, the progress of the reaction was monitored using thin-layer chromatography (TLC) and by observing the consumption of the starting material. After the reaction reached completion, the mixture was cooled. Then, 50 mL of CH_2Cl_2 was added to the mixture. The resulting mixture was washed with water, dried using magnesium sulfate, filtered, and evaporated. The residue obtained after evaporation was a mixture of **2a-36** and **2a-35**. To separate these products, silica gel flash chromatography was performed using a solvent system consisting of CH_2Cl_2/pentane (2:1) as the eluent.

Fraction 1: The compound 3,3-dinitrodifurazanyl ether (referred to as **2a-36**) was obtained as colorless crystals. The melting point of the crystals was determined to be in the range of 63–64 °C. Through analysis using techniques such as infrared spectroscopy (IR) and thin-layer chromatography (TLC), it was confirmed that the properties and characteristics of this substance matched those of the previously described compound.

Fraction 2: Bis-3,3′-(nitro-NNO-azoxy)difurazanyl ether (**2a-35**) was afforded as colorless crystals. T_m, 66 °C; ^{13}C NMR ($CDCl_3$): d.147.2 ($C \pm N_3O_3$), 154.3 ($C \pm O$); ^{14}N NMR ($CDCl_3$): 46.8 (NO_2), 71.5 (N=O); $C_4N_{10}O_9$ (332.11): calcd C 14.46, N 42.18; found C 14.42, N 42.05.

3-Amino-4-(5-methyl-1,2,4-oxadiazol-3-yl)furazan (**AMOF**): To prepare **AMOF** (4-amino-N′-hydroxy-1,2,5-oxadiazole-3-carboximidamide), a mixture of **AAOF** (4-amino-N′-hydroxy-1,2,5-oxadiazole-3-carboximidamide) (2.003 g, 14.0 mmol) and acetamide (1.654 g, 28.0 mmol) was heated at 180 °C for a duration of 7 h. After the heating process, the resulting mixture was poured into 100 mL of water, forming a suspension. The suspension was then heated to its boiling point and subsequently filtered while still hot. The filtrate obtained from the filtration was allowed to cool, leading to the precipitation of **AMOF**. The precipitate was then filtered off and dried at 50 °C. The yield of the final product, **AMOF**, was found to be 1.731 g (74% yield).

4-(5-Methyl-1,2,4-oxadiazol-3-yl)-N-(2,2,2-trinitroethyl)-1,2,5-oxadiazol-3-amine (**2a-37**): At ambient temperature, 3-amino-4-(5-methyl-1,2,4-oxadiazol-3-yl)furazan (0.334 g, 2.0 mmol) was dissolved in a 20 mL solution of dilute hydrochloric acid with a concentration of 0.1 M. Following this, 2,2,2-trinitroethanol (0.905 g, 5.0 mmol) was added to the reaction solution. The reaction mixture was then heated to 50 °C and stirred for a period of 2 h. Upon completion of the reaction, the mixture was allowed to cool to room temperature, resulting in the formation of a yellow viscous liquid. To purify the compound, the viscous liquid was washed with a large amount of water and subsequently dried. This process yielded pure compound **2a-37** with a weight of 1.078 g (85.3% yield). T_m, 125 °C; T_d, 200 °C; ^1H NMR (300 MHz, $CDCl_3$): d = 6.36 (t, J = 6.0 Hz, 1H, –NH–), d = 5.11 (d, J = 6.0 Hz, 2H, $–CH_2–$) ppm, 2.77 (s, 3H, $–CH_3$); ^{13}C NMR (75 MHz, acetone-d_6): d = 178.93, 158.92, 155.39, 136.75, 125.98, 48.10, 11.28 ppm; IR: n = 3009, 1590, 1546, 1464, 1398, 1350, 1276, 1193, 1113, 1022, 1084, 977, 929, 854, 824, 806, 743, 698 cm^{-1}; elemental analysis calcd (%) for $C_7H_6N_8O_8$ (330.17): calcd C 25.46, H 1.83, N 33.94; found: C 24.68, H 1.92, N 33.24.

N-(4-(5-methyl-oxadiazol-3-yl)-oxadiazol-3-yl)-N-(trinitroethyl)nitramide (**2a-38**): To initiate the nitration reaction, **2a-37** (0.660 g, 2.0 mmol) was added to a vigorously stirred mixture of nitric acid (10 mL, 100%) and acetic anhydride (10 mL) at a temperature of − 5 °C. The resulting reaction mixture was then stirred at 0 °C for a duration of 1 h. After completion of the reaction, the mixture was poured into ice water (50 mL). A white solid formed, which was immediately filtered and washed with cold water. Within several minutes, the white solid transformed into a yellow viscous solid, identified as **2a-38**. The weight of the obtained product was 0.625 g, corresponding to a yield of 83.3%. T_m, 25 °C; T_d, 172 °C; ^1H NMR (300 MHz, $CDCl_3$): d = 5.93 (s, 2H, $-CH_2-$), 2.74 (s, 3H, $-CH_3$) ppm; ^{13}C NMR (75 MHz, $CDCl_3$): d = 179.21, 157.39, 150.88, 142.86, 121.63, 53.07, 12.29 ppm; IR: n = 3669, 2978, 2901, 2378, 2351, 2323, 1992, 1587, 1462, 1339, 1349, 1277, 1194, 1111, 1079, 1047, 975, 928, 855, 803, 725, 608, 563 cm^{-1}; elemental analysis calcd (%) for $C_7H_5N_9O_{10}$ (375.17): calcd C 22.41, H 1.34, N 33.60; found: C 22.09, H 1.26, N 32.59.

N-(4-(5-methyl-1,2,4-oxadiazol-3-yl)-1,2,5-oxadiazol-3-yl)nitramide (**2a-39**): To initiate the reaction, nitric acid (100%, 20 mL) was placed in a 50 mL two-necked round bottom flask and cooled to − 10 °C. Slowly, 3-amino-4-(5-methyl-1,2,4-oxadiazol-3-yl)furazan (0.334 g, 2.0 mmol) was added to the cooled nitric acid while maintaining the temperature below − 10 °C. The resulting reaction mixture was stirred for 2 h and then poured into ice water (80 mL). The mixture was transferred to a refrigerator and left overnight to allow the precipitation of **2a-39**. The precipitate was then filtered, washed with cold water, and dried in ambient air. The final yield of **2a-39** was determined to be 0.642 g, corresponding to an 87.2% yield. T_d, 159 °C; ^1H NMR (300 MHz, $CDCl_3$): d = 11.55 (br, 1H, –NH–), 2.85 (s, 3H, $-CH_3$) ppm; ^{13}C NMR (75 MHz, $CDCl_3$): d = 179.10, 157.87, 147.45, 137.29, 12.41; IR: n = 3122, 3075, 3007, 2972, 2845, 2741, 1623, 1610, 1583, 1544, 1485, 1452, 1369, 1323, 1300, 1267, 1152, 1040, 1010, 987, 968, 931, 915, 889, 839, 760, 740, 707, 632, 592 cm^{-1}; elemental analysis calcd (%) for $C_5H_4N_6O_4$ (212.13): calcd C 28.31, H 1.90, N 39.62; found: C 28.62, H 2.01, N 39.16.

4-Amino-3-(5-acetic acid methyl ester-1,2,4-oxadiazolyl)- furazan (**2a-41**): Method A: In this procedure, a suspension of 4-amino-1,2,5-oxadiazole-3-carboxamidoxime (1.43 g, 10 mmol) in 20 mL of acetonitrile was prepared. Methyl malonyl chloride (1.37 g, 10 mmol) was added to the suspension, resulting in the formation of a clear solution. After a few minutes, a solid precipitated from the solution. The reaction mixture was stirred for 10 min, and then potassium carbonate (2.76 g, 20 mmol) was added all at once. The reaction mixture was heated to 50 °C and maintained at this temperature for a duration of 24 h. After the reaction was complete, the mixture was cooled to room temperature and the solvent was removed under vacuum. The remaining solid residue was added to 80 g of ice water, and the resulting precipitate was collected by filtration and air dried. The yield of compound 1 obtained from this process was 1.63 g, corresponding to a yield of 72.4%.

Method B: A mixture containing 4-amino-1,2,5-oxadiazole-3-carboxamidoxime (4.29 g, 30 mmol) and dimethyl malonate (19.8 g, 150 mmol) was subjected to heating at a temperature of 150 °C. During the reaction, the H_2O and MeOH byproducts were collected using a Dean-Stark apparatus. The heating process was continued for a duration of 24 h. After completion of the heating, the reaction mixture was allowed to cool to room temperature. Excess dimethyl malonate was removed by blowing air over the mixture. Following this, water (50 mL) was added to the remaining solid residue. The mixture was then stirred for 1 h, and subsequently, the solid was collected via filtration. The resulting product, referred to as compound **2a-41**, had a yield of 4.58 g, corresponding to a yield of 67.8%. Notably, the ^{1}H and ^{13}C NMR spectra of this product exhibited characteristics identical to those obtained using method A. T_m: 73.4; T_d: 270.9 °C; IR (cm^{-1}) n 3429, 3335, 2989, 2958, 1735, 1697, 1637, 1608, 1590, 1555, 1463, 1437, 1408, 1391, 1350, 1291, 1207, 1184, 1164, 1149, 1014, 993, 969, 944, 902, 865, 810, 766, 734, 703, 648, 575, 524; ^{1}H NMR (CD_3CN): 5.46 (2H, s, NH_2); 4.21 (2H, s, CH_2); 3.75 (3H, s, CH_3); ^{13}C NMR (CD_3CN): d 175.97, 167.32, 161.00, 156.54, 138.20, 53.72, 33.65. EA ($C_7H_7N_5O_4$, 225.16): calcd, C, 37.34; H, 3.13; N, 31.10; found, C: 37.31; H: 3.08; N: 31.07.

4-Nitramino-3-(5-(2,2′-dinitro)-acetic acid methyl ester-1,2,4-oxadiazolyl)-furazan (**2a-42**): To begin the reaction, compound **2a-41** (0.6 g, 2.7 mmol) was suspended in 6 mL of concentrated H_2SO_4 at a temperature of − 5 °C. Then, 100% HNO_3 (2.3 mL) was added dropwise to the suspension. The resulting reaction mixture was stirred at 0 °C for a duration of 30 min. Afterward, the reaction mixture was slowly warmed to room temperature and stirred for an additional 2 h. A white solid precipitated from the mixture, which was collected by filtration. The collected solid was washed with trifluoroacetic acid (3 × 2 mL) and subsequently air dried. This process resulted in the formation of compound **2a-42**, with a weight of 0.84 g, corresponding to a yield of 87.5%. T_d: 88.2 °C; IR (cm^{-1}) n 3273, 2971, 2924, 1776, 1622, 1594, 1471, 1438, 1317, 1303, 1260, 1167, 1082, 1044, 1029, 972, 960, 920, 842, 807, 750, 720, 622, 611, 509. ^{1}H NMR (CD_3CN): 11.76 (1H, br, NH); 4.16 (3H, s, CH_3); ^{13}C NMR(CD_3CN): d 166.73, 160.34, 155.68, 149.33, 142.64, 110.76, 58.68. EA ($C_7H_4N_8O_{10}$, 360.15): C, 23.34; H, 1.12; N, 31.11; found, C: 23.16; H: 1.12; N: 28.45.

3-Amino-4-(5-amino-1,2,4-oxadiazol-3-yl)furazan (**2a-43**): A solution of 8 g (80 mmol) of potassium bicarbonate ($KHCO_3$) in 30 mL of water was added to a suspension containing 10 g (70 mmol) of amide oxime **AAOF** in 50 mL of alcohol. Subsequently, 7.42 g (70 mmol) of cyanogen bromide was added to the mixture in portions. The resulting mixture was stirred for a duration of 24 h at a temperature ranging between 15 and 20 °C. After completion of the reaction, the precipitate formed was filtered off, washed with water, and dried. The final yield of the product obtained was determined to be 10.5 g, corresponding to an 89% yield, $T_m > 275$ °C (decomp.; from water). ^{1}H NMR spectrum (DMSO-d_6), d, ppm: 8.35 s (2H, NH_2), 6.34 s (2H, NH_2). ^{13}C NMR spectrum (DMSO-d_6), d, ppm: 173.2, 160.0, 156.19, 138.4. Found, %: C 28.51; H 2.48; N 50.08. $C_4H_4N_6O_2$. Calculated, %: C 28.58; H 2.40; N 5.00.

3-(4-Nitro-1,2,5-oxadiazol-3-yl)-1,2,4-oxadiazol-5-amine (**2a-44**): In an ice-salt bath at − 10 °C, a three-necked round-bottomed flask equipped with a stirrer and dropping funnel was charged with sodium tungstate dihydrate (2.20 g, 6.7 mmol). Next, 50% hydrogen peroxide solution (20.00 g, 0.29 mol) was added to the flask. Under stirring at − 5 °C, sulfuric acid (20.00 g, 0.21 mol) was added drop-wise to the reaction mixture. The temperature was then allowed to warm to room temperature. Slowly, compound **2a-43** (0.84 g, 5.0 mmol) was added to the reaction mixture. Once **2a-43** was completely added, the reaction mixture was stirred at room temperature for an additional 24 h. The resulting mixture was then subjected to extraction with ethyl acetate (50 mL × 4). The combined organic extracts were subsequently washed with ice-water. The white precipitate obtained was concentrated using a rotary evaporator, resulting in the formation of solid **2a-44** (0.79 g, 4 mmol) with a yield of 80%. T_d: 89 °C. ^{1}H NMR (500 MHz, DMSO-d_6): δ 7.91 (s, 2H) ppm. ^{13}C NMR (500 MHz, DMSO-d_6): δ 139.58, 155.40, 158.01, 172.06 ppm. IR (KBr): $\tilde{\nu}$ 3431, 3353, 3257, 3159, 1673, 1607, 1557, 1537, 1468, 1386, 1353, 1309, 1137 cm^{-1}. Elemental analysis for $C_4H_2N_6O_6$ (198.10): calcd C 24.25, H 1.02, N 42.42%. Found: C 24.23, H 1.01, N 42.45%.

4-Amino-3-(5-amino-1,2,4-oxadiazolyl) furoxan (**2a-45**): A solution of acetic acid (4.00 mL, 70.00 mmol) was added to a mixture of malononitrile (3.00 g, 45.4 mmol) and $NaNO_2$ (3.70 g, 53.6 mmol) in 50 mL of water at a temperature range of − 5 to 0 °C. The resulting mixture was stirred at 20 °C for a duration of 4 h. Afterward, a solution containing $NH_2OH{\cdot}HCl$ (3.15 g, 45.4 mmol) in 15 mL of water was added to the reaction mixture at 25 °C. The mixture was then stirred for an additional 4 h at 25 °C. The resulting precipitate was filtered off, washed with ice-water, and air-dried. This process yielded a yellowish solid known as 2-cyano-N-hydroxy-2-(hydroxyimino)ethanimidamide (68.2% yield). The crude product was used directly without undergoing further purification. Next, a solution of 2-cyano-N-hydroxy-2-(hydroxyimino)ethanimidamide (5.30 g, 41.30 mmol) in 125 mL of diethyl ether was prepared. Acetic acid (10 mL, 17.49 mmol) was added to this solution. In portions, PbO_2 (35.00 g, 146.00 mmol) was added to the mixture. The reaction mixture was stirred for 4 h at room temperature. The resulting precipitate was filtered off, and the filtrate was washed with water (100 mL × 3) and dried using magnesium sulfate. The yellow precipitate obtained from the drying process was concentrated using a rotary evaporator, resulting in the formation of a yellow solid weighing 3.12 g. Subsequently, a 50% hydroxylamine solution was added under stirring to a suspension of the yellow solid (3.12 g) in 30 mL of water. The reaction mixture was stirred for 2 h at room temperature. The precipitate was filtered off, washed with water, and air-dried. This process yielded a light-yellow powder known as 4-amino-3-aminooxim furoxan (5.24 g) with a yield of 79.6%. ^{1}H NMR (500 MHz, DMSO-d_6): δ 10.52 (1H, s) 6.45 (2H, s) 6.13 (2H, s) ppm, ^{13}C NMR (50.69 MHz, DMSO-d_6): δ155.87, 141.94, 104.42 ppm. Finally, pdotassium bicarbonate (1.00 g, 10 mmol) was dissolved in 7 mL water and 4-amino-3-aminooxim furoxan (1.27 g, 8 mmol) was added. The cyanogen bromide (0.85 g, 8 mmol) was added in portions to the mixture. The reaction mixture was stirred for 72 h and the precipitate was filtered off and washed extensively with water and diethyl ether. Compound **2a-45** was obtained as yellowish-brown powder (1.13 g) in the yield of 76.8%. ^{1}H NMR (500 MHz, DMSO-d_6): δ 8.36 (2H, s) 6.45 (2H, s) ppm, ^{13}C NMR (50.69 MHz, DMSO-d_6): 171.91, 157.26, 155.92, 101.38 ppm.

4-Nitro-3-(5-amino-1,2,4-oxadiazol-3-yl)furoxan (**2a-46**): Ammonium persulphate (1.53 g, 6.7 mmol) was added slowly to 30% hydrogen peroxide solution (32.87 g, 0.29 mol) at 0 °C. Sulfuric acid (20.00 g, 0.21 mol) was added drop-wise to the solution under stirring at 5 °C. Then, allow the reaction mixture warm to room temperature and compound **2a-45** (0.92 g, 5.0 mmol) were added slowly. Then the reaction mixture was stirred at room temperature for another 4 h and extracted with ethyl acetate (50 mL × 4). The combined extracts were washed with ice-water and dried on magnesium sulfate. The solvent was removed under reduced pressure to leave a yellow solid (0.94 g) in a yield of 88.6%. ^{1}H NMR (500 MHz, DMSO-d_6): δ 7.15 (2H, s) ppm, ^{13}C NMR (50.69 MHz, DMSO-d_6): δ 102.99, 156.26, 158.68, 173.90 ppm. IR (KBr): n 3168, 3141, 2953, 2927, 2855, 1678, 1634, 1572, 1503, 1386, 1342, 1314, 1261, 1229, 1141, 1097, 1070, 1049, 991, 960, 908, 847, 790, 760, 730, 692, 663 cm^{-1}. Elemental analysis for $C_4H_2N_6O_5$ (214.10): calcd C 22.44, H 0.94, N 39.25. Found: C 22.48, H 0.91, N 39.21.

3,3′-Bis(1,2,4-oxadiazole)]-5′-diylbis(methylene) Dinitrate (**2a-47**): A 2 L round-bottom flask holding a stir bar was filled with 1 L of acetonitrile, compound **2a-46** (15.0 g, 0.127 mol), and K_2CO_3 (35.1 g, 0.254 mol) in that order. Next, a pressure-equalizing liquid addition funnel was attached to the flask and a solution containing acetoxy acetyl chloride (35.4 g, 27.9 mL, 0.254 mol) in 200 mL of CH_3CN was slowly introduced over two hours. Stirring of the mixture was maintained overnight at room temperature. Subsequently, the flask was fitted with a reflux condenser and was heated until it reached boiling point. After stirring for 4 h, the reaction mixture was cooled to ambient temperature. The resultant solid was isolated by Buchner filtration and discarded. The remaining liquid was shifted into a separate 2 L round-bottom flask and condensed under vacuum conditions to yield a crude solid product. This raw product underwent purification by washing it with 500 mL of H_2O. The washed solids were gathered again using Buchner filtration and dried in an oven at 60 °C throughout the night. This process yielded 24.4 g of [3,3′-bis(1,2,4-oxadiazole)]-5,5′-diylbis(methylene) diacetate (68% yield), appearing as a white powder. T_m = 107.5 °C; ^{1}H NMR (90 MHz, DMSO-d_6): 5.53 (s, 4H), 2.18 (s, 6H); ^{13}C NMR (90 MHz, DMSO-d_6): 176.91, 169.84, 159.37, 56.45, 20.09; IR (neat) cm^{-1} 1745.23 (s), 1577.13 (m), 1208.08 (s).

A 250 mL round-bottom flask fitted with a stir bar was filled in order with 50 mL of MeOH, 3,3′-bis(1,2,4-oxadiazole)]-5,5′-diylbis(methylene) diacetate (24.0 g, 85.1 mmol), and K_2CO_3 (1.18 g, 8.51 mmol). The mixture was stirred at room temperature for a period of 48 h. Subsequently, it was concentrated under vacuum which produced a crude solid. This solid underwent purification by rinsing with 200 mL of H_2O. The cleansed solid was isolated using Buchner filtration and then dried in an oven set to 60 °C throughout the night. This procedure yielded 16.0 g of [3,3′-bis(1,2,4-oxadiazole)]-5,5′-diyldimethanol (95% yield) which was identified as a white powdery substance. T_m = 197.8 °C; T_d = 231.5 °C (onset), 278.2 °C (peak); ^{1}H NMR (90 MHz, DMSO-d_6): 6.16 (s, 2H), 4.87 (s, 4H); ^{13}C NMR (90 MHz, DMSO-d_6): 178.50, 156.94, 52.74; IR (neat) cm^{-1} 3339.57 (w), 1575.77 (m), 1431.04 (m), 1202.56 (m), 1081.27 (s).

A 250 mL round-bottom flask set in an ice bath was filled with 100 mL of 100% HNO_3. After the acid cooled to 0 °C, [3,3′-bis(1,2,4-oxadiazole)]-5,5′-diyldimethanol (20.0 g, 101.0 mmol) was gradually added in four equal segments over an hour. Following the completion of addition, the reaction mixture was stirred for four hours, allowing the ice bath to naturally thaw and the reaction mixture to attain room temperature. Subsequently, this mixture was introduced onto crushed ice while stirring continuously. After an hour, the solid product was separated through Buchner filtration and allowed to air dry in a well-ventilated fume hood. The final product was 24.7 g of dinitrate **2a-47** (85% yield) as a white powder. $T_m = 84.5$ °C; $T_d = 183.4$ °C (onset), 214.6 °C (peak); ^{1}H NMR (90 MHz, DMSO-d_6): 6.16 (s, 4H); ^{13}C NMR (90 MHz, DMSO-d_6): 175.48, 159.43, 64.26; IR (neat) cm^{-1} 2950.51 (w), 1653.16 (s), 1574.20 (m), 1415.61 (m), 1350.42 (m), 1282.23 (s), 1212.83 (s), 1059.18 (m).

3,3′-Bis(1,2,4-oxadiazolyl)-5,5′-bis(2,2′-dinitro)diacetic acid diethyl ester: Compound 3,3′-bis(1,2,4-oxadiazolyl)-5,5′-diacetic acid diethyl ester (**2a-48**, 12 mmol) was dissolved in a solution of 96% H_2SO_4 (34 mL) and cooled to 0 °C. Further chilling brought the temperature down to – 5 °C, at which point a solution of 100% nitric acid (14 mL) was incrementally introduced. The system's temperature was then raised to 10 °C, and stirring persisted for 3 h within the range of 10–15 °C. Post this, the mixture was set to cool to – 15 °C, filtered, and rinsed with trifluoroacetic acid as well as chilled water. Eventually, a colorless yield was obtained as a solid substance. The quantity achieved was 5.70 g (97%), forming a colorless solid. T_m: (DSC, 5 °C/min^{-1}) 72 °C, T_d: 286 °C; ^{1}H NMR (400 MHz, [D_6]DMSO, 25 °C): d = 1.06 (t, J = 6.87 Hz, 6H, CH_3), 3.38 ppm (q, J = 6.87, 7.15 Hz, 4H, CH_2CH_3); ^{13}C NMR (101 MHz, [D_6]DMSO, 25 °C): d = 19.0 (s, 2C, CH_3), 56.5 (s, 2C, CH_2CH_3), 124.7 (s, 2C, $C(NO_2)_2$), 160.4 (s, 2C, CC), 172.9 (s, 2C, $CC(NO_2)_2$), 173.2 ppm (s, 2C, CO); ^{14}N NMR (29 MHz, [D_6]DMSO, 25 °C): d = – 18 ppm (s, 4N, NO_2); IR: ~n = 3007 (w), 2948 (w) 1780 (vs), 1739 (w), 1605 (vs), 1592 (vs), 1559 (m), 1472 (w), 1456 (w), 1445 (w), 1415 (w), 1368 (w), 1351 (w), 1299 (m), 1280 (m), 1236 (s), 1220 (vs), 1159 (m), 1107 (w), 1087 (m), 1025 (m), 987 (m), 972 (s), 915 (m), 849 (s), 840 (vs), 817 (s), 795 (s), 724 (m), 693 (m), 665 cm^{-1} (m); elemental analysis calcd (%) for $C_{12}H_{10}N_8O_{14}$ (490.25): C 29.40, H 2.06, N 22.86; found: C 28.91, H 1.98, N 22.40.

5,5′-Bis(fluorodinitromethyl)-3,3′-bi(1,2,4-oxadiazole) (**2a-50**): Diammonium 5,5′-bis(dinitromethanide)-3,3′-bi(1,2,4-oxadiazole) (0.40 g, 0.95 mmol) was mixed with anhydrous acetonitrile (100 mL) and Selectfluor (1.34 g, 3.79 mmol) was introduced at room temperature. The result was a yellow mixture that was stirred until the color disappeared, which took around 2 h. The solvent was then evaporated under reduced pressure. The residual yellowish substance was combined with anhydrous diethyl ether (150 mL) and this new mixture was warmed to 40 °C while being stirred in a sealed container for half an hour. After filtration of the solid and condensing the solvent under reduced pressure to about 5 mL, a crystalline solid formed overnight at a temperature of 8 °C. Upon complete removal of the solvent, the residue yielded compound **2a-50** (240 mg, 0.63 mmol, 66%) as colorless crystals appropriate for X-ray diffraction. DSC (5 °C min^{-1}, T_{onset}): $T_m = 67$ °C, $T_d = 151$ °C; IR: n = 1620 (s), 1572 (m), 1445 (w), 1363 (w), 1334 (w), 1297 (m), 1259 (m), 1241 (m), 1212 (m), 1065 (m), 1016 (s), 954 (m), 917 (w), 831 (m), 793 (s), 744 (w), 716 (m), 700 (m), 661 cm^{-1} (w); Raman (1064 nm, 300 mW): n = 1628 (15), 1597 (100), 1586 (44), 1424 (31), 1335 (10), 1244 (5), 1016 (9), 951 (24), 922 (20), 833 (24), 627 (5), 493 (15), 404 (16), 368 (13), 340 (9), 318 cm^{-1} (9); ^{13}C NMR (CD_2Cl_2): d = 164.8 (d, 2JC, F = 28.8 Hz, OCN), 159.9 (CC), 111.9 ppm (dquint, JC,F = 297.1 Hz, JC,

N = 9.6 Hz, $C(NO_2)_2F$); ^{14}N NMR (CD_2Cl_2): d = – 34 ppm (d, 2JN, F = 9.9 Hz, NO_2); ^{19}F NMR (CD_2Cl_2): d = – 98.93 ppm (quint, 2JF, N = 9.8 Hz, CF); MS (DCI+): m/z: 383.2 [M+H]+; elemental analysis calcd (%) for $C_6N_8O_{10}F_2$: C 18.86, N 29.33; found: C 19.25, N 28.18.

5,5′-Bis(trinitromethyl)-3,3′-bi(1,2,4-oxadiazole) (**2a-51**): NaOH (500 mg, 12.05 mmol) was combined with a suspension of 3,3′-bis(1,2,4-oxadiazolyl)-5,5′-bis(2,2′-dinitro)diacetic acid diethyl ester (500 mg, 1.61 mmol) dispersed in water (25 mL). This combination was stirred for an hour at 30 °C. The concoction was subsequently made acidic to reach a pH level of 3 using aqueous HNO_3 (60%), and then condensed under reduced pressure. The mixture was chilled through placement in an ice bath, leading to the formation of a precipitate. This solid material was separated out by filtration which resulted in compound 3,3′-bi(1,2,4-oxadiazolyl)-5,5′-diacetic acid. It appeared as a colorless solid and its yield was estimated to be 310 mg (1.22 mmol, 76%). DSC (5 °C min^{-1}): T_m = 160 °C, T_d = 236 °C; IR: n = 2986 (w), 2617 (w), 1707 (s), 1585 (m), 1432 (m), 1404 (m), 1330 (m), 1300 (m), 1272 (w), 1233 (s), 1202 (m), 1161 (m), 1022 (m), 956 (w), 933 (m), 904 (m), 825 (m), 713 (m), 676 cm^{-1} (s); ^1H NMR ([D_6]DMSO): d = 13.33 (s, 2H, OH), 4.31 ppm (s, 4H, CH_2); ^{13}C NMR ([D_6]DMSO): d = 176.6 (OCN), 168.0 (COO), 160.1 (CC), 33.4 ppm (CH_2); elemental analysis calcd (%) for $C_8H_6N_4O_6$: C 37.81, N 22.04, H 2.38; found: C 37.83, N 22.34, H 2.41.

Method 1: Diammonium 5,5′-bis(dinitromethanide)-3,3′-bi(1,2,4-oxadiazole) (250 mg, 0.66 mmol) was incrementally added over a span of 10 min to a solution of NO_2BF_4 (200 mg, 1.50 mmol) which was already dissolved in anhydrous acetonitrile (100 mL) and cooled down to – 10 °C. The resultant mixture was stirred for an hour at 0 °C followed by the removal of the solvent under reduced pressure. The yellowish solid that persisted had Et_2O (50 mL) introduced to it. This mixture was then heated to 40 °C while stirring within a sealed container for half an hour. Post filtration of the solid and concentration of the solvent under reduced pressure to around 3 mL, a crystalline solid was formed after leaving it overnight at 8 °C. After complete evaporation of the solvent, the product obtained was compound **2a-51** (80 mg, 0.18 mmol, 27%). It manifested as colorless crystals.

Method 2: Highly concentrated H_2SO_4 (6.6 mL, 96%) was chilled to –10 °C, and then HNO_3 (2.0 mL, 100%) was slowly introduced. Following this, the compound 3,3′-bi(1,2,4-oxadiazolyl)-5,5′-diacetic acid (250 mg, 0.98 mmol) was dispersed in small segments over half an hour. The flask was fitted with a drying tube and stirring of the reaction mixture persisted for 16 h while it naturally warmed up to room temperature. This concoction was then transferred onto ice (40 g), leading to the formation of a solid precipitate which was separated via filtration and rinsed with icy water. After drying under reduced pressure, the residue yielded compound **2a-51** (380 mg, 0.87 mmol, 89%), a colorless solid. Recrystallizing from Et_2O or CH_2Cl_2 resulted in single crystals appropriate for X-ray diffraction. DSC (5 °C min^{-1}, Tonset): T_m = 119 °C, T_d = 124 °C; IR: n = 1710 (w), 1608 (s), 1561 (m), 1405 (w), 1347 (w), 1291 (m), 1271 (s), 1227 (s), 1139 (s), 1072 (s), 1038 (s), 975 (m), 962 (m), 933 (m), 904 (s), 839 (s), 792 (s), 725 (m), 668 (m), 656 cm^{-1} (m); Raman (1064 nm, 300 mW, 25 °C): n = 1621 (23), 1595 (100), 1573 (45), 1415 (32), 1349 (8), 1294 (10), 1275

(11), 1097 (4), 1023 (11), 960 (29), 933 (16), 844 (45), 799 (4), 769 (5), 627 (7), 518 cm^{-1} (12); ^{13}C NMR (CD_2Cl_2): d = 163.6 (OCN), 159.9 (CC), 117.5 ppm (sept, JC, N = 9.4 Hz, $C(NO_2)_3$); ^{14}N NMR (CD_2Cl_2): d = – 43 ppm (NO_2); MS (DCI+): m/z: 437.3 [M+H]+; elemental analysis calcd (%) for $C_6N_{10}O_{14}$: C 16.52, N 32.12; found: C 16.52, N 31.02.

5,5′-Dimethyl-3,3′-bi(1,2,4-oxadiazole) (**2a-52**): The compound 3,3′-Bi(1,2,4-oxadiazolyl)-5,5′-diacetic acid was mixed with DMF and warmed up to 60 °C for an hour duration. The emergence of CO_2 was verified using a $Ba(OH)_2$ solution. Overnight crystallization at 8 °C resulted in the quantitative formation of compound **2a-52**. DSC (5 °C min^{-1}): T_m = 166 °C, T_d = 266 °C; IR: n = 3020 (w), 2932 (w), 2621 (w), 1931 (w), 1576 (s), 1544 (w), 1438 (m), 1413 (m), 1395 (w), 1374 (m), 1265 (s), 1218 (s), 1113 (w), 1047 (m), 1036 (m), 976 (m), 949 (m), 908 (s), 701 (s), 680 cm^{-1} (m); ^1H NMR (CD_3CN): d = 2.67 ppm (s, 6H, CH_3); ^{13}C NMR (CD_3CN): d = 179.0 (OCN), 160.1 (CC), 11.7 ppm (CH_3); MS (DEI+): m/z: 166.2 [M]+; elemental analysis calcd (%) for $C_6H_6N_4O_2$: C 43.38, N 33.72, H 3.64; found: C 43.27, N 33.44, H 3.54.

Bis(5-(fluorodinitromethyl)-1,2,4-oxadiazol-3-yl) methane (**2a-54**): Diethyl 2,2′-(methylenebis(1,2,4-oxadiazole-3,5-diyl))diacetate (324 mg, 1 mmol) was added into a 96% H_2SO_4 solution (2.5 mL) at 0 °C. Following further cooling to 0–5 °C, concentrated HNO_3 (1.0 mL) was introduced bit by bit. This reaction mixture then gradually warmed up to room temperature and was stirred for approximately 12 h. Post stirring, the concoction was neutralized using ice water and an extraction process was carried out thrice using ethyl acetate. The unified organic phase was cleaned with aqueous $NaHCO_3$ and brine, desiccated over anhydrous sodium sulphate, and then evaporated under diminished pressure to acquire crude product as a yellow oil. Without any additional purification, the oil was mixed with an ammonia solution in methanol at ambient temperature. After being stirred for 5 h, the resulting yellow precipitate was gathered via filtration, air dried to yield ammonium (methylenebis(1,2,4-oxadiazole-3,5-diyl))bis(di-nitromethanide) (341 mg, 86.5%), which appeared as a yellow solid. T_d (onset) = 137 °C. ^1H NMR (400 MHz, DMSO-d_6): δ = 4.42 (s, 2H, CH_2), 7.07 (s, 8H, NH_4) ppm. ^{13}C NMR (100 MHz, DMSO-d_6): δ = 24.16 (s, 1C, CH_2), 122.51 (s, 2C, $C(NO_2)_2$), 166.78 (s, 2C, NCN), 171.47 (s, 2C, OCN) ppm. IR: 3226 (s), 3007 (w), 1749 (w), 1591 (s), 1489 (w), 1419 (m), 1247 (vs), 1134 (vs), 1109 (s), 1022 (w), 979 (m), 829 (m), 781 (s), 750 (vs) cm^{-1}; HRMS: calc. for $C_7H_2N_8O_{10}$ $[M-N_2H_8]^{2-}$: 178.9952, found: 178.9959, calc. for $C_7H_3N_8O_{10}$ $[M-N_2H_7]^-$: 358.9978, found: 358.9973.

Ammonium (methylenebis(1,2,4-oxadiazole-3,5-diyl))bis(di-nitromethanide) compound (47 mg, 0.12 mmol) was mixed with dry CH_3CN (3.0 mL), and then Selectfluor (170 mg, 0.48 mmol) was introduced at a temperature of 25 °C. The mixture initially yellow, turned colorless and clear following a stirring process that lasted for around 20 min and an additional 3 h. Post solvent evaporation, the residue was extracted using CH_2Cl_2. After purification through column chromatography, a pure yellow oil known as **2a-54** (40 mg, 84.2%) was acquired. T_d (onset) = 166. ^1H NMR (400 MHz, DMSO-d_6): δ = 4.89 ppm (s, 2H, CH_2). ^{13}C NMR (100 MHz, DMSO-d_6): δ = 23.50(s, CH_2), 110.70(d, 1 JC, F = 292.0 Hz, $C(NO_2)_2F$), 163.33

(d, 2 JC, F = 27.0 Hz, OCN), 166.13 (NCN) ppm. ^{19}F NMR (376 MHz, DMSO-d_6): δ = – 103.11 ppm. (s, CF). IR: 2893 (w), 1610 (vs), 1577 (m), 1517 (w), 1415 (m), 1394 (w), 1340 (w), 1274 (s), 1240 (m), 1212 (m), 1188 (m), 1099 (s), 966 (m), 840 (vs), 796 (vs), 750 (w), 711 (m), 657 (w) cm^{-1}; HRMS: Calc. for $C_7HF_2N_8O_{10}$ [M-H]$^-$: 394.9789, found: 394.9794.

Bis(5-(trinitromethyl)-1,2,4-oxadiazol-3-yl)methane (**2a-55**): Dihydroxymalonimidamide (6.61 g, 50 mmol) was suspended in 100 mL of CH_3CN, after which methyl malonyl chloride (15.81 g, 105 mmol) was added. As the mixture stirred at a temperature between 0 and 10 °C, it transitioned from a suspension to a yellow solution containing solids. DBU (22.84 g, 150 mmol) was incorporated in a single addition, after which the reaction mixture's temperature was increased to 50 °C and this heat was sustained for a 6 h period. Once the mixture had cooled down to room temperature, the solvent was evacuated under vacuum. The remaining residue was then put through column chromatography, resulting in the production of diethyl 2,2′-(methylenebis(1,2,4-oxadiazole-3,5-diyl)) diacetate (3.21 g, 19.8%) as a colorless solid. T_m (peak) = 44 °C. ^{1}H NMR (400 MHz, DMSO-d_6): δ = 4.38 (s, 2H, CH_2), 4.28 (s, 4H, CH_2CO), 4.13 (q, 4H, CH_2CH_3), 1.19 (t, 6H, CH_2CH_3). ^{13}C NMR (100 MHz, DMSO-d_6): δ = 14.36 (s, 1C, CH_2CH_3), 23.45 (s, 1C, CCH_2C), 32.94 (s, 1C, CCH_2CO), 61.95 (s, 1C, CH_2CH_3), 166.50 (s, 2C, OCN), 166.74 (s, 2C, COO), 174.46 (s, 2C, NCN) ppm. IR: 2991 (m), 2945 (w), 2906 (w), 1747 (vs), 1589 (vs), 1519 (w), 1475 (m), 1404 (w), 1375 (m), 1301 (w), 1280 (w), 1195 (s), 1172 (s), 1112 (w), 1047 (m), 1025 (m), 939 (s), 893 (m), 813 (m), 761 (m), 719 (w), 677 (w), 628 (w), 580 (m) cm^{-1}.

NaOH (3.0 M, 4.0 mL) was introduced into a solution containing diethyl 2,2′-(methylenebis(1,2,4-oxadiazole-3,5-diyl)) diacetate (648 mg, 2 mmol) in water (6.0 mL). Stirred for 5 h at room temperature, this mixture was then acidified to pH 3 using hydrochloric acid and subjected to three rounds of extraction with ethyl acetate. The collected organic phases were rinsed with aqueous sodium bicarbonate and brine, desiccated over anhydrous Na_2SO_4 and concentrated under diminished pressure. This produced the intermediate compound 2,2′-(methylenebis(1,2,4-oxadiazole-3,5-diyl))diacetic acid (516 mg, 96.3%), which did not undergo further purification. Subsequently, it (295 mg, 1.10 mmol) was added gradually over 15 min to a cooled mixture (0 °C) of fuming HNO_3 (2.0 mL) and 96% H_2SO_4 (5.0 mL). The mixture stirred for 6 h at room temperature before being poured into ice water (40 mL), forming a precipitate that was filtered off and rinsed with chilled water to yield compound **2a-55** (156 mg, 31.5%) as a white powder. T_m (peak) = 52 °C. T_d (onset) = 117 °C. ^{1}H NMR (400 MHz, $ClCD_3$): δ = 4.65 (s, 2H, CH_2). ^{13}C NMR (100 MHz, $ClCD_3$): δ = 23.9 (s, 1C, CH_2), 116.4 (s, 2C, $C(NO_2)_3$), 162.4 (s, 2C, NCN), 166.2 (s, 2C, OCN) ppm. IR: 2916 (w), 1504 (vs), 1516 (w), 1415 (w), 1338 (w), 1274 (s), 1097 (s), 960 (m), 840 (s), 798 cm^{-1} (vs); HRMS: calc. for $C_7HN_{10}O_{14}$ [M-H]$^-$: 448.9679, found: 448.9672, calc. for $C_7H_2N_9O_{12}$ [M-NO_2]$^-$: 403.9828, found: 403.9827.

5,5′-Diamino-3,3′-azo-1,2,4-oxadiazole (**2a-56**): Sodium dicyanamide (10 g, 112 mmol) was dispersed in 180 mL of ethanol. This was followed by a segmented addition of hydroxylamine hydrochloride (7.78 g, 112 mmol). The blend was stirred at room temperature for a duration of 12 h. The inorganic salt that precipitated out was then filtered off. The filtrate went through an evaporation process until completely dried, resulting in the formation of 8 g (71%) of the compound 3,5-diamino-1,2,4-oxadiazole as white solid; ^{1}H NMR: 10.1 (s, 2H) 7.2 (s, 2H). ^{13}C NMR: 171.4, 169.8.

The compound 3,5-Diamino-1,2,4-oxadiazole (10 g, 100 mmol) was scattered in a concoction comprising of 150 mL of acetonitrile and 80 mL of concentrated hydrochloric acid. A solution of potassium permanganate (15.8 g, 100 mmol), dissolved in the least amount of water, was gradually added to this mixture. The reaction blend was stirred for a duration of 8 h, after which it was diluted with 400 mL of water. Hydrogen peroxide (at 5% concentration) was then incorporated until the solution became colorless. The resulting precipitate was filtered out and dried, yielding 5.5 g (55%) of the product identified as compound **2a-56** as yellow solid; T_d. 355 °C (onset); IR (KBr): n 3491, 3172, 1680, 1514, 1403, 1347, 1215, 1111, 974, 929, 792, 769, 669, 5883 cm^{-1}; ^{1}H NMR: 8.33 (s, 4H). ^{13}C NMR: 176.5, 174.0. ^{15}N NMR: 136.19, − 28.94, − 209.69, 314.06; Elemental analysis: ($C_4H_6N_8O_3$, 214) calcd: C, 22.43, H, 2.82, N, 52.33. Found: C, 22.87, H, 2.70, N, 51.73.

3,3′-Azo-1,2,4-oxadiazol-5,5′-dione (**2a-57**): Compound **2a-56** (2 g, 10.2 mmol) was introduced into a nitration blend. This blend was prepared by gradually adding 100% nitric acid (4.3 mL, 102 mmol) to acetic anhydride (15.6 mL, 153 mmol) while keeping the temperature at 0 ºC. Following this, the reaction mixture was stirred at room temperature over a period of 8 h. It was then poured onto 80 g of ice and extracted with ethyl acetate (60 mL), after which the solvent was evaporated. The resulting crude product was rinsed with dichloromethane (20 mL) and dried, leading to the production of 1.1 g (54%) of compound **2a-57**.

3-Amino-4-(5-amino-1,3,4-oxadiazol-2-yl)furazan (**2a-59**): A solution was prepared by dissolving 8 g (80 mmol) of potassium hydrogen carbonate in 30 ml of water. This solution was then added to a mixture containing 10 g (70 mmol) of hydrazide suspended in 50 ml of alcohol. Cyanogen bromide (7.42 g, 70 mmol) was subsequently introduced in segments. The combined mixture was stirred for a period of 24 h while maintaining a temperature between 15 and 20 °C. Following this, the formed precipitate was filtered out, rinsed with water and dried to give a yield of 9.9 g (84%), T_m, 225 °C (decomp.; from water). ^{1}H NMR spectrum (DMSO-d_6), d, ppm: 7.78 s (2H, NH2); 6.49 s (2H, NH2). ^{13}C NMR spectrum (DMSO-d_6), d, ppm: 165.38, 155.56, 143.12, 135.57. Found, %: C 28.53; H 2.44; N 50.03. $C_4H_4N_6O_2$. Calculated, %: C 28.58; H 2.40; N 50.00.

3-Nitroamino-4-(5-amino-1,3,4-oxadiazol-2-yl)furazan monohydrate (**2a-60**): First, nitric acid (100%, 2 mL) was placed in a two-necked flask with a 10 mL capacity and then cooled down to 0 °C. Following this, **2a-59** (1.0 g, 5.95 mmol) was gradually added to the chilled nitric acid, ensuring that the reaction temperature remained under 5 °C. Once the addition was complete, the resulting mixture was stirred for an hour at 0 °C. As it was poured onto 50 g of ice-water, a white solid emerged which was subsequently separated, rinsed with cool water, and air-dried to yield a pale-yellow powder identified as pure **2a-60**. Yield: 0.90 g (3.89 mmol,

65.4%). T_d: 132 °C. ^{1}H NMR (300 MHz, CD_3CN): d 6.19 ppm (s, 1H), 4.71 ppm (s, 2H). ^{13}C NMR (300 MHz, DMSO-d_6): d 166.3, 149.2, 149.0, 140.4 ppm. IR (KBr): n 3504, 3423, 3315, 1730, 1666, 1622, 1498, 1450, 1412, 1298, 1119, 1032, 937, 876, 814, 746, 714, 654, 584 cm^{-1}. Elemental analysis for $C_4H_5N_7O_5$ (231.13): calcd C 20.79, H 2.18, N 42.42%. Found: C 20.51, H 2.15, N 41.78%.

5-(4-nitro-1,2,5-oxadiazol-3-yl)-1,3,4-oxadiazol-2-amine (**2a-61**): In a three-necked round-bottomed flask equipped with a stirrer and dropping funnel, sodium tungstate dihydrate (2.20 g, 6.7 mmol) and 50% hydrogen peroxide solution (30.00 g, 0.44 mol) were placed. This setup was cooled in an ice-salt bath at − 10 °C. Following this, sulfuric acid (30.00 g, 0.31 mol) was incorporated drop-wise at − 5 °C while the mixture was stirred. The reaction blend was then allowed to reach room temperature, subsequent to which compound **2a-59** (0.84 g, 5.0 mmol) was slowly added. Once all of **2a-61** was incorporated, the mixture was left stirring at room temperature for an additional 24 h. Post this, the mixture was extracted using ethyl acetate (50 mL × 4). The pooled extracts were rinsed with ice-water and dried over magnesium sulfate. Concentration on a rotary evaporator yielded a white solid (0.79 g, 4 mmol) with an 80% yield. T_m: 139 °C. T_d: 155 °C. ^{1}H NMR (500 MHz, DMSO-d_6): δ 7.84 (s, 2H) ppm. ^{13}C NMR (500 MHz, DMSO-d_6): δ 137.55, 144.24, 158.03, 164.81 ppm. IR (KBr): $\tilde{v}$ 3323, 3128, 1652, 1568, 1511, 1480, 1457, 1354, 1081, 1030, 910, 891 cm^{-1}. Elemental analysis for $C_4H_2N_6O_4$ (198.12): calcd C 24.25, H 1.02, N 42.42%. Found: C 24.23, H 1.01, N 42.45%.

5,5′-Bis(trinitromethyl)-2,2′-bi(1,3,4-oxadiazole) (**2a-62**): The process started with a combination of 0.15 mol of carbethoxyethylacetimidate chlorohydrate and 0.1 mol acid hydrazide in 30 mL of ethanol at a temperature of 60 ºC for an hour. Following this, the blend was evaporated and 50 mL of water was incorporated. The resulting solution underwent extraction with chloroform, after which the extract was evaporated as well. The product was then chromatographed on silica gel L 40/100 using a hexane-ether (1:1) eluent. In a separate procedure, 10 g of P_2O_5 were combined with 40 mL of dichloroethane at a temperature range of 50–60 ºC. Introducing 0.02 mol of the aforementioned product to this mixture, it was boiled for an hour. The dark fluid was decanted from the residual P_2O_5 and evaporated. The oily end-product obtained was combined with MeOH and left for a day at a temperature between 5 and 10 ºC. This resulted in the precipitation of 2,2′-bi(1,3,4-oxadiazolyl)-5,5′-diacetic acid diethyl ester which was subsequently filtered out.

A solution of 2,2′-bi(1,3,4-oxadiazolyl)-5,5′-diacetic acid diethyl ester (0.5 g, 1.61 mmol) was prepared in acetonitrile (10 mL). To this solution, NaOH (6 M, 2 mL) was incorporated. The mixture was stirred for a couple of hours at room temperature. Post filtration, a yellow intermediate compound was formed. The moist powder was then dissolved in water. Acidification of the solution occurred with the addition of hydrochloric acid and it was stirred for half an hour. Following further filtration, the remaining precipitate was washed with water to yield 0.296 g of 2,2′-bi(1,3,4-oxadiazolyl)-5,5′-diacetic acid. This white powdered substance was obtained with a yield of 72.3%. ^{1}H NMR (300 MHz, DMSO-d_6): d = 4.28 (s, 4H) ppm. ^{13}C NMR (75 MHz, DMSO-d_6): d = 167.99, 163.43, 153.47, 31.70 ppm. IR (KBr): 3403, 3203,

2625, 2175, 1695, 1566, 1514, 1425, 1348, 1278, 1220, 1160, 1076, 991, 915, 866, 806, 687. Elemental analysis calcd (%) for $C_8H_6N_4O_6$ (254.16): C 37.81, H 2.38, N 22.04; found: C 37.63, H 2.51, N 21.10%.

Solid 2,2′-bi(1,3,4-oxadiazolyl)-5,5′-diacetic acid (0.22 g, 1.04 mmol) was gradually added over a quarter of an hour to a blend of 100% nitric acid (2.0 mL) and 96% sulfuric acid (6.0 mL), which was chilled to 0 ºC. This mixture was then left stirring for 18 h at room temperature before being poured into ice-cold water (50 mL). The solid that precipitated was filtered, rinsed with cold water, and air-dried, yielding 0.280 g of **2a-62**. The end product was a white powder obtained with a yield of 74.1%. ^{13}C NMR (75 MHz, Acetone-d_6): d = 157.28, 154.33, 118.10 ppm. IR (KBr): 2888, 2613, 1620, 1464, 1414, 1354, 1273, 1165, 1078, 956, 841, 796, 707, 641; Elemental analysis calcd (%) for $C_6N_{10}O_{14}$ (436.12): C 16.52, N 32.12; found: C 16.81, N 32.06%.

5,5′-Bis(fluorodinitromethyl)-2,2′-bi(1,3,4-oxadiazole) (**2a-63**): The process began with dissolving compound 2,2′-bi(1,3,4-oxadiazolyl)-5,5′-diacetic acid (2.0 g, 6.45 mmol) in 96% sulfuric acid (10 mL). To the resulting solution, 100% HNO_3 (2.0 mL) was gradually added at a temperature of 0 ºC, and the blend was stirred for two hours at room temperature. To quench the reaction, the mixture was poured onto crushed ice (50 g). The formed precipitate was rinsed with ice-cold water and air-dried. This resulted in the production of 2.600 g of 2,2′-bi(1,3,4-oxadiazolyl)-5,5′-bis(2,2′-dinitro)-diacetic acid diethyl ester, a white solid, with an 82.3% yield. ^{1}H NMR (300 MHz, $CDCl_3$): d = 4.65 (q, 6H), 1.45 (t, 4H) ppm. ^{13}C NMR (75 MHz, $CDCl_3$): d = 155.76, 154.52, 153.65, 109.29, 68.12, 13.45 ppm. IR (KBr): 2993, 2902, 1782, 1611, 1575, 1541, 1463, 1352, 1302, 1227, 1172, 1152, 1095, 1067, 1031, 994, 951, 837, 794, 777, 695, 677, 643, 611. Elemental analysis calcd (%) for $C_{12}H_{10}N_8O_{14}$ (490.25): C 29.40, H 2.96, N 22.86; found: C 29.21, H 3.08, N 22.99%.

5,5′-Dinitromethyl-2,2′-bis(1,3,4-oxadiazolate): An initial solution was prepared by dissolving 2,2′-bi(1,3,4-oxadiazolyl)-5,5′-bis(2,2′-dinitro)-diacetic acid diethyl ester (0.5 g, 1.02 mmol) into methanol (10 mL). To this mixture, a base solution (0.2 mL) was added gradually and stirred for 2 h. The resulting precipitate was filtered and rinsed with methanol to acquire 5,5′-Dinitromethyl-2,2′-bis(1,3,4-oxadiazolate). Subsequently, the obtained salt (0.12 mmol) was diluted in dry CH_3CN (3.0 mL), followed by the addition of Selectfluor (0.48 mmol) at 25 °C. The yellow blend turned colorless and transparent over a stirring period of 20 min, plus an additional 3 h. After removing the solvent through evaporation, the resultant was extracted with CH_2Cl_2. Post column chromatography purification, compound **2a-63** was procured.

5-(Tetrazol-1-yl)-2H-tetrazole (**2b-1**): A blend of 5-aminotetrazole (8.50 g, 100 mmol), sodium azide (7.80 g, 120 mmol) and triethyl orthoformate (22.2 g, 150 mmol) was prepared at 60 °C under intense stirring. To this mixture, glacial acetic acid (50 mL) was gradually introduced. After refluxing at 90 °C for 4 h, a transparent solution was formed which was then cooled to room temperature. Hydrochloric acid (60 mL) was added to the cooled mixture and stirred for an hour. Evaporation under vacuum resulted in removal of the solvent, following which, the residue was suspended in toluene. Further evaporation removed any remaining acetic acid traces. This step was repeated to ensure complete removal. The dry residue was then dissolved in hydrochloric acid (200 mL) and extracted using ethyl acetate (4 × 50 mL). The combined organic phases were dried over magnesium sulfate. After a final round of solvent evaporation under vacuum, the residue was thoroughly rinsed with diethyl ether, yielding anhydrous 5-(tetrazol-1-yl)-2H-tetrazole as a colorless powder (up to 5.80 g, 42.0 mmol, 42%). To obtain the less sensitive monohydrate form, recrystallization from water was performed. ^{1}H NMR (DMSO-d_6): d = 14.29 (s, NH), 10.21 ppm (s, CH); ^{13}C NMR (DMSO-d_6): d = 155.0, 143.8 ppm; ^{15}N NMR (DMSO-d_6): d = 13.4 (d, N_7), – 19.0 (Nc6), – 35.2 (N_2), – 51.3 (d, N_8), – 85.2 (N_1), – 153.8 ppm (d, N_5); IR (ATR): n = 3163 (m), 3130 (m), 3034 (m), 2971 (m), 2865 (m), 2766 (m), 2702 (m), 2649 (m), 1697 (w), 1582 (vs), 1455 (m), 1385 (m), 1372 (m), 1272 (s), 1208 (m), 1178 (m), 1148 (m), 1114 (m), 1086 (s), 1021 (s), 993 (m), 957 (m), 913 (m), 891 (m), 871 (m), 742 (m), 649 cm^{-1} (m); elemental analysis calcd (%) for $C_2H_2N_8$ (138.1): C 17.40, H 1.46, N 81.14; found C 18.02, H 1.55, N 79.29; DSC (5 °C min^{-1}): T_d = 145 °C; sensitivities (grain size < 100 mm): IS 1 J, FS 5 N, ESD 130 mJ.

2-Methyl-5-(tetrazol-1-yl)tetrazole (**2b-2**): An initial mixture was prepared by combining 5-amino-2-methyltetrazole (1.98 g, 20.0 mmol), sodium azide (1.56 g, 24.0 mmol), and triethyl orthoformate (4.44 g, 30.0 mol). To this mixture, glacial acetic acid (8.40 g, 140 mmol) was gradually incorporated. The blend was then heated to reflux at 100 °C for a duration of three hours. After heating, the crude substance precipitated as a colorless powder. Hydrochloric acid (1 m, 20 mL) was subsequently added. The solvent was extracted and the remaining residue was dissolved in boiling benzene, followed by filtration. The end product, referred to as **2b-2**, was procured through crystallization from the filtered solution. It appeared as colorless needles, weighing 1.10 g (7.23 mmol), accounting for a yield of 36%. ^{1}H NMR (DMSO-d_6): d = 10.26 (s, 1H, NH), 4.51 ppm (s, 3H, CH_3); ^{13}CNMR (DMSO-d_6): d = 156.6, 144.2, 41.4 ppm; ^{15}N NMR (D_2O): d = 13.9 (d, N7), 1.9 (q, N_3), – 19.3 (N_6), – 50.8 (d, N_8), – 66.6 (N_4), – 89.7 (q, N_1), – 100.9 (q, N_2), – 154.3 ppm (d, N_5); IR (KBr): n = 3427 (w), 3142 (s), 3051 (w), 3967 (w), 2930 (w), 2886 (w), 2777 (w), 2710 (w), 2385 (w), 2342 (w), 2254 (w), 2129 (w), 1978 (w), 1781 (w), 1752 (w), 1588 (vs), 1534 (m), 1476 (m), 1467 (s), 1433 (m), 1414 (s), 1363 (w), 1351 (m), 1291 (m), 1266 (s), 1260 (s), 1198 (m), 1178 (m), 1134 (m), 1088 (vs), 1056 (s), 1038 (w), 1018 (w), 1001 (s), 944 (m), 893 (m), 880 (m), 808 (w), 743 (m), 721 (s), 713 (s), 684 (w), 652 (s), 501 cm^{-1} (w); elemental analysis calcd (%) for $C_3H_4N_8$ (152.1): C 23.69, H 2.65, N 73.66; found C 23.64, H 2.67, N 73.56; DSC (5 °C min^{-1}): T_m = 97 °C, T_d = 135 °C; sensitivities (grain size 100–500 mm): IS 9 J, FS 240 N, ESD 300 mJ.

5,5′-Bistetrazole-1,1′-diolate (**TKX-50**): Method A: 5,5′-Bistetrazole (3.00 g, 21.7 mmol) was dissolved in 200 mL of water to form a clear solution. To this, Oxone (80.0 g, 109 mmol, 5 eq.) was added and the solution was buffered with trisodium phosphate till pH 7 was reached. The solution turned pink when the pH exceeded a certain level. The mixture was stirred at room temperature for 5 h before being acidified with concentrated sulfuric acid. It was then extracted using diethyl ether. Upon solvent evaporation, an unrefined product was obtained as a slightly yellow solid. Recrystallization from methanol helped remove any remaining sulfates or phosphates. The reaction resulted in a mixture of three isomers **1,1′-BTO**, **2,2′-BTO**, and **1,2′-BTO** with a collective yield of 71% (2.60 g, 15.3 mmol). Among these, **2,2′-BTO** was the primary product. This isomer blend (1.70 g, 10 mmol) was then dissolved in 20 mL of hot water. Hydroxylamine (50% w/w, 1.32 g, 20 mmol) was added in aqueous solution, leading to immediate formation of a colorless precipitate. Warming the mixture redissolved the precipitate, before it started to reprecipitate. This product was the dihydroxylammonium salt of the 1,1′-isomer (**1,1′-BTO**). After filtration and recrystallization from water, to remove the salt of the more water-soluble 2,2′-isomer (**2,2′-BTO**), the dihydroxylammonium salt of the 1,1′-isomer (**1,1′-BTO**) was obtained. However, this product was obtained only in minor yields (0.31 g, 1.3 mmol, 13%) due to the dominant formation of the 2,2′-isomer (**2,2′-BTO**) in the oxidation process with Oxone.

Method B: First, NaOH (27.5 g, 0.69 mol) was dissolved in 75 mL of water. The solution was then cooled to 0 °C using a salt-ice bath. While stirring the solution, hydroxylammonium chloride (69.5 g, 1.00 mol) was added. The next step involved adding glyoxal (72.5 g, 0.50 mol, 40% in H_2O) to the resulting solution, making sure to maintain the temperature below 10 °C. Once all of the glyoxal was incorporated, the solution was further cooled in the salt-ice bath until glyoxime started precipitating. This solid substance was separated through suction filtration and washed with a small amount of ice-water to eliminate residual sodium chloride. ^{1}H NMR (270 MHz, DMSO-d_6, 25 °C, ppm) δ: 7.73, 11.61; ^{13}C NMR (270 MHz, DMSO-d_6, 25 °C, ppm) δ: 145.9; EA (found, calc. for $C_2H_4N_2O_2$, MW = 88.07): C (27.07, 27.28), H (4.69, 4.58), N (31.45, 31.81) %.

Initially, glyoxime (17.6 g, 200 mmol) was suspended in 200 mL of ethanol. Chlorine gas was then introduced into this suspension and the process was carried out at − 20 °C until the suspension changed from green to yellowish. Thereafter, the solution was allowed to gradually return to room temperature, letting the dissolved chlorine be released simultaneously. Subsequent to this, the solvent was extracted under vacuum conditions. The remaining solid material was redistributed in 50 mL of chloroform. After stirring this new mixture for about 15 min at room temperature, it was filtered to obtain the final product—a colorless substance weighing 26.6 g, which accounted for an 85% yield. ^{1}H NMR (270 MHz, DMSO-d_6, 25 °C, ppm) δ: 13.10; ^{13}C NMR (270 MHz, DMSO-d_6, 25 °C, ppm) δ: 131.2; EA (found, calc. for $C_2H_2Cl_2N_2O_2$, MW = 156.96): C (15.65, 15.30), H (1.25, 1.28), N (17.49, 17.85) %.

Dichloroglyoxime (784 mg, 5 mmol) was initially dissolved in 10 mL of dimethyl formamide. Sodium azide (841 mg, 12.93 mmol) was then introduced at a temperature of 0 °C. The resulting suspension was stirred for around 20 min, still maintaining the temperature at 0 °C. Following this, water (100 mL) was added to the mixture. A precipitate formed which was subsequently filtered and washed with an additional 20 mL of water. After air-drying, the colorless final product was obtained, weighing 713 mg, corresponding to an 84% yield. DSC (5 °C min^{-1}): 170 °C (T_d); IR (atr, cm^{-1}): n = 3209 (w), 2170 (w), 2123 (w), 1622 (w), 1400 (w), 1361 (w), 1286 (m), 1013 (vs), 930 (m), 920 (s), 855 (s), 731 (s); Raman (1064 nm, 300 mW, 25 °C,

cm^{-1}): n = 2166 (8), 2129 (5), 2091 (3), 1621 (100), 1457 (14), 1390 (12), 1216 (19), 1034 (3), 882 (20), 672 (3), 442 (6); ^{1}H NMR (270 MHz, DMSO-d_6, 25 °C, ppm) δ: 12.08; ^{13}C NMR (270 MHz, DMSO-d_6, 25 °C, ppm) δ: 136.5; EA (found, calc. for $C_2H_2N_8O_2$, MW = 170.09): C (14.38, 14.12), H (1.46, 1.19), N (66.01, 65.88) %;

Initially, diazidoglyoxime (850 mg, 5 mmol) was suspended in 40 mL of diethyl ether. Then, gaseous HCl was introduced into this reaction mixture while stirring for a duration of two hours at a temperature ranging between 0 and 5 °C. Subsequently, the flask was sealed and left to stir overnight at room temperature. Afterwards, the solution was allowed to rest in order to enable crystallization. This process resulted in the formation of 5,5′-bistetrazole-1,1′-diol dihydrate, which appeared as pale-yellow crystals and weighed 760 mg, accounting for a 73% yield. DSC (5 °C min^{-1}): 214 °C (dec.); IR (atr, cm^{-1}): n = 3229 (m), 1665 (m), 1411 (w), 1375 (w), 1302 (w), 1208 (w), 1144 (m), 995 (s), 714 (w), 662 (w); Raman (1064 nm, 300 mW, 25 °C, cm^{-1}): n = 1608 (100), 1270 (26), 1157 (46), 1133 (38), 1019 (22), 766 (31), 738 (13), 693 (4), 597 (6), 402 (29); ^{1}H NMR (400 MHz, DMSO-d_6, 25 °C, ppm) δ: 6.80; ^{13}C NMR (400 MHz, DMSO-d_6, 25 °C, ppm) δ: 135.8; EA (found, calc. for $C_2H_6N_8O_4$, MW = 206.12): C (12.01, 11.65), H (2.81, 2.93), N (54.04, 54.36) %; BAM drophammer: > 40 J; friction tester: 216 N; ESD: 0.5 J.

Firstly, 5,5′-bistetrazole-1,1′-diol dihydrate (2.06 g, 10 mmol) was dissolved in 50 mL of warm water. Then Hydroxylamine (1.32 g, 20 mmol, 50% w/w in H_2O) was added to the solution while stirring. As the solution cooled down to room temperature, the dihydroxylammonium salt started to crystallize. The resulting **TKX-50** was separated through suction filtration and air-dried, with a yield of 82%. DSC (5 °C min^{-1}, °C): 221 °C (T_d); IR (KBr, cm^{-1}): n = 3425 (m), 3219 (s), 3081 (s), 3050 (s), 2936 (s), 2689 (s), 2513 (m), 1599 (m), 1577 (m), 1527 (s), 1426 (s), 1413 (s), 1389 (m), 1352 (m), 1338 (m), 1316 (w), 1236 (vs), 1174 (m), 1145 (w), 1095 (w), 1046 (w), 1029 (w), 1011 (m), m997 (m), 800 (m), 723 (m), 676 (w), 612 (w), 579 (w), 539 (w), 498 (w); Raman (1064 nm, 300 mW, 25 °C, cm^{-1}): n = 1616 (100), 1469 (3), 1278 (2), 1239 (25), 1173 (2), 1143 (6), 1116 (10), 1014 (7), 1004 (12), 763 (4), 612 (3), 409 (4), 335 (2), 257 (2), 199 (2); ^{1}H NMR (400 MHz, DMSO-d_6, 25 °C, ppm) δ: 9.66 (s, 8H, NH_3OH); ^{13}C NMR (400 MHz, DMSO-d_6, 25 °C, ppm) δ: 135.5 ($(CN_4)_2$); m/z (FAB^+): 34.0 [NH_3OH^+]; m/z (FAB^-): 169.1 [$C_2HN_8O_2^-$]; EA (found, calc. For $C_2H_8N_{10}O_4$, MW = 236.15): C (10.50, 10.17), H (3,63, 3.41), N (59.31, 59.31) %; BAM drophammer: 20 J; friction tester: 120 N; ESD: 0.11 J.

1,1′-Diamino-5,5′-bistetrazole (**2b-3**): First, 5,5′-Bistetrazole (1.00 g, 7.2 mmol) was dissolved in 20 mL of distilled water. This was followed by the addition of a CsOH solution (4.34 g, 14.5 mmol, 50 wt%). The mixed solution was then evaporated, leaving behind a residue which was dissolved in a mixture of 250 mL of acetonitrile and 100 mL of distilled water. Separately, ethyl O-p-tolylsulphonylacetohydroximate (4.4 g, 17.1 mmol) was added to 40 mL of perchloric acid. This new blend was stirred for 2 h at room temperature before being poured onto 100 mL of iced water, which was then stirred until all the ice had melted. This resulting mixture was extracted with dichloromethane in 5 iterations of 10 mL each. The extract collected from this process was combined with the solution of bistetrazole, after which it was stirred overnight under regular atmospheric conditions. The final solution was evaporated and its extract was collected using ethyl acetate. After filtration, the filtrate was evaporated again and purified through column chromatography (using a ratio of ethyl acetate:benzene:petroleum ether = 3:1:1). This resulted in the production of 124 mg (10% yield) of 1,1′-diamino-5,5′-bistetrazole. DSC (5 °C min^{-1}): 145 °C (T_m), 185 °C (T_d); IR (cm^{-1}) n = 3292 (w), 3249 (w), 3140 (m), 2361 (w), 2200 (w), 1612 (m), 1459 (w), 1406 (w), 1385 (w), 1330 (w), 1303 (w), 1274 (w), 1213 (w), 1193 (w), 1131 (m), 1065 (w), 1032 (s), 1008 (m), 973 (s), 956 (s), 924 (s), 816 (w), 716 (w), 705 (w), 681 (w), 666 (w); Raman (1064 nm): n = 3330 (1), 3256 (1), 3177 (4), 2937 (1), 2859 (2), 2026 (1), 1626 (100), 1613 (10), 1502 (2), 1392 (1), 1334 (1), 1278 (27), 1206 (1), 1181 (2), 1130 (1), 1082 (26), 1044 (11), 983 (5), 927 (1), 752 (4), 742 (22), 706 (1), 579 (1), 550 (1), 406 (4), 388 (7), 324 (1), 290 (1), 256 (10); ^{1}H NMR (DMSO-d_6) (ppm) = 7.40 (s, 4H); ^{13}C NMR (DMSO-d_6) (ppm) = 141.7 (s, 2C); m/z: (DCI+) 169.2 ($C_2H_5N_{10}$); EA ($C_2H_4N_{10}$, 168.12) calcd: C 14.29, H 2.40, N 83.31%, found: C 14.70, H 2.33, N 51.03%.

1,2′-Diamino-5,5′-bistetrazole (**2b-4**): Initially, 5,5′-bistetrazole (1.00 g, 7.2 mmol) was dissolved in 20 mL of distilled water. Then a CsOH solution (4.34 g, 14.5 mmol, 50 wt%) was added. After evaporation, the resulting residue was re-dissolved in a mixture of 250 mL acetonitrile and 100 mL distilled water. Separately, ethyl O-p-tolylsulphonylacetohydroximate (4.4 g, 17.1 mmol) was introduced into 40 mL of perchloric acid. This mixture was stirred for 2 h at room temperature and then poured onto 100 mL of iced water. The stirring continued until the ice had completely melted. The resultant mixture was then extracted using dichloromethane in five increments of 10 mL. This extract was integrated with the previously prepared bistetrazole solution, which was then stirred overnight at room temperature. The final mixture was evaporated and further extracted with ethyl acetate. Post filtration, the filtrate was evaporated again and purified using column chromatography (with the ratio of ethyl acetate:benzene:petroleum ether being 3:1:1). This process yielded 72 mg (6% yield) of 1,2′-diamino-5,5′-diaminobistetrazole. DSC (5 °C min^{-1}): 90 °C (T_m), 170 °C (T_d); IR (cm^{-1}) n = 3299 (w), 3224 (w), 3143 (m), 1634 (w), 1602 (m), 1591 (m), 1499 (w), 1429 (w), 1398 (w), 1348 (w), 1326 (w), 1216 (w), 1184 (m), 1114 (w), 1093 (w), 1064 (m), 993 (s), 953 (s), 822 (w), 733 (w), 718 (w), 704 (m), 682 (w); Raman (1064 nm): n = 3229 (5), 3147 (3), 2944 (1), 2859 (2), 1635 (100), 1609 (4), 1501 (2), 1460 (10), 1434 (4), 1395 (7), 1364 (1), 1348 (1), 1327 (7), 1221 (21), 1189 (10), 1126 (13), 1092 (1), 1062 (8), 1030 (43), 993 (3), 950 (4), 755 (5), 736 (1), 708 (7), 682 (1), 543 (1), 454 (2), 417 (3), 366 (9), 320 (5), 269 (4); ^{1}H NMR (DMSO-d_6) (ppm) = 8.59 (s, 2H), 7.35 (s, 2H); ^{13}C NMR (DMSO-d_6) (ppm) = 150.5 (s, 1C), 143.7 (s, 1C); m/z: (DCI^+) 169.2 ($C_2H_5N_{10}$); EA ($C_2H_4N_{10}$, 168.12) calcd: C 14.29, H 2.40, N 83.31%, found: C 14.70, H 2.33, N 51.03%.

Dipotassium dinitraminobistetrazolate (**K_2DNABT**): Firstly, a 40% aqueous glyoxal solution was mixed with methanol and combined with the suitable amount of hydrazine derivative also in methanol. A small volume of glacial acetic acid was introduced, and the concoction was stirred until precipitation occurred. After several hours, the end product was collected via filtration and rinsed with methanol. Yield improvement or secondary crop isolation was not pursued. The methodology is exemplified through the preparation of the methyl carbazate derivative of glyoxal or glyoxal derivatives. To begin, a 40% aqueous glyoxal solution (14.5 g, 0.1 M) dissolved in 50 mL of methanol, was combined with 18.9 g of methyl carbazate (0.21 M) dissolved in 150 mL of methanol. Glacial acetic acid (five drops) was added thereafter. A white precipitate appeared after 5 min; the reaction was ceased after one hour and the product filtered to yield 17.75 g (88%). Then, glyoxal derivatives (10.1 g, 50 mmol) were suspended in 300 mL DMF before adding NCS (20 g, 150 mmol). This blend was stirred overnight at room temperature, then filtered, and washed with water, ethanol, and ether. The resulting dry product, dichlorodimethoxycarbonylglyoxal bishydrazone, weighed 9.79 g (72%). In the next step, dichlorodimethoxycarbonylglyoxal bishydrazone (960 mg, 3.452 mmol) was suspended in 10 mL DMF and cooled to 0–5 °C. Sodium azide (650 mg, 10 mmol) was added, and the mixture was stirred overnight at room temperature. The blend was then diluted with 30 mL of ice-water, filtered, and washed with water, ethanol, and ether to yield 387 mg (38%) diazidodimethoxycarbonylglyoxal bishydrazone. Finally, diazidodimethoxycarbonylglyoxal bishydrazone (1 g) was suspended in 50 mL of diethyl ether. This suspension was cooled with an ice bath and saturated at 5 °C with gaseous HCl. The flask was sealed and stirred at room temperature for three days. After the removal of the solvent, the remaining solid was recrystallized from methanol, resulting in dimethoxycarbonyl diaminobistetrazole as colorless plates. DSC (5 °C min^{-1}): 205 °C (T_{dec}); ^{1}H NMR (DMSO-d_6) δ: 12.86 (s), 3.78; ^{13}C NMR (DMSO-d_6 25 °C, ppm) δ: 155.3, 142.0, 54.4; IR (atr, cm^{-1}): 3257 (w), 1773 (w), 1746 (vs), 1520 (m), 1458 (m), 1435 (w), 1387(w), 1282 (s), 1261 (s), 1192 (w), 1114 (w), 1063 (w), 1012 (w), 965 (w), 818 (w), 765 (m), 721 (w); EA ($C_6H_8N_{10}O_2$, 284.19): calc.: C 25.36, H 2.84, N 49.29%; found: C 25.95, H 2.90, N 48.32%; BAM drophammer: 3 J; friction tester: 192 N; ESD: 0. 5 J.

Initially, dimethoxycarbonyl diaminobistetrazole (1 g, 3.52 mmol) was suspended in 50 mL of dry acetonitrile and temperatures were dropped to between 0 and 5 °C. Following this, N_2O_5 (1.5 g, 13.9 mmol) was introduced into the mixture which was then stirred at the same temperature. Once all starting material had dissolved after approximately 1–2 h, the solution was further stirred for an additional half an hour. Next, KOH (14 mL) was added while stirring the blend vigorously at ice bath temperatures. More KOH solution was introduced until the pH of the aqueous part remained consistently at least 12. Finally, the resulting solid that precipitated was gathered through filtration and re-suspended in 20 mL of ice water, stirred for another 5 min, and filtered once more. This process resulted in the yield of 720 mg (61%) of a finely powdered, colorless **K_2DNABT**. T_d: 200 °C; IR (KBr): n = 3450 (m), 1635 (w), 1440 (s), 1382 (w), 1360 (w), 1300 (vs), 1256 (m), 1163 (w), 1124 (w), 1031 (w), 998 (w), 872 (w), 773 (w), 729 cm^{-1}(w); EA ($C_2K_2N_{12}O_4$,334.30): C 7.19, N 50.28%; found: C 7.62, N 47.95%;m/z: 257 ($C_2HN_{12}O_4^-$); BAM drophammer: 1 J; friction tester: < 5N; ESD: 3 mJ.

1,1′-Dinitramino-5,5′-bistetrazole (**2b-5**): To start, K_2DNABT (522 mg, 1.56 mmol) was mixed in 60 mL of 2 m HCl and the resulting solution was extracted six times using ethyl acetate in quantities of 30 mL each time. Post extraction, the combined organic layers were dehydrated over magnesium sulfate and then concentrated under reduced pressure without additional heating. This process yielded an orange oil. This orange oil was meticulously dried under high vacuum, resulting in an ochre-colored residue. This residue was then dissolved in a small amount of diethyl ether and left to crystallize. After a few hours, colorless single crystals were obtained that were suitable for X-ray analysis in almost quantitative yield. DTA (5 °C min^{-1}): 107 °C (T_d); IR (atr): n = 3397 (w), 3092 (m), 3031 (m), 2925 (m), 2854 (m), 2803 (m), 2613 (w), 1712 (w), 1625 (vs), 1460 (w), 1450 (w), 1398 (s), 1316 (s), 1290 (s), 1194 (s), 1116 (s), 1012 (m), 969 (m), 957 (m), 863 (s), 756 (s), 734 (s), 706 cm^{-1} (m); ^{1}H NMR (DMSO-d_6): d = 5.67 ppm; ^{13}C NMR (DMSO-d_6): d = 140.5 ppm; ^{15}N NMR (DMSO-d_6): 5.6 ppm (N_6); ^{15}N NMR (DMSO-d_6): d = 3.9 (N3$_)$, − 5.6 (N_6), − 9.0 (N_2), − 49.2 (N_4), − 117.0 (N_1), − 122.0 ppm (N_5); BAM drop-hammer: < 1 J; friction tester: < 5 N; ESD: 0.03 J.

General procedure for alkoxy aminotetrazoles (**2b-(6-13)**): Starting at 0 °C, cyanogen bromide (10 mmol) was mixed in 50 mL of dry acetonitrile. To this mixture, dry sodium azide (40 mmol) was added. This reaction blend was stirred at 0 °C for a duration of 4 h. Post stirring, the inorganic salt was extracted through filtration and quickly dissolved in cold water. Following that, the cyanogen azide solution was introduced to a solution containing 5 mmol alkoxy amine [or 2.5 mmol for bis(oxyamine)]. For alkoxy ammonium hydrochloride, either 5 or 2.5 mmol NaOH was necessary for neutralization in 20 mL of water at 0 °C. After stirring overnight at room temperature, the solvent was evaporated by blowing air over the liquid surface. The end product was cleaned by rinsing with both water and acetonitrile. For the synthesis of target compounds, it is recommended that twice the initial amount of cyanogen bromide, sodium azide, and methylene diamine hydrochloride be utilized.

1,1′-Ethylenebis(oxy)bis(5-aminotetrazole) (**2b-9**): Light brown solid; Yield 20%; T_d, 175 °C (DSC: 10 °C min^{-1}); IR (KBr) n 3331, 3150, 1661, 1603, 1449, 1314, 1263, 1106, 1032, 1014, 858; ^{1}H NMR δ 7.10 (s, 4H, NH_2), 4.65 (s, 4H, CH_2); ^{13}C NMR δ 149.3, 76.4; ^{15}N NMR δ 1.3 (N_3), − 25.0 (N_2), − 87.7 (N_4), − 127.2 (N_1), − 328.7 (NH_2); EA ($C_4H_8N_{10}O_2$, 228.17) calcd: C, 21.06; H, 3.53; N, 61.39%. Found: C, 21.37; H, 3.55; N, 60.34%.

1,1′-Methylenebis(oxy)bis(5-aminotetrazole) (**2b-10**): Light brown solid; Yield 7%; T_d, 123 °C (DSC: 10 °C min^{-1}); IR (KBr) n 3320, 3146, 2924, 1668, 1333, 1215, 1154, 1096, 986, 638; ^{1}H NMR δ 7.29 (s, 4H, NH_2), 6.03 (s, 2H, CH_2); ^{13}C NMR δ 149.8, 105.1; ^{15}N NMR δ 2.7 (N_3), − 26.2 (N_2), − 88.5 (N_4), − 129.6 (N_1), − 327.2 (NH_2); MS (EI) m/z: 214 (2.1), 56 (89), 43 (100); EA ($C_3H_6N_{10}O_2$, 214.15) calcd: C, 16.83; H, 2.82; N, 65.41%. Found: C, 17.04; H, 2.92; N, 63.15%.

1,3-dimethylureabis(5-aminotetrazole) (**2b-11**): colorless crystal; IR (KBr): n = 3331, 3146, 1707, 1659 cm^{-1}; ^{1}H NMR (DMSO-d_6): d = 7.01 (s, 4H; NH_2), 11.09 ppm (br s, 2H; NH); ^{13}C NMR (DMSO-d_6): d = 152.5 (s), 154.8 ppm (s); ^{15}N NMR (DMSO-d_6): d = _331.9 (t, NH_2), − 269.3 (br s; NH), − 172.8 (N_1), − 94.6 (N_4), − 9.4 (N_3), 4.2 ppm (N2); elemental analysis (%): calcd for $C_3H_6N_{12}O$ (226.16): C 15.93, H 2.67, N 74.32; found C 16.27, H 2.77, N 72.71.

N1,N2-dimethyloxalamidebis(5-aminotetrazole) (**2b-12**): white solid; IR (KBr): n = 3402, 3339, 3283, 3223, 3183, 1738, 1661, 1452, 1115, 1075, 672 cm^{-1}; ^{1}H NMR

(DMSO-d_6): d = 7.13 (s, 4H; NH_2), 12.71 ppm (s, 2H; NH); ^{13}C NMR (DMSO-d_6): d = 154.3 (s), 156.9 ppm (s); elemental analysis (%): calcd for $C_4H_6N_{12}O_2$ (254.17): C 18.90, H 2.38, N 66.13; found C 19.21, H 2.36, N 65.65.

N1,N2-dimethylhydrazine-1,2-dicarboxamidebis(5-aminotetrazole) (**2b-13**): white solid; IR (KBr): n = 3412, 3327, 3291, 3213, 3000, 1717, 1649, 1584, 1518, 1331, 1254, 598 cm^{-1}; ^{1}H NMR (DMSO-d_6): d = 6.80 (s, 4H; NH_2), 9.13 (s, 2H; NH), 10.75 ppm (br s, 2H; NH); ^{13}C NMR (DMSO-d_6): d = 154.9 (s), 155.4 ppm (s); ^{15}N NMR (DMSO-d_6): d = − 332.9 (t, NH_2), − 276.0 (d, NH), − 270.4 (br s, NH), − 172.1 (N_1), − 96.2 (N_4), − 18.2 (N_3), 1.7 ppm (N_2); elemental analysis (%): calcd for $C_4H_8N_{14}O_2$ (284.20): C 16.90, H 2.84, N 69.00; found C 16.88, H 2.95, N 67.21.

For the creation of alkoxy nitroiminotetrazole derivatives (**2b-(14-18)**): Initially, at 0 °C, oxy aminotetrazole (2 mmol) was gradually integrated into 5 mL of 100% HNO_3. This blend was then stirred at room temperature for a length of 18 h. Following that, the solution was deposited into approximately 20 g of ice. Post stirring for a further 3 h, the end product was precipitated (if this doesn't occur, water should be left to evaporate in the open air), extracted through filtration, cleaned slightly with water, and then left to dry in ambient conditions at room temperature.

1,2-Bis(4,5-dihydro-5-nitroimino-1H-tetrazol-1-yl)ethane (**2b-14**): Colorless crystal; T_d: 194 °C; IR (KBr) n = 3430, 3196, 3107, 3031, 1581, 1489, 1308, 1261, 1025; ^{1}H NMR (DMSO-d_6): δ = 4.66 (s, 4H), 11.65 (br. s, 2H); ^{13}C NMR (DMSO-d_6): δ = 44.6, 150.9; ^{15}N NMR (55 °C, DMSO-d_6) δ = − 170.6 (s), − 156.8 (s), − 154.3 (s), − 24.7 (s), − 20.1 (s), − 15.5 (s, NO_2); C, H, N analysis (%): calcd for $C_4H_6N_{12}O_4$ (286.17): C 16.79, H 2.11, N 58.73; found C 16.81, H 2.10, N 57.37.

1,2-Bis(4,5-dihydro-5-nitroimino-1H-tetrazol-1-yl)propane (**2b-15**): White solid; T_d: 172 °C; IR (KBr): ν = 3575, 3469, 3064, 1577, 1499, 1427, 1449, 1339, 1299, 1246, 1043, 988; ^{1}H NMR (DMSO-d_6): δ = 2.43 (quin, 2H), 4.27 (t, 4H), 8.94 (br. s, 4H); ^{13}C NMR (DMSO-d_6): δ = 25.5, 44.1, 150.1; C,H,N analysis (%): calcd for $C_5H_{10}N_{12}O_5$ (318.21): C 18.87, H 3.17, N 52.82; found C 18.82, H 3.07, N 52.50.

1,2-Bis(4,5-dihydro-5-nitroimino-1H-tetrazol-1-yl)butane (**2b-16**): White solid; mp.: 145 °C (dec.); IR (KBr): n = 3428, 3179, 3117, 2966, 1578, 1491, 1430, 1318, 1242, 1208, 1036, 988, 779, 716; ^{1}H NMR (DMSO-d_6): δ = 1.85 (s, 4H), 4.19 (s, 4H), 7.25 (br. s, 2H); ^{13}C NMR (DMSO-d_6): δ = 25.5, 44.0, 150.2; C, H, N analysis (%): calcd for $C_6H_{10}N_{12}O_4$ (314.22): C 22.93, H 3.21, N 53.49; found C 22.79, H 3.21, N 52.47.

1,1′-Methylenebis(oxy)bis(5-nitroiminotetrazole) (**2b-17**): White solid; Yield 62%; T_d, 157 °C (DSC: 5 °C min^{-1}); IR (KBr) n 3436, 3109, 1598, 1498, 1412, 1319, 1246, 1046, 976, 924, 777, 700; ^{1} H NMR δ 13.96 (s, 2H, H-tetrazole), 6.23 (s, 2H, CH_2); ^{13}C NMR δ 147.0, 104.7; Caution! during elemental analysis the compound exploded and was not determinable.

1,1′-Ethylenebis(oxy)bis(5-nitroiminotetrazole) (**2b-18**): White solid; T_d, 134 °C (DSC: 10 °C min^{-1}); IR (KBr) n 3421, 3288, 3027, 1593, 1494, 1426, 1322, 1216, 1045, 1026, 1001, 978, 878, 842, 779, 694, 675, 631; ^{1}H NMR δ 10.49 (s, 2H, H-tetrazole), 4.77 (s, 4H, CH_2); ^{13}C NMR δ 147.0, 76.9; ^{15}N NMR δ − 10.6 (N_2), −

17.8 (N_6), − 24.5 (N_3), − 117.0 (N_1), − 124.3 (N_4), − 154.7 (N_5); MS (EI) m/z: 319 (0.2), 68 (8), 63 (2.4), 46 (100), 44 (61); Caution! during elemental analysis the compound exploded and was not determinable.

General procedure for synthesis of (**2b-(19-22)**): Initially, mono-, bis-, or trisubstituted 1,5-diaminotetrazoles were produced following a known method. Cyanogen bromide (3.15 g, 30 mmol) was dissolved in dry acetonitrile (50 mL) at 0 °C and dry sodium azide (9.3 g, 0.143 mol) was then added. This reaction blend was stirred for 4 h between 0 and 5 °C. The resulting inorganic salt was removed through filtration (Caution! it is crucial to quickly dissolve the salt in cold water after filtration). Cyanogen azide was combined with an aqueous monoamine or diamine solution at 0 °C. The monoamines included methanamine (0.47 g), methylhydroxylamine (0.71 g), and 2-aminoethanol (0.92 g) all at 15 mmol. Diamines encompassed ethane-1,2-diamine (0.42 g), propane-1,3-diamine (0.52 g), and butane-1,4-diamine (0.62 g), each at 7 mmol. Following two days of stirring at room temperature, solids were filtered out and rinsed with a bit of water (5 mL). Mono- and disubstituted 1,5-diaminotetrazoles in the form of white or light-yellow solids were acquired in high purity levels. Subsequently, water (2 mL) was added to these mono- and disubstituted compounds (1 mmol for each), forming a suspension while stirred. Then, an aqueous formalin solution (30%; 1.1 mmol for monoamine, 2.2 mmol for diamine) was introduced. The combo was heated to 70 °C in a water bath until it clarified. Afterward, this solution was thinned with another 5 mL of water and cooled to 20 °C. The right quantity of trinitromethane was included (1.1 mmol for monoamine, 2.2 mmol for diamine), and the blend was maintained at a temperature between 15 and 20 °C. In about 0.5–2 h, a yellow oil formed which completely hardened over an approximate span of two to three days. The solid mass was then separated through filtration, rinsed with water three times (each time with 5 mL), and left to dry in open air overnight. The objective compounds were ultimately achieved as white solids or light yellow powders, with productivity rates ranging between 43 and 82%.

1,1′-(Ethane-1,2-diyl)bis(N-(2′,2′,2′-trinitroethyl)-tetrazol-5-amine) (**2b-19**): Pale yellow solid (yield 76%); T_d 183 °C; ^{1}H NMR (300 MHz, CD_3CN): d = 6.07–6.05 (t, 2H, –NH–), 5.25–5.23 (d, 4H), 4.54 ppm (s, 4H, CH_2); ^{13}C NMR (300 MHz, CD_3CN): d = 156.5, 126.7, 49.1, 45.3; ^{15}N NMR (500 MHz, CD_3CN) d = 5.5 (N_2), − 19.0 (N_3), − 30.8 (t, NO_2), − 88.8 (N_4), − 176.24 (N_1), − 340.3 (t, NH), − 342.1 ppm (t, NH); IR (KBr): n = 3273, 3121, 3036, 2955, 2597, 1063, 1530, 1462, 1417, 1295, 1108, 1077, 889, 855, 804, 769, 736, 667 cm^{-1}; elemental analysis (%) calcd for $C_8H_{10}N_{16}O_{12}$ (522.26): C 18.40, H 1.93, N 42.91; found C 18.86, H 2.06, N 42.60.

1,1′-(Propane-1,3-diyl)bis(N-(2′,2′,2′-trinitroethyl)-tetrazol-5-amine) (**2b-20**): pale yellow solid (yield 61%); T_d, 143 °C; ^{1}H NMR (300 MHz, CD3CN): d = 6.14 (s, 2H, –NH–), 5.28–5.26 (d, 4H), 4.13 (t, 4H, CH_2), 2.30 ppm (m, 2H, CH_2); ^{13}C NMR (300 MHz, CD_3CN): d = 156.5, 126.9, 63.9, 49.3, 44.1 ppm; ^{15}N NMR (500 MHz, CD_3CN): d = 4.6 (N_2), − 18.4 (N_3), − 30.6 (t, NO_2), − 88.8 (N_4), − 160.6 (N_1), − 340.7 (NH), − 342.5 ppm (NH); IR (KBr): n = 3410, 3252, 3115, 3050, 2953, 1590, 1466, 1425, 1339, 1304, 1110, 893, 853, 804, 742, 550 cm^{-1};

elemental analysis (%) calcd for $C_9H_{12}N_{16}O_{12}$ (536.29): C 20.16, H 2.26, N 41.79; found C 20.15, H 2.62, N 40.02.

1,1′-(Butane-1,4-diyl)bis(N-(2′,2′,2′-trinitroethyl)-tetrazol-5-amine) (**2b-21**): pale yellow solid (yield 47%); T_d, 166 °C; ^1H NMR (300 MHz, CD_3CN): d = 5.96 (s, 2H, –NH–), 5.26–5.24 (d, 4H,), 4.12–4.10 (t, 4H, CH_2), 1.77–1.73 ppm (m, 4H, CH_2); ^{13}C NMR (300 MHz, –CD_3CN): d = 156.3, 127.4, 49.3, 46.4, 26.6 ppm; IR (KBr): n = 3263, 3126, 3049, 2952, 1609, 1468, 1418, 1377, 1298, 1246, 1112, 889, 855, 806, 773, 602, 536 cm^{-1}; elemental analysis (%) calcd for $C_{10}H_{14}N_{16}O_{12}$ (550.32): C 21.83, H 2.56, N 40.72; found C 22.04, H 2.63, N 39.82.

1,1′-(Ethane-1,2-diylbis(oxy))bis(N-(2′,2′,2′-trinitroethyl)-tetrazol-5-amine) (**2b-22**): Formalin solution (37% concentration, 2 mmol) was incrementally added to an aqueous mixture of **2b-9** (0.23 g, 1 mmol) suspended in water (2 mL). This solution was heated up to 70 °C and mixed until it clarified, then was cooled down to room temperature. The liquid portion was evaporated by air drying. Later, methanol (10 mL) was introduced to dissolve the resulting solid, followed by adding trinitromethane (0.17 g, 1.1 mmol). This concoction was warmed up to 65 °C and stirred for four hours. Upon adding more water (10 mL), a solid started to form which was then separated via filtration, rinsed with water three times (5 mL each time), and finally dried out to yield the pure product **2b-22** (a pale yellow solid with a 36% yield). T_d, 147 °C; ^1H NMR (300 MHz, CD_3CN): d = 6.42 (s, 2H, –NH–), 5.26–5.23 (d, 4H, –CH_2–), 4.69 ppm (s, 4H, –CH_2–CH_2–); ^{13}C NMR (300 MHz, CD_3CN): d = 150.6, 127.3, 78.5, 48.9 ppm; IR (KBr): n = 3272, 3132, 2924, 1600, 1454, 1415, 1299, 1233, 1164, 1102, 1034, 806, 729, 631, 538 cm^{-1}; elemental analysis (%) calcd for $C_8H_{10}N_{16}O_{14}$ (550.32): C 17.34, H 1.82, N 40.43; found C 17.72, H 2.10, N 39.21.

2,2′-(Oxybis(methylene))bis(5-nitro-1H-tetrazole) (**2b-23**): A mixture of sodium 5-nitrotetrazolate dihydrate1 (0.87 g, 5.03 mmol) suspended in acetone (15 mL) was combined with bis(chloromethyl) ether (0.30 g, 2.61 mmol) and sodium bromide (0.52 g, 5.03 mmol). This combination was mixed at room temperature for three days. The resulting white solid was separated by filtration, and the filtrate was then concentrated. Diluting the oily remains with water induced the formation of another white solid. After filtering and washing this precipitate with dichloromethane (10 mL, three times), product **2b-23** was obtained, yielding a 60% as a white solid (0.41 g). T_m: 148 °C. T_d: 203 °C. ^1H NMR: d 6.32 ppm (s, 4H). ^{13}C NMR: d 166.6, 81.5 ppm. IR (KBr): n 3456, 1637, 1563, 1487, 1450, 1414, 1328, 1233, 1192, 1110, 1070, 1018, 991, 945, 845, 779, 689, 660, 660, 543 cm^{-1}. Elemental analysis for $C_4H_4N_{10}O_5$ (272.14): calcd C 17.65, H 1.48, N 51.47%. Found: C 17.68, H 1.32, N 50.51%. IS: 4 J. FS: 40 N.

2,2′-(Oxybis(methylene))bis(5-(trinitromethyl)-1H-tetrazole) (**2b-24**): A quantity of 5-(trinitromethyl)-1H-tetrazole1 (1.00 g, 4.57 mmol) was introduced to pre-chilled sulfuric acid at 0 °C (98% concentration, 5 mL). Then, paraformaldehyde (0.50 g) was gradually added within a span of five minutes. The reaction environment was slowly warmed up to 25 °C and held at room temperature for 12 h. The mixture was then discharged into 100 g of crushed ice, resulting in the formation of white solid. This solid was isolated through filtration and washed with water and dichloromethane three times (10 mL each time). After drying the residue under vacuum, product **2b-24** was obtained as a white solid (0.44 g, yielding 40%). T_m: 75 °C. T_d: 130 °C. ^{1}H NMR: d 6.38 ppm (s, 4H). ^{13}C NMR: d 154.3, 118.5, 82.9 ppm. IR (KBr): n 1628, 1599, 1445, 1337, 1280, 1219, 1189, 1127, 1094, 1030, 993, 943, 843, 799, 770, 682 cm^{-1}. Elemental analysis for $C_6H_4N_{14}O_{13}$ (480.18): calcd C 15.01, H 0.84, N 40.84%. Found: C 14.82, H 1.03, N 42.87%. IS: < 2 J. FS: 20 N.

5,5′-Bis(1H-tetrazolyl)amine (**H_2BTA**): A 2000 mL three-necked flask holding a boiling mixture of sodium dicyanamide (44.5 g, 0.5 mol), sodium azide (65.0 g, 1.0 mol), ethanol (400 mL) and water (250 mL) was combined with 2 M HCl (750 mL) over a duration of five hours. The mixture was then boiled for an additional 48 h. Once cooled to 0 °C in an ice bath and mixed with concentrated HCl (80 mL), 5,5′-Bis(1H-tetrazolyl)amine monohydrate manifested as a fine colorless solid matter. After filtering this precipitate and washing it with small volumes of ethanol and diethyl ether, it was dried under vacuum conditions. The yield produced was 75 g (0.43 mol, 88%). This product can be recrystallized from diluted acids like HCl or HClO4. 250 °C (T_d), (DSC, 5 °C min^{-1}); IR (KBr, cm^{-1}): n 3456 (s), 3028 (s), 2932 (s), 2858 (s), 2671 (m), 2438 (m), 1796 (w), 1656 (vs), 1611 (s), 1556 (s), 1454 (m), 1352 (m), 1337 (m), 1282 (m), 1263 (m), 1154 (w), 1110 (m), 1072 (s), 1501 (s), 1036 (m), 1003 (m), 899 (m), 819 (m), 790 (m), 738 (m), 690 (m), 503 (m), 406 (w). ^{1}H NMR (DMSO-d_6, 25 °C) d: 11.92 (s, br), 9.53 (s, br); ^{13}C NMR (DMSO-d_6, 25 °C) d: 154.7 (C); ^{15}N NMR (DMSO-d_6, 25 °C) d: − 17.9 (N_2), − 123.8 (N_1), − 315.7 (NH); m/z (DEI+): 153 (M+H_2O, 66), 128 (18), 127 (9), 97 (16), 96 (20), 70 (35), 69 (34), 68 (36), 57 (33), 53 (23), 43 (33), 42 (81), 41 (39), 29 (39), 28 (100); EA ($C_2H_5N_9O$, 171.15) calcd: C 14.04, H 2.95, N 73.67%; found: C 14.10, H 2.61, N 73.19%; impact sensitivity: > 100 J; friction sensitivity: > 360 N.

5,5′-Bis(1H-tetrazolyl)amine monohydrate was dehydrated under vacuum at elevated temperature (120 °C). 250 °C (T_d); IR (KBr, cm^{-1}) n 3267 (m), 3136 (m), 3080 (m), 2934 (m), 1654 (vs), 1618 (s), 1607 (s), 1570 (s), 1547 (m), 1481 (m), 1445 (w), 1384 (w), 1321 (w), 1288 (w), 1228 (w), 1174 (w), 1133 (m), 1078 (w), 1057 (m), 1042 (m), 1014 (w), 835 (w), 806 (w), 749 (w), 719 (w), 600 (w); EA ($C_2H_3N_9$, 181.17) calcd: C 15.69, H1.97, N 82.34%; found: C 15.61, H 2.07, N 81.62%; impact sensitivity: > 30 J; friction sensitivity: > 300 N.

1,1′-Azobistetrazole (**2b-25**): A quantity of 1-aminotetrazole1 (0.17 g, 2.0 mmol) was dissolved in acetonitrile (4 mL) and then cooled to 0 °C. This was followed by the addition of an oxidizing mixture, which was prepared by combining glacial acetic acid (0.5 mL) with water (8 mL) containing sodium dichloroisocyanurate (0.45 g, 2 mmol). The solution immediately turned yellow and was stirred at 0 °C for half an hour. Afterwards, this mixture was poured into a low-temperature (5 °C) 2% sodium bicarbonate solution (100 mL) and stirred briefly for five minutes. Filtering this substance produced a white solid. The yield is approximately 72% (0.12 g). (Due to sensitivity, the exact yield cannot be determined accurately). DSC (5 °C min^{-1}): 80 °C (dec); IR (cm^{-1}) ν% = 3134 (m), 1757 (w), 1465 (m), 1424 (m), 1405 (w), 1364 (w), 1328 (w), 1318 (w), 1305 (w), 1211 (w), 1185 (w), 1173 (m), 1155 (s), 1071 (m), 1054 (s), 982 (m), 951 (s), 910 (s), 878 (m), 776 (w), 720 (w), 702 (s), 640 (s); Raman (1064 nm): ν % = 3135 (12), 1899 (1), 1838 (1), 1688 (1), 1530 (100),

1465 (66), 1405 (11), 1372 (25), 1312 (77), 1216 (11), 1152 (26), 1066 (11), 1015 (6), 982 (13), 936 (43), 922 (4), 872 (4), 706 (1), 627 (1), 375 (3), 291 (8), 245 (7); ^{1}H NMR (DMSO d_6) δ (ppm) = 9.33 (s, 2H, C-H); ^{13}C NMR (DMSO d_6) δ (ppm) = 143.2 (s, 2C, CN4). m/z: (DCI+) 167.1 (M+H); EA ($C_2H_2N_{10}$, 166.10 g mol^{-1}) calcd: C 14.46, N 84.32, H 1.21%, found: too sensitive for measurement.

1,1′-Azobis(5-methyltetrazole) (**2b-26**): A combination of benzaldehyde hydrazone (60 g, 0.5 mol), NaN_3 (39 g, 0.6 mol), and triethyl orthoacetate (150 mL, 0.9 mol) was mixed together. Acetic acid (250 mL) was gradually introduced into this mixture while stirring and the resulting solution was heated to 80 °C for a duration of 6 h. This reaction concoction was then slowly poured into 1.2 L of water accompanied by continuous stirring. The formed solid substance was separated through filtration and rinsed with water. It was thereafter exposed to a mixture of water (200 mL) and concentrated hydrochloric acid (100 mL), with benzaldehyde being eliminated from the mixture via steam distillation. After neutralizing the remaining distillate with an aqueous ammonia solution and evaporating it under vacuum to eliminate water, the residue was put through an extraction process using ethyl acetate. The extract was dried over $MgSO_4$ and the remaining ethyl acetate was eliminated by vacuum evaporation. This resulted in the production of 8 g of 5-methyl-1-aminotetrazole as a slightly yellowish, transparent liquid (yielding 16%). ^{1}H NMR (500 MHz, DMSO-d_6, TMS): d 6.85 (s, 2H), 2.43 ppm (s, 3H). ^{13}C NMR (126 MHz, $CDCl_3$, TMS): d 150.5, 7.4 ppm.

The oxidizing agent used in this procedure is trichloroisocyanuric acid (TCICA). The method begins with dissolving 5-methyl-1-aminotetrazole (0.99 g, 10 mmol) in 20 mL of CH_3CN. After cooling the solution to 0 °C, TCICA (4.65 g, 20 mmol) was incrementally added. This mixture was then stirred at the same temperature, 0 °C, for half an hour. Solid Na_2CO_3 was used to neutralize the solution before filtration to remove undissolved solids. The residual solid on the filter paper was rinsed multiple times with acetonitrile, and then combined filtrates were concentrated through evaporation under vacuum. One should note that even after repeated washing with acetonitrile, the drying filter cake spontaneously exploded. Following the solvent's removal under vacuum, the raw product was purified using silica chromatography with 1:1 ethyl acetate: petroleum ether as the eluent. This process produced 0.65 g of **2b-26**, a faint yellow solid, yielding a 67%. ^{1}H NMR (500 MHz, $CDCl_3$, TMS): d 2.91 ppm (s, CH_3). ^{13}C NMR (126 MHz, l_3, TMS): d 153.1, 9.1 ppm. IR: 2832 (vw), 1695 (m), 1542 (m), 1452 (m), 1385 (vs), 1094 (m), 1044 (m), 988 (m), 937 (m), 738 (m), 684 (m) cm^{-1}. ESI–MS: m/z 193.00 [M-H]. Anal. calcd for $C_4H_6N_{10}$ (194): C, 24.74; H, 3.11; N, 72.14. Found: C, 24.76; H, 3.10; N, 72.13. Impact sensitivity and friction sensitivity: too sensitive for measurement; electrostatic sensitivity discharge (ESD): 5 mJ.

Sodium dichloroisocyanurate (SDIC) as an oxidizing agent: A solution of sodium dichloroisocyanurate (SDIC, 2.2 g, 10 mmol) in water (4 mL) was stirred vigorously while acetic acid (0.5 mL) was added at room temperature for a duration of 30 min. This oxidizing mixture was then introduced to a solution of 5-methyl-1-aminotetrazole (0.99 g, 10 mmol) in 20 mL of CH_3CN. The combined reaction content was further stirred at a low temperature of 0 °C for 30 min. The solution underwent neutralization with solid Na_2CO_3 and filtration to eliminate undissolved substances. Acetonitrile was used to wash the filtered solid several times, followed by concentration of the filtrate. Removal of the solvent under vacuum resulted in a crude product which was subsequently purified using silica chromatography with a 1:1 ratio of ethyl acetate: petroleum ether as the eluent. This purification process yielded 0.51 g of a slightly yellow solid known as **2b-26** (yield: 53%).

Tert-butyl hypochlorite as an oxidizing agent: The process commenced with the dissolution of 5-Methyl-1-aminotetrazole (0.99 g, 10 mmol) in 20 mL CH_3CN. After cooling this solution to 0 °C, tert-butyl hypochlorite (1.1 mL, 1.1 g, 10 mmol) was incrementally added. The reaction content was stirred at the same temperature of 0 °C for an additional 30 min. The mixture was neutralized using solid Na_2CO_3 and filtered to remove any un-dissolved substances. Acetonitrile was used to wash the filter residue several times, after which the filtrate was concentrated. Following the removal of solvent under vacuum conditions, the resulting crude product was purified through silica chromatography using a 1:1 ethyl acetate: petroleum ether mix as the eluent. This resulted in 0.57 g of **2b-26**, a faint yellow solid with an overall yield of 59%.

2,2′-Azobis(5-nitrotetrazole) (**2b-27**): The process began with the addition of ethyl O-ptolylsulphonylacetohydroximate (0.95 g, 3.69 mmol), freshly prepared and pulverized, to perchloric acid (7.5 mL, 60%) at room temperature. This was stirred for 2 h under normal conditions. Following this, the resultant tosylhydroxylamine suspension was poured into an ice/water slurry (80 mL) and after the ice had melted, the mixture underwent extraction using several portions (7 × 10 mL) of dichloromethane. The dichloromethane extracts were combined, dried over sodium sulfate, and then added all at once to a solution of ammonium nitrotetrazolate hemihydrate (0.50 g, 3.54 mmol) in acetonitrile (250 mL). This mixture was allowed to stir for two days under normal conditions before being evaporated to dryness and resuspended in ethyl acetate. After filtration, the filtrate was evaporated again and purified via silica chromatography using a 4:4:2 mix of ethyl acetate–hexane–benzene as the eluent. This resulted in a yield of 0.230 g (50%) of 2-amino-5-nitrotetrazole. DSC (5 °C min^{-1}): 102 (T_m), 140 (T_d); IR (cm^{-1}) n = 3340 (m), 3255 (w), 3191 (w), 2945 (w), 2921 (w), 1729 (w), 1678 (w), 1622 (m), 1563 (s), 1529 (m), 1495 (m), 1473 (s), 1454 (w), 1418 (m), 1340 (w), 1319 (s), 1249 (m), 1227 (w), 1190 (m), 1174 (m), 1154 (m), 1144 (m), 1119 (m), 1049 (m), 1073 (m), 1030 (m), 954 (s), 912 (m), 893 (m), 840 (s), 711 (w), 696 (w), 652 (w). Raman (1064 nm, cm^{-1}) n = 3343 (5), 3258 (3), 3195 (8), 2985 (1), 2939 (3), 2915 (1), 1722 (2), 1679 (1), 1617 (4), 1567 (14), 1528 (2), 1497 (20), 1475 (31), 1457 (6), 1420 (100), 1394 (14), 1322 (16), 1252 (3), 1159 (19), 1074 (10), 1030 (86), 842 (8), 772 (11), 762 (3), 746 (17), 663 (1), 569 (8), 455 (8), 419 (6), 332 (3), 229 (6); ^1H NMR (DMSO-d_6) 8.98; ^{13}C NMR (DMSO-d_6) 163.1 (s, 1C); m/z: (DCI+) 131.07 (M+H) ($CH_2O_2N_6$); EA ($CH_2O_2N_6$, 130.07 g mol^{-1}) calcd: C 9.23%, N 64.61%, H 1.55%; found: too sensitive for measurement.

A quantity of 2-amino-5-nitrotetrazole (0.13 g, 1.00 mmol) was dissolved in acetonitrile (4 mL) and cooled to 0 °C. Following this, a solution of sodium dichloroisocyanurate dehydrate (0.225 g, 0.88 mmol) in water (8 mL) with acetic acid (0.25 mL) was introduced gradually. The solution immediately took on a yellow hue and was allowed to stir in the ice bath for another 30 min. Afterwards, this solution was transferred into an aqueous sodium bicarbonate solution (100 mL, 2%), resulting in a precipitate. Upon filtering the solution, the filtercake spontaneously detonated. Raman (1064 nm): n = 3073 (7), 2970 (3), 2910 (1), 2875 (1), 1604 (3), 1575 (4), 1530 (100), 1495 (44), 1395 (37), 1296 (23), 1257 (5), 1225 (3), 1176 (48), 1113 (3), 1062 (7), 1005 (27), 932 (62), 889 (4), 833 (3), 776 (8), 707 (3), 638 (2), 554 (1), 464 (1), 372 (1), 341 (4), 249 (2) (very wet measurement) m/z: (FAB–) 256 (M–), 114 (CN_5O_2–).

Synthesis of 5,5′-Hydrazinebistetrazole (HBT). The procedure began with the dissolution of sodium 5,5′-azotetrazolate pentahydrate (50.00 g, 166.63 mmol) in water (1.4 L). Excess magnesium powder (28.71 g, 1.18 mol) was then introduced into the yellow solution. This mixture underwent a reflux process for 6 h and was then allowed to cool to room temperature under a nitrogen flow. The leftover magnesium powder along with magnesium hydroxide that precipitated upon cooling were filtered. To avoid oxidation of sodium 5,5′-hydrazinobistetrazolate, this filtration step was done swiftly. The filtrate was sent to a flask containing 6 M HCl (160 mL, 1.58 mol). The contact with HCl prevented any further oxidation of the hydrazino-bridged compound. An immediate formation of a white solid was noticed. To encourage precipitation, the flask was thoroughly shaken. The resulting slightly yellow powder was filtered, washed with ample water and acetone, yielding the desired product as a white powder (26.64 g, 95.2%). Crystals adequate for structure determination were grown by dissolving a small quantity of the solid in boiling water and allowing it to cool slowly. $C_2H_4N_{10}$ (calcd/found, %): C, 14.29/14.27; H, 2.40/2.58; N, 83.31/ 82.68. T_m, 206.9–208.2 °C. ^{1}H NMR (DMSO-d_6): 9.66 (4H, s, NH). ^{13}C NMR (DMSO-d_6); 159.68 (s, CN_3). IR (Gas): 3332 (w, NH_3), 3336 (s, HCN), 3282 (s, HCN), 2168 (w, HCN), 2102 (w, HCN), 1621 (m, NH_3), 1431 (w, HCN), 1380 (w, HCN), 965 (vs, NH_3), 926 (vs, NH_3), 732 (s, HCN), 710 (vs, HCN), 679 (s, HCN).

Sodium 5,5′-azoxybistetrazolate pentahydrate: The process began with the dissolution of potassium permanganate (17.1 g, 108 mmol) in water (150 mL). Subsequently, 5-aminotetrazole (12.4 g, 146 mmol) was incrementally added at 0 °C. The ensuing brown mixture was stirred at different temperatures: first at 0 °C for an hour, then at room temperature for another hour, and finally at 50 °C for three hours. After this, ethanol (30 mL, 514 mmol) alongside a solution of sodium hydroxide in water (15%, 150 mL, 562 mmol) were introduced. This resulted in a thick, brown slurry that was stirred at 90 °C for three hours. The mixture was filtered over diatomaceous earth and washed with boiling water until the filtrate was no longer yellow. Rotary evaporation was used to remove the solvent, leaving behind a light-yellow solid. This residue was extracted with hot ethanol, after which the ethanol was removed using rotary evaporation. The remaining yellow solid went through two cycles of recrystallization from EtOH-H_2O, eventually yielding clear, yellow crystals of Sodium 5,5′-azoxybistetrazolate pentahydrate (2.80 g, 8.80 mmol, 16%). DSC (5 °C min^{-1}): 222 °C (T_d); IR (KBr, cm^{-1}): n = 3835 (w), 3422 (vs), 1632 (m), 1500 (m), 1473 (vw), 1450 (m), 1396 (m), 1275 (w), 1207 (w), 1160 (w), 1061 (w), 1034 (w), 913 (w), 768 (m), 559 (w); ^{1}H NMR (DMSO-d_6): 3.40 (H_2O); ^{13}C NMR (DMSO-d_6):

168.4 (CN_4), 164.0 (CN_4); EA ($C_2H_{10}N_{10}O_6Na_2$, 316.14) calcd: N 44.30, C 7.60, H 3.10; found: N 43.57, C 7.78, H 3.06; impact sensitivity: 40 J; friction sensitivity: 324 N.

1,1′-Dihydroxyazotetrazole (**2b-28**): The process began with the dissolution of cyanogen bromide (2.12 g, 20 mmol) in acetonitrile (MeCN, 100 mL). This solution was then cooled to a temperature between 0 and 5 °C before sodium azide (1.40 g, 21.5 mmol) was introduced. Following 4 h of stirring in an ice bath, the suspension was filtered and rinsed with MeCN (10 mL). Importantly, the filter paper must not be allowed to dry out during this step. The solution was then chilled to − 20 °C and a 50% solution of hydroxylamine (2.64 g) in MeCN (20 mL) was added all at once. Subsequently, the resulting suspension was stirred at room temperature (RT) for an hour and filtered. After washing the solid with ether (Et_2O, 30 mL) and air drying, compound hydroxylammonium aminotetrazole-1-oxide was obtained (1.29 g, 48% yield). After an additional 3 h at RT, pure hydroxylammonium aminotetrazole-1-oxide was isolated from the filtrate (0.58 g, 22% yield).^{1}H NMR (DMSO-d_6): d = 8.46, 8.22 ppm; ^{13}C NMR (DMSO-d_6): d = 149.3 ppm; IR (ATR): n = 2464 (m), 1739 (w), 1644 (w), 1557 (m), 1531 (m), 1479 (w), 1308 (m), 1269 (w), 1257 (w), 1204 (w), 1164 (w), 1144 (w), 1127 (w), 1102 (m), 1002 (m), 830 (s), 801 (w), 781 (vs), 721 (m), 668 cm^{-1} (w); DSC (5 °C min^{-1}): 155 °C (T_d); elemental analysis calcd (%) for $CH_6N_6O_2$ (134.1): C 8.96, H 4.51, N 62.67; found: C 9.46, H 4.33, N 62.39; BAM drophammer: 10 J; friction tester: 360 N; ESD: 0.3 J.

The process kicked off with the dissolution of crude hydroxylammonium aminotetrazole-1-oxide (2.165 g, 16.1 mmol) in water (50 mL), followed by the addition of 2 M KOH (11 mL). The solution was then heated to 70 °C. Subsequently, a hot water solution of $KMnO_4$ (2.543 g, 16 mmol) was incorporated into the mixture dropwise. After maintaining the stirring for 30 min at the same temperature, methanol (MeOH, 10 mL) was added, and the solution was stirred for an additional 10 min at 70 °C. The resulting suspension was then hot-filtered through celite and concentrated until the red product began to precipitate. The solution was allowed to crystallize at room temperature (RT). Precipitation was completed by introducing ethanol (EtOH). The precipitate was filtered and rinsed with both EtOH and ether (Et_2O), leading to the production of pure azotetrazole-1,1′-dioxide dipotassium salt (1.726 g, 78% yield) as an orange-red solid. IR (ATR): n = 1738 (w), 1471 (w), 1428 (m), 1378 (w), 1271 (m), 1216 (m), 1143 (m), 1066 (w), 1003 (w), 784 (vs), 740 (w), 717 cm^{-1} (w); Raman (1064 nm, 200 mW, 25 °C): n = 1388 (100), 1205 (9), 1157 (5), 1085 (8), 1018 (3), 922 cm^{-1} (15); DSC (5 °C min^{-1}): 285 °C (T_d); elemental analysis calcd (%) for $K_2C_2N_{10}O_2$ (274.28): C 8.76, N 51.07; found: C 9.21, N 50.96; BAM drophammer: 20 J; friction tester: 360 N; ESD: 0.35 J.

The process began with the dissolution of azotetrazole-1,1′-dioxide dipotassium salt (2.74 g, 10 mmol) in 2 M HCl (50 mL). The solution was then subjected to multiple extractions with ethyl acetate (EtOAc, 30 mL portions) until a faint yellow color was noticeable in the aqueous phase. Following this, the organic phase was dried over magnesium sulfate. Upon removal of the solvent under vacuum, the free acid **2b-28** was obtained as an orange solid, with a yield of 91% (1.80 g). ^{1}H NMR (DMSO-d_6): d = 10.62 ppm; ^{13}C NMR (DMSO-d_6): d = 154.3 ppm; ^{15}N NMR

(DMSO-d_6): d = 98.6 (N_5), − 5.2 (N_3), − 15.1 (N_2), − 68.1 (N_4), − 105.6 ppm (N_1); IR (ATR): n = 2196 (m), 1529 (m), 1402 (m), 1240 (s), 1162 (m), 1077 (s), 867 (vs), 770 (s), 738 (s), 725 (s), 708 (s), 681 cm^{-1} (s); Raman (1064 nm, 200 mW, 25 °C): n = 1628 (4), 1423 (100), 1224 (52), 1163 (7), 1115 (34), 1036 (3), 925 (22), 724 (2), 549 cm^{-1} (9); DSC (5 °C min^{-1}): 170 °C (T_d); BAM drophammer: < 1 J; friction tester: < 5 N; ESD: 0.01 J.

Bistetrazolylhydrazinedioxide (**2b-29**): The procedure started with dissolving **K_2AzTO** (10 mmol) in hot water (50 mL), to which powdered magnesium (3 g) was added. This solution was boiled until the color vanished within a few minutes. Following this, the solution was quickly cooled over ice, filtered, and acidified to pH 1 using concentrated hydrochloric acid. Subsequent extraction took place five times with ethyl acetate (EtOAc, 20 mL). The organic phase was then dried over magnesium sulfate and evaporated. The remaining residue was crystallized from water, resulting in the formation of complex **2b-29** as colorless crystals, with a yield of 70% (1.59 g). ^{1}H NMR (DMSO-d_6): d = 7.38, 13.37, 9.49 ppm; ^{13}C NMR (DMSO-d_6): d = 151.1 ppm; ^{15}N NMR (DMSO-d_6): d = − 11.1 (N_3), − 25.9 (N_2), − 91.9 (N_4), − 147.0 (N_1), − 306.5 ppm (N_5); IR (ATR): n = 3307 (w), 2896 (w), 1621 (s), 1539 (m), 1514 (m), 1466 (w), 1435 (w), 1353 (w), 1312 (w), 1245 (vs), 1212 (w), 1157 (w), 1143 (m), 1115 (w), 1090 (m), 1036 (w), 1004 (w), 973 (w), 862 (m), 789 (w), 716 cm^{-1} (m); Raman (1064 nm, 200 mW, 25 °C): n = 3305 (14), 3110 (15), 1630 (17), 1545 (5), 1523 (4), 1389 (100), 1319 (16), 1268 (17), 1214 (10), 1145 (13), 1098 (10), 1006 (9), 980 (5), 929 (10), 840 (14), 689 (6), 532 cm^{-1} (4); DSC (5 °C min^{-1}): 120 °C (T_d).

1,1′-Diamino-5,5′-azotetrazole (**2b-30**): The procedure initiated with the dissolution of **DAT** (1.0014 g, 10 mmol) in acetone (20 mL), to which 4 drops of H_2SO_4 were added. This mixture was heated under reflux at 58 °C for an hour. Following this, the precipitate was filtered and rinsed with acetone, yielding a white solid. The yield was 66.3%. (0.9301 g). T_m: 154 °C. ^{1}H NMR (400 MHz, DMSO-d_6): d 6.71 (2H, s, NH_2), d 2.39 (3H, s, CH_3), d 2.08 (3H, s, CH_3). IR (KBr): 3342 (s), 3243 (m), 3174 (s), 2997 (w), 1661 (vs), 1637 (vs), 1581 (vw), 1462 (vw), 1424 (vw), 1363 (vw), 1105 (vw), 1056 (vw), 709 (vw), 563 (vw) cm^{-1}. EI-MS: m/z 140 [M+]. Anal. caclcd for $C_4H_8N_6$ (140): C, 28.10; H, 6.30; N, 65.60. Found: C, 28.00; H, 6.35; N, 65.65.

The process began with the addition of acetic acid (1 mL) to a solution containing sodium dichloroisocyanurate (SDIC, 0.8653 g, 3 mmol) in methanol (20 mL). This mixture was stirred energetically at 30 °C for 20 min. Following this, 5-amino-1-isopropylidene-tetrazole (0.4236 g, 3 mmol) dissolved in methanol (10 mL) was incorporated dropwise. The reaction mixture was then stirred at 30 °C for an additional 20 min. Lastly, the precipitate was filtered off and rinsed three times with hot water, yielding an orange solid with a 64.7% yield (0.2701 g). T_d: 151 °C, ^{1}H NMR (500 MHz, DMSO-d_6): d 2.42 (3H, s, CH_3), d 2.10 (3H, s, CH_3). IR (KBr): 3434 (vs), 2363 (w), 1705 (vw), 1628 (m), 1418 (vw), 1375 (vw), 1213 (vw), 1078 (vw), 612 (vw) cm^{-1}. EI-MS: The solubility is too low to allow for measurement, in addition the compound is heat sensitive. Anal. caclcd for $C_8H_{12}N_{12}$ (276.30): C, 34.78; H, 4.35; N, 60.87. Found: C, 34.69; H, 4.41; N, 60.90.

The procedure initiated with the suspension of 1,1′-diisopropylideneamino-5,5′-azotetrazole (0.2851 g, 2 mmol) in 80% acetic acid (2 mL) and heated at 60 °C for 30 min. The mixture was then cooled to room temperature and filtered. After washing the product with water, a light yellow compound **2b-30** was obtained. This product was further purified by recrystallization from DMSO. The overall yield was 83.5%. T_d: 179 °C. ^{1}H NMR (500 MHz, DMSO-d_6): d 7.93 (4H, s, NH_2). ^{13}C NMR (500 MHz, DMSO-d_6): d 156.8 ppm. IR (KBr): 3297 (vs), 3248 (vw), 3199 (vw), 2147 (w), 1613 (m), 1497 (vw), 1462 (vw), 1320 (vw), 1277 (vw), 1156 (vw), 1072 (vw), 1019 (vw), 971 (vw), 753 (vw), 699 (vw), 566 (vw), 505 (vw) cm^{-1}. EI-MS: The solubility is too low to allow for measurement, in addition the compound is heat sensitive. Anal. calcd for $C_2H_4N_{12}$ (196.17): C, 12.24; H, 2.04; N, 85.71. Found: too sensitive for measurement.

1,1′-Dinitramino-5,5′-azobitetrazole (**2b-31**): The process started with the suspension of 1,1′-diamino-5,5′-azobitetrazole (490 mg, 2.5 mmol) in dry acetonitrile (50 mL). This mixture was cooled using an ice bath, and a solution of N_2O_5 (540 mg, 5 mmol) in cold acetonitrile (30 mL) was added gradually. As the diaminoazotetrazole dissolved, a red color indicative of the product formation was seen. The reaction was then neutralized by adding 7 mL of 2 M KOH. The resulting red precipitate was filtered off and air-dried. The final product was 550 mg (a yield of 62%) of crystalline dipotassium 1,1′-dinitramino-5,5′-azobitetrazolate. DSC (5 °C min^{-1}): 180 °C (T_d). IR (atr): n = 1635 (m), 1451 (s), 1434 (s), 1297 (s), 1242 (m), 1226 (w), 1119 (m), 1080 (vs), 1037 (s), 875 (w), 770 (w), 751 (w), 740 cm^{-1} (w).

Bis(tetrazol-5-yl)-triazene (**2b-32**): The procedure began with dissolving barium 5,5′-bitetrazolate (1.15 g, 4.2 mmol) in hot water (30 mL). A solution of about 1 g (1.38 mmol) of the corresponding rare earth metal sulfate hydrate in water (10 mL) was then added. The resulting barium sulfate precipitate was filtered out. After evaporating the water, a solid was left, which was dried under vacuum at room temperature.

Bis(1-methyl-tetrazol-5-yl)-triazene (**2b-33**): The procedure started with suspending 1-methyl-5-aminotetrazole (19.82 g, 200 mmol) in water (100 mL) and cooling it down to 0 °C. Upon the addition of concentrated HCl (21.24 g, 200 mmol), the 1-methyl-5-aminotetrazole completely dissolved. Following this, a solution of $NaNO_2$ (6.90 g, 10 mmol) in water (40 mL) was added drop by drop over 20 min. The solution was then stirred for 24 h. The resulting colorless solid was filtered and rinsed with water. Single crystals were acquired from a methanol solution (11.0 g, yield 52.6%). T_m: 160 °C, T_d, 182 °C; IR (KBr, cm^{-1}): 3375 (m), 3270 (m), 3081 (w), 3031 (w), 2662 (m), 2554 (m), 1664 (m), 1631 (s), 1530 (m), 1502 (m), 1458 (m), 1437 (s), 1396 (m), 1308 (m), 1268 (s), 1247 (m), 1226 (m), 1194 (s), 1102 (w), 1036 (m), 1003 (w), 983 (w), 936 (m), 806 (m), 747 (m), 729 (m), 699 (s), 666 (m), 641 (m), 605 (m), 482 (w); ^{1}H NMR (DMSO-d_6) d: 4.04 (s, 6H, CH3); ^{13}C NMR (DMSO-d_6) d: 34.9 (s, CH_3), 155.2 (tetrazole); ^{15}N NMR (DMSO-d_6) d: – 265.0 (s, N_5, N_7), – 168.7 (d, N_1), – 76.59 (s, N_4), – 12.72 (d, N_2), 5.54 (s, N_3), 85.20 (s, N_6); EA ($C_4H_9N_{11}O$, 227.19): calc.: C, 21.15; H, 3.99; N, 67.82; O, 7.04. Found: C, 21.18; H, 3.89; N, 67.48%; BAM drophammer: 3 J; friction tester: 360 N.

1,1′-(triaz-1-ene-1,3-diyl)bis(1H-tetrazol-5-amine) (**2b-34**): The process began by adding a solution of sodium nitrite (0.69 g, 10 mmol) in water (5 mL) drop by drop to a stirred solution of **DAT** (1.0 g, 10 mmol) in concentrated HCl (5 mL). This reaction mixture was maintained at 0 °C for 30 min. After the solvent was removed under vacuum, resultant crude residues were treated with hot ethanol (20 mL) and then filtered to eliminate insoluble salt. The filtrate was subsequently evaporated, dried, and recrystallized with ethanol, yielding 0.5 g of needle-like crystals. ^{1}H NMR (500 MHz, DMSO-d_6, 25 °C, TMS): δ 7.93 ppm. ^{13}C NMR (126 MHz, DMSO-d_6, 25 °C, TMS): δ 153.8, 152.5, 150.1 ppm. IR: 3403 (m), 3193 (m), 3157 (m), 3056 (s), 1692 (s), 1640 (vs), 1578 (w), 1472 (vw), 1447 (w), 1274 (w), 1052 (w), 984 (w), 893 (w), 773 (w), 728 (w), 673 (w) cm^{-1}. Raman: 1684, 1582, 1390, 1284, 1062, 746 cm^{-1}. BAM impact: 10 J; BAM friction: 160 N; Electrostatic sensitivity discharge (ESD): 0.32 J.

1,4-Bis-[1-methyltetrazol-5-yl]-1,4-dimethyl-2-tetrazene (**2b-35**): The process initiated by cooling a solution of 1-methyl-5-(methylamino)-1H-tetrazole (2.83 g, 25 mmol) in 2 M HCl to 273 K. Sodium nitrite (1.73 g, 0.25 mol) dissolved in 20 mL water was then slowly added, maintaining the temperature between 273 and 278 K. Once diazotization finished, the pH was adjusted to 8 using anhydrous potassium carbonate, followed by extraction with dichloromethane (3 × 100 mL). The combined organic layers were rinsed twice with water (70 mL each time), and dried over $MgSO_4$. Post solvent removal under vacuum, the yellow residue of N-nitrosamino-1H-tetrazole was dissolved in acetic acid (10 mL). This solution was rapidly added dropwise to a mix of zinc dust (5 g) suspended in water (50 mL), while keeping the temperature at 0–25 °C. Afterward, the reaction mixture was warmed to 10 °C and filtered. Liquid bromine was added dropwise to the filtrate until a persistent color was seen. After resting overnight at 25 °C, the precipitate was gathered and recrystallized from acetone/chloroform, yielding colorless rods of **2b-35** (720 mg, 5.7 mmol, 22.8% yield), suitable for single-crystal X-ray structure analysis. IR (KBr): n 3026 (vw), 1622 (m), 1568 (m), 1468 (s), 1411 (m), 1396 (w), 1382 (vw), 1319 (vw), 1274 (vw), 1217 (w), 1117 (s), 1101 (m), 1039 (vw), 1006 (w), 962 (vw), 730 (w), 716 (w), 681 (vw), 656 (m), 468 (vw) cm^{-1}; ^{1}H NMR (400 MHz, DMSO-d_6): 3.63 (s, 3H, –CH_3), 4.13 (s, 3H, –CH_3); ^{13}C NMR (400 MHz, DMSO-d_6): 35.7 (CH_3), 39.8 (CH_3), 154.9 (C); $C_6H_{12}N_{12}$ Calc. C, 28.6; H, 4.8; N, 66.6%; Found: C, 28.5; H, 5.0; N, 66.8%.

Di(1H-tetrazol-5-yl)methanone oxime (**2b-36**): The process kicked off with suspending potassium dicyanonitrosomethanide (2.79 g, 21 mmol), sodium azide (2.66 g, 42 mmol), and zinc chloride (7 g, 51 mmol) in water (200 mL). It was then heated under reflux for 6 h, leading to the formation of a yellow precipitate that gradually turned brown. After cooling the reaction mixture to room temperature, it was acidified with 2 N HCl, causing the brown precipitate to dissolve and a new one to form upon stirring. This newly formed precipitate was filtered, washed with water, and dried under vacuum to yield a white solid with an 81.5% yield. This technique can be scaled up for larger quantities. T_m, 143.9 °C, T_d (onset), 288.7 °C, IR (KBr) n 3552, 3407, 2966, 2879, 1831, 1627, 1572, 1523, 1476, 1418, 1328, 1242, 1134, 1077, 945, 761, 474 cm^{-1}; ^{13}C NMR d 151.1, 145.5, 132.1; elemental analysis (%)

calcd for $C_3H_3N_9O$ (199.13): C, 18.09; H, 2.53; N, 63.31; found C, 18.08; H, 2.21; N, 63.20.

5,5′-(Hydrazonomethylene)bis(1H-tetrazole) (**2b-37**): The procedure started with suspending di(1H-tetrazol-5-yl)methanone oxime in water. Four equivalents of hydrazine hydrate were then slowly added while stirring continuously. The resultant solution was heated under reflux overnight. After letting it cool down to room temperature, concentrated HCl was added to form a white precipitate. This solid was filtered off and rinsed with water. The filtrate was checked to ensure complete precipitation. The combined precipitate was dried under vacuum, resulting in an almost quantitative yield. T_d (onset), 247.6 °C, IR (KBr) n 3400, 3362, 3164, 1641, 1617, 1581, 1551, 1415, 1327, 1247, 1197, 1136, 1069, 1040, 933, 873, 742, 582 cm^{-1}; ^1H NMR d 16.42 (s, 2H, NH), 10.24 (s, 2H, NH_2); ^{13}C NMR d 152.83, 147.26, 109.26; elemental analysis (%) calcd for $C_3H_4N_{10}$ (181.13): C, 20.00; H, 2.24; N, 77.76; found C, 19.65; H, 2.22; N, 76.95.

3,3′-Diamino-5,5′-bis(1H-1,2,4-triazole) (**DABT**): The process initiated with the addition of hydrochloric acid (60 mL) to a stirred blend of oxalic acid (20.0 g, 159 mmol) and aminoguanidinium bicarbonate (45.4 g, 332 mmol). The reaction mixture was maintained at 70 °C with stirring for an hour, followed by filtration to collect the precipitate. This colorless solid was then dissolved in water (240 mL), and its pH was adjusted to 14 using sodium hydroxide. After heating it under reflux for an hour, the reaction mixture was acidified to pH 4 using acetic acid. The resulting precipitate was gathered by filtration, rinsed with water (200 mL), and left to air dry, yielding **DABT** (18.6 g, 112 mmol, 70% yield) as a colorless solid. ^1H NMR (400 MHz, DMSO-d6): d = 6.46 ppm (s, 2H; NH_2); ^{13}C NMR (100 MHz, DMSO-d6): d = 157.3, 149.3 ppm; IR: n = 3325 (m), 3116 (m), 2863 (m), 2784 (m), 1706 (s), 1668 (s), 1654 (s), 1618 (m), 1606 (m), 1484 (m), 1457 (m), 1267 (m), 1104 (vs), 1061 (s), 987 (w), 956 (w), 769 (w), 721 cm^{-1}(s); MS (DEI+): m/z: 166.1 $[C_4H_7N_8]^+$; elemental analysis (%) calcd for $C_4H_6N_{10}$: C 28.92, H 3.64, N 67.44; found: C 28.72, H 3.58, N 66.1.

3,3′-Dinitro-5,5′-bis(1H-1,2,4-triazole) (**DNBT**): The procedure started by gradually adding a solution of **DABT** (11.9 g,72 mmol) in 20% sulfuric acid (140 mL) to a solution of sodium nitrite (98.8 g, 1.4 mol) in water (140 mL) at 40 °C. This mix was stirred at 50 °C for an hour. Once cooled to room temperature, the mix was further acidified with 20% sulfuric acid until no more nitrogen dioxide was produced. The resultant precipitate was collected through filtration and dissolved in boiling water. The hot filtrate was allowed to cool down to room temperature, which led to the precipitation of a pale green solid. This precipitate was collected to obtain **DNBT**·$2H_2O$ (15.5 g, 59 mmol, 82% yield) as a crystalline solid. ^1H NMR (400 MHz, DMSO-d6): d = 9.68 ppm (s, 2H); ^{13}C NMR (100 MHz, DMSO-d6): d = 162.7, 145.6 ppm; ^{14}N NMR (28.9 MHz, DMSO-d6): d = 26 ppm (NO_2); ^{15}N NMR (40.5 MHz, DMSO-d6): d = – 27.8 (N_4), – 88.8 (N_2), – 141.7 (N_3), – 156.1 ppm (N_1); IR: n = 3599 (m), 3499 (m), 3052 (w), 2849 (w), 2747 (w), 2670 (m), 2621 (m), 2574 (m), 2530 (m), 2488 (m), 2419 (m), 1844 (w), 1609 (m), 1532 (vs), 1466 (w), 1416 (vs), 1314 (vs), 1245 (m), 1183 (m), 1024 (m), 953 (s), 837 (s), 690 (w), 690 cm^{-1} (w); elemental analysis (%) calcd for $C_4H_2N_8O_4$: C 21.25, H 0.89, N

49.56; found: C 21.44, H 0.95, N 49.19;sensitivities (grain size < 100 mm): friction = 360 N, impact = 10 J, ESD = 0.1 J; DSC (onset, 58C min-1): $T_d = 251$ °C.

Synthesis of 5,5′-Bis(3-diazo-1,2,4-triazole) (**BDTZ**): The process began with carefully adding a suspension of **DABT** (0.50 g, 3.01 mmol) in 20% sulfuric acid (15 mL) to a solution of sodium nitrite (0.63 g, 9.13 mmol) in water (5 mL), maintaining the temperature at 0–5 °C. The reaction mixture was then stirred at room temperature for an hour. Following this, the resulting white precipitate was filtered and rinsed with water before being air dried to obtain 0.4 g of a white solid, which corresponds to a yield of 70.6%. ^{13}C NMR (125 MHz, DMSO-d6): 156.411, 136.762; IR (KBr): 3441, 2277, 2232, 1721, 1402, 1326, 1311, 1275, 1205, 1074, 1016, 970, 706, 655. Anal. calcd for C_4N_{10} (%): C 25.53, N 74.47; Found: C 25.42, N 73.29.

3,3′-Dinitro-5,5′-bis-1H-1,2,4-triazole-1,1′-diol (**2c-1**): The process commenced by dissolving 3,3′-dinitro-5,5′-bis(1H-1,2,4-triazole) (5.0 g, 19 mmol) in a blend of water (125 mL) and potassium acetate (25.0 g, 0.25 mol), while maintaining the temperature at 40 °C. Oxone (83.0 g, 0.27 mol) was added in portions over 2 h, with pH carefully adjusted to 4–5 by introducing potassium acetate (38.0 g, 0.38 mol) dissolved in water (50 mL) drop by drop. This mix was then stirred at 40 °C for 48 h. Afterward, the solution was acidified using sulfuric acid (50 wt %, 150 mL) and then extracted with ethyl acetate (4 × 100 mL). The combined organic layers were dried over magnesium sulfate and the solvent was evaporated under vacuum to yield **2c-1** (4.0 g, 16 mmol, 81% yield) as a colorless solid. ^{1}H NMR (DMSO-d6): 9.01 (s, 2H, OH) ppm; ^{13}C NMR (DMSO-d6): 154.9 (C–NO_2), 134.4 (C–C); ^{14}N NMR (DMSO-d6): – 27 (NO_2); ^{15}N NMR (DMSO-d6): – 28.4 (N_4), – 90.7 (N_2), – 121.5 (N_1), – 140.6 (N_3); IR: 3502 (m), 3462 (s), 3346 (s), 1628 (m), 1550 (vs), 1461 (s), 1407 (m), 1346 (w), 1314 (m), 1232 (m), 1177 (m), 1041 (w), 1006 (w), 872 (w), 831 (m), 803 (m), 760 (w), 753 (w), 732 (w), 669 (w). EA ($C_4H_2N_8O_6$): calcd: C 18.61, H 0.78, N 43.41; found: C 18.77, H 0.94, N 42.13 m/z (FAB–): 257.0 [$C_4N_8O_6H^-$]. Sensitivities (grain size: < 100 μm): BAM friction: 360 N, BAM impact: 10 J, ESD: 0.4 J; DSC (onset, 5 °C min^{-1}): T_d: 191 °C.

(5,5′-Dinitro-2H,2′H-[3,3′-bi(1,2,4-triazole)]-2,2′-diyl)bis(methylene) dinitrate (**2c-2**): The process started with adding the ring-coupled precursor **2c-1** (0.68 g, 3 mmol) to a blend of water (2 mL) and formaldehyde solution (37%, 5 mL). For **DNBT**, a drop of 10% sulfuric acid was added to the mix. After stirring for a day, the resulting solid was filtered off, rinsed with water, and air-dried to yield the crude diol product, which can be used directly in the next step. For the nitration mixture, fuming nitric acid (2 mL) was slowly added to acetic anhydride (3 mL) at 0 °C. Then the resulting diol product was gradually introduced at 0 °C. The reaction mix was allowed to stir for an additional hour at 0–5 °C after complete addition. Subsequently, this mixture was poured onto ice, leading to a precipitate that was filtered, washed with cold ethanol, and air-dried to obtain **2c-2**, a white solid (767 mg, 68% yield). ^{1}H NMR d: 6.98 (s, 4H, CH_2) ppm; ^{13}C NMR d: 161.1, 143.7, 76.5 ppm; IR (KBr): n 1686, 1561, 1422, 1317, 1281, 986, 833 cm^{-1}; anal. calcd for $C_6H_4N_{10}O_{10}$ (376.01): C 19.16, H 1.07, N 37.24; found C 19.69, H 1.06, N 35.81.

3,3′-Dinitrimino-5,5′-bis(1H-1,2,4-triazole) (**2c-3**): The procedure began with the slow addition of nitric acid (100%, 3.0 mL) to a solution of **DABT** (1.0 g, 6.0 mmol) in concentrated sulfuric acid (9.0 mL) at 0 °C. The mix was left to warm up to room temperature and stirred for an hour. Subsequently, the clear solution was poured onto ice, leading to a precipitate which was gathered by filtration. It was recrystallized from boiling water to yield **2c-3**·$2H_2O$ (1.35 g, 4.6 mmol, 77%) as a yellow crystalline solid. ^{1}H NMR (400 MHz, DMSO-d6): d = 5.52 ppm (s, 2H); ^{13}C NMR (100 MHz, DMSO-d6): d = 153.1, 142.1 ppm; ^{14}N NMR (28.9 MHz, DMSO-d6): d = − 21 ppm (NO_2); IR: n = 3165 (m), 3154 (m), 1565 (vs), 1508 (s), 1463 (s), 1446 (s), 1380 (m), 1298 (vs), 1229 (vs), 1140 (m), 1085 (m), 1054 (s), 989 (m), 947 (s), 849 (m), 778 (s), 766 (s), 751 (s), 708 cm^{-1}(vs); elemental analysis (%) calcd for $C_4H_2N_8O_4$: C 16.44, H 2.76, N 47.94; found: C 16.73, H 2.69, N 47.73; sensitivities (water-free compound, grain size < 100 mm): friction = 108 N, impact = 3 J, ESD = 0.5 J;DSC (onset, 5 °C min^{-1}): T_d = 194 °C.

3,3′-Diazido-5,5′-bis(1H-1,2,4-triazole) (**2c-4**): The process started by gradually adding a solution of sodium nitrite (0.37 g, 5.4 mol, 3 equiv) in water (2.0 mL) to a suspension of **DABT** (0.3 g, 1.8 mmol) in 20% sulfuric acid (20 mL) at 0 °C. The mix was allowed to warm to room temperature before being stirred at 40 °C for an hour. After cooling to room temperature, a solution of sodium azide (5.9 g, 9.0 mmol) in water (2.0 mL) was slowly added in. The suspension was then stirred overnight at room temperature to eliminate excess sodium azide, followed by extraction with ethyl acetate (3 × 20 mL). Post evaporation of the solvent, the resulting solid was recrystallized from water. This led to the formation of a colorless precipitate, which was collected to yield **2c-4**·$2H_2O$ (0.25 g, 1.0 mmol, 56% yield), appearing as crystalline needles.^{1}H NMR (400 MHz, DMSO-d6): d = 14.86 ppm (s, 2H); ^{13}C NMR (100 MHz, DMSO-d6): d = 157.7, 145.9 ppm; ^{14}N NMR (28.9 MHz, DMSO-d6): d = − 145 ppm (N3); ^{15}N NMR (40.5 MHz, DMSO-d6): d = − 115.3 (N_2), − 141.9 (N_3), − 146.2 (N_5), − 153.0 (N_6), − 173.2 (N_1), − 295.2 ppm (N_4); IR: n = 3141 (s), 3042 (s), 2876 (s), 2710 (s), 2655 (s), 2630 (m), 2571 (m), 2435 (m), 2362 (m), 2240 (m), 2228 (m), 2156 (vs), 2138 (vs), 2137 (vs), 1541 (vs),1518 (s), 1483 (vs), 1457 (s), 1420 (vs), 1418 (vs), 1391 (s), 1333 (s), 1299 (m), 1299 (m), 1275 (s), 1241 (m), 1218 (m), 1188 (s), 1142 (s), 1122 (s),1033 (vs), 1014 (m), 980 (vs), 958 (m), 846 (m), 799 (s), 780 (m), 729 (s),714 (m), 661 (m), 532 cm^{-1} (m); MS:(DEI+):m/z: 218.1 [$C_4H_2N_{12}$]+; elemental analysis (%) calcd for $C_4H_6N_{12}O_2$: C 18.90, H 2.38, N 66.13; found: C 19.41, H 2.14, N 65.75; sensitivities (grain size < 100 mm): friction = 48 N, impact = 3 J, ESD = 0.04 J; DSC (onset, 5 °C min^{-1}): T_d = 202 °C.

5,5′-Dinitro-2H,2′H-3,3′-bi(1,2,4-triazole)-2,2′-diamine (**2c-6**): The procedure began with the slow addition of a dichloromethane solution (100 mL) containing O-tosylhydroxylamine (15 mmol) to an icy suspension of diammonium 3,3′-dinitro-bis(1,2,4-triazolate) (781 mg, 3 mmol) in dimethylformamide (30 mL). Following which, potassium carbonate was introduced, and the resulting reaction mixture stirred at room temperature for 12 h. Afterward, dichloromethane was eliminated by rotary evaporation, and ethyl acetate (50 mL) was added, followed by heating this blend at reflux for 10 min. The hot suspension was then filtered using Celite and rinsed with hot ethyl acetate. Post evaporation of the filtrate, the crude product underwent purification via chromatography using hexane/ethyl acetate/dimethylformamide (100:50:5), yielding compound **2c-6** as a white solid. 45% yield. ^{1}H NMR (DMSO-d6): 7.51

(s, 4H); ^{13}C NMR (DMSO-d6): 157.8, 140.4; ^{15}N NMR (DMSO-d6): – 28.23, – 79.38, – 132.78, – 148.51, – 292.47 (t, J = 73.52 Hz); IR (KBr pellet): 3333, 3300, 3237, 3165, 1594, 1561, 1418, 1315, 1018, 934, 837, 677 cm^{-1}; elemental analysis (%) calcd for $C_4H_4N_{10}O_4$ (312.16): C, 18.76; H, 1.57; N, 54.68; found: C, 19.16; H, 1.70; N, 53.18.

N,N′-(5,5′-dinitro-2H,2′H-3,3′-bi(1,2,4-triazole)-2,2′-diyl)dinitramide (**2c-7**): The process initiated with dissolving compound **2c-6** (512 mg, 2 mmol) in concentrated sulfuric acid (20 mL) at 70 °C, which was then cooled to – 15 °C using an ice-salt bath. Fuming nitric acid (5 mL) was gradually introduced into this cold solution, and the resultant reaction mixture was maintained at – 15 °C for 2 h. Subsequently, the precipitate was filtered, rinsed with trifluoroacetic acid, and vacuum-dried to obtain **2c-7** as a white solid, with an 85% yield. ^{1}H NMR (DMSO-d6): 6.20 (br, 2H); ^{13}C NMR (DMSO-d6): 158.6, 140.6; ^{15}N NMR (DMSO-d6): – 4.31, – 27.06, – 79.28, – 113.52, – 131.76, – 134.12; IR (KBr pellet): 3578, 3186, 3019, 2872, 1628, 1492, 1459, 1421, 1385, 1357, 1295, 1224, 1052, 1017, 955, 868, 836, 767, 736, 673, 640 cm^{-1}; elemental analysis (%) calcd for $C_4H_2N_{12}O_8$ (346.13): C, 13.88; H, 0.58; N, 48.56; found: C, 13.59; H, 0.73; N, 46.15.

4,4′,5,5′-Tetraamino-3,3′-bi-1,2,4-triazole (**2c-8**): The protocol commenced by slowly dissolving phosphorus pentoxide (10 g, 70.4 mmol) in preheated (to 50 °C) phosphoric acid (30 g, 306 mmol). A finely pulverized blend of oxalic acid dihydrate (3.15 g, 25.0 mmol) and diaminoguanidine monohydrochloride (8.29 g, 66 mmol) was subsequently introduced into this warmed solution. Upon complete addition, the viscous mixture was gradually heated to 120 °C, during which gas evolution of HCl was noted. The mixture was heated at 120 °C for 4 h, followed by cooling to room temperature under continuous stirring. On adding 150 mL of ice water into this mix, a white precipitate materialized. Around 75 mL of 10 M NaOH was utilized to neutralize the reaction blend, causing the suspension color to shift from white to brown. The precipitate was filtered out, rinsed several times with water, and air-dried to yield an unrefined compound **2c-8** in the form of a brownish solid (Yield: 1.39 g, 7.09 mmol, 28%). For purification purposes, the crude product underwent recrystallization using hydrochloric acid or glacial acetic acid. Compound **2c-8** (1000 mg, 5.10 mmol, 1.00 eq.) was gradually added to glacial acid, then heated until complete dissolution of **2c-8**. Post removal from the heating bath, the mixture was allowed to cool to room temperature. After filtration and repeated water washes, the residue was dried under nitrogen flow, followed by oven drying at 100 °C overnight. Subsequently, the solid was suspended in 50 mL of water and basified using 1 mL of 10 M NaOH. The suspension was filtered, and the residue air-dried, yielding the pure form of compound **2c-8** as a white solid (Yield: 696 mg, 3.55 mmol, 70%). DTA (5 °C min^{-1}) onset: 342 °C (T_d); IR (ATR, cm^{-1}): n = 3782 (vw), 3400 (m), 3341 (m), 3278 (m), 3142 (w), 2348 (vw), 1626 (vs), 1537 (s), 1478 (m), 1421 (w), 1317 (w), 1252 (vw), 1232 (vw), 1085 (m), 1022 (m), 987 (vs), 935 (m), 799 (vw), 778 (vw), 723 (m), 678 (vw), 664 (vw); ^{1}H NMR (DMSO-d6): 5.91 (br s, 4H), 5.81 ppm (br s, 4H); ^{13}C NMR (DMSO-d6): δ = 155.4 (s, C–NH_2), 139.4 ppm (s, C–C); m/z (DEI$^+$): 196.2 ($C_4H_8N_{10}$); EA ($C_4H_8N_{10}$, 196.17): C 24.49, H 4.11, N 71.40; found: C 25.04, H 4.04, N 69.93; BAM impact: 40 J, BAM friction: 360 N, ESD: 1.5 J.

5,5′-Diamino-4,4′-dinitramino-3,3′-bi-1,2,4-triazole (**2c-9**): The procedure initiated with the placement of HNO_3 (100%, 80 mL) in a 500 mL three-neck round-bottom flask, which was cooled to − 10 °C. Subsequently, compound **2c-8** (10.0 g, 50.98 mmol) was gradually introduced while ensuring that the temperature stayed below 0 °C. Upon complete addition, the resulting orange solution was stirred for 90 min at − 5 °C. Next, the solution was transferred into 300 g of ice and stirred for around 10–15 min. As the stirring continued, **2c-9** started precipitating as a colorless solid out of the ice-water and HNO_3 mixture. The precipitate was isolated by filtration, followed by washes with ice-cold water, ethanol, and ether (each 30 mL, 3 times), yielding a slightly yellowish powder of pure **2c-9** (Yield: 10.1 g, 35.29 mmol, 69%). Recrystallization in boiling water resulted in colorless crystals of **2c-9** suitable for X-Ray diffraction experiments. DTA (5 °C min^{-1}) onset: 259 °C (T_d); IR (ATR): n = 3434 (w), 3214 (w), 2715 (w), 1688 (s), 1515 (w), 1440 (m), 1273 (vs), 1237 (m), 1093 (w), 1047 (w), 997 (s), 886 (m), 782 (m), 705 cm^{-1} (m); ^1H NMR (DMSO-d_6): d = 14.43 (br s, 2H, N–H), 8.43 ppm (s, 4H, –NH_2); ^{13}C NMR (DMSO-d_6): d = 149.1 (s, C-NH2), 136.5 ppm (s, C=C); ^{14}N NMR (DMSO-d6): d = − 4.7 ppm (s, –NO_2); m/z (ESI–): 285.1 ($C_4H_5N_{12}O_4^-$) m/z (ESI+): 287.1 ($C_4H_7N_{12}O_4^+$); EA ($C_4H_6N_{12}O_4$, 286.17): C 16.79, H 2.11, N 58.74, found: C 16.72, H 2.38, N 57.03; BAM impact: 9 J, BAM friction: 120 N, ESD: 0.15 J.

3,3′-Dinitromethyl-5,5′-bis(1,2H-1,2,4-triazole) (**2c-10** or **2c-11**): The process began with the dissolution of 3,3′-diethyl-acetate-5,5′-bis(1H-1,2,4-triazole) (0.5 g, 1.62 mmol) in concentrated sulfuric acid (6 mL), which was then cooled to 0 °C. Following this, concentrated nitric acid (4 mL) and 60% oleum (SO_3, 2 mL) were introduced gradually, after which the mixture was allowed to reach room temperature before being stirred for another 2 h. Next, the mixture was transferred into ice, neutralized with sodium hydroxide, and stirred until complete dissolution of solids. The solution's pH was then adjusted to 1 by adding concentrated sulfuric acid. The precipitate that formed was isolated through filtration and recrystallized from water, resulting in the extraction of compound **2c-11** as a yellow solid (0.27 g, 0.79 mmol, 49%). ^1H NMR (400 MHz, DMSO-d_6, 60 °C): d = 8.94 (s, 2H; NH), 3.91 ppm (s, 2H; CH); ^{13}C NMR (100 MHz, DMSO-d6, 60 °C): 151.1, 149.3, 106.5 ppm; ^{14}N NMR(28.9 MHz, DMSO-d6, 60 °C): − 23 ppm (NO_2); IR: n = 3544 (m), 3467 (m), 3319 (m), 2360 (vw), 2094 (vw), 1643 (w), 1511 (s), 1484 (m),1467 (m), 1436 (s), 1388 (m), 1364 (m), 1300 (m), 1286 (m), 1271 (s),1200 (vs), 1146 (m), 1128 (s), 1085 (vs), 987 (vs), 976 (vs), 832 (s), 779 (w), 779 (w), 745 (s), 722 (s), 666 cm^{-1} (m); sensitivities: (grain size < 100 mm): friction = 360 N, impact > 20 J, ESD = 0.2 J; DSC (onset, 5 °C min^{-1}): T_d = 130 °C.

5,5′-Dinitro-4,4′-bi-l,2,3-triazole (**2c-13**): In a suspension of NaN_3 (0.65 g, 0.01 mol) in methanol (15 mL), diacetate 1,1,4,4-tetranitrobutan-2,3-diol (1.3 g, 0.0037 mol) was incrementally added at 20 °C with stirring. After letting the mixture react for 20 h at 20 °C under periodic stirring, it was cautiously acidified using 20% H_2SO_4, until cessation of gas evolution. Subsequently, methanol was removed through rotary evaporation and the remaining aqueous solution was subjected to ether extraction (3 × 10 mL). The pooled ether extracts were dried using anhydrous $MgSO_4$ and evaporated via rotary evaporation. The resulting residue was treated with a slow addition of 30% KOH solution in methanol, adjusting the pH to 7–8, after which dipotassium salt **2c-13** was precipitated out with ether, yielding 0.75 g of light-brown powdered **2c-13** salt, T_d, 210 °C. Found, %: K 26.00. $C_4K_2N_8O_4$. Calculated, %: K 25.9. IR: 1380, 1517 (NO_2), 985, 1020 (triazole). To extract NH acid **2c-13**, the dipotassium salt was first dissolved in a minimal quantity of water. Following this, 30 mL of ether was added and the solution was acidified with 10% H_2SO_4. The ether layer was then separated, and the remaining aqueous layer underwent an additional extraction with 20 mL of ether. These ether extracts were dried using anhydrous $MgSO_4$, reduced to 10 mL using a rotary evaporator, and then precipitated out with dry dichloroethane. This process yielded 0.65 g (67%) of colorless or light-brown crystalline NH acid **2c-13**. T_d, 155 °C. Found, %: C 21.80; H 1.02; N 48.80. $C_4H_2N_80_4$. Calculated, %: C 21.25; H 0.89; N 49.56. Mass spectrum: 226 (M^+).

General procedure for **2c-14 (i-iv)**: Ammonia gas was bubbled through an ether solution (50 mL) containing **2c-13** (1.13 g, 5 mmol) for 5 min while stirring. Following the removal of the organic solvent, the residual solid was rinsed with ethanol (3 × 5 mL) and air-dried. This resulted in the formation of a brown solid-diammonium 4, 4′-bis(5-nitro-1, 2, 3-2H-triazolate), weighing 1.20 g (92% yield). T_d: 241 °C; ^{1}H NMR: 7.00 (8H, br, NH_4^+); ^{13}C NMR: δ 134.1 (C_4, C_4'); 151.5 (C_5, C_5'); IR (cm^{-1}): ^{13}C NMR [D_2O]: 133.7 (C_4, C_4'); 152.2 (C_5, C_5'). Then, the compound diammonium 4, 4′-bis(5-nitro-1, 2, 3-2H-triazolate) (0.52 g, 2 mmol) was dissolved in 5 mL of DMF. Next, dimethyl sulfate (1.26 g, 10 mmol) was added to the solution while stirring. The reaction mixture was then heated to 80 °C and maintained at this temperature for 2 h. After the reaction time, the mixture was carefully poured into a solution of ice-water (50 mL). As a result, a precipitate formed in the solution. This precipitate was collected by filtration to separate it from the liquid phase. To purify the obtained compound, silica chromatography was employed. In this process, a mixture of 2:3 ethyl acetate-hexane was used as the eluent. This eluent helped separate the desired compounds, which are labeled as compounds **2c-14 (i-iv)** in the study.

1,2′-Dimethyl-5,5′-dinitro-4,4′-bis(1,2,3-triazole) (**2c-14i**): Needle crystals. T_m: 139 °C, T_d: 229 °C; IR (cm^{-1}) n 3034, 2969, 2921, 2892, 1548, 1480, 1447, 1409, 1369, 1354, 1288, 1205, 1182, 1135, 1045, 1006, 844, 832, 760, 722, 699, 662, 601. ^{1}H NMR (CD_3CN): 4.38 (3H, s, CH_3); 4.42 (3H, s, CH_3).

2,2′-Dimethyl-5,5′-dinitro-4,4′-bis(1,2,3-triazole) (**2c-14ii**): Colorless crystals. T_m: 165 °C, IR: 3034, 2967, 1545, 1451, 1421, 1360, 1298, 1199, 1057, 1007, 850, 831, 772, 718, 662, 610, 446; ^{1}H NMR (CD_3CN): 4.37 (6H, s, CH_3); ^{13}C NMR (CD_3CN): δ 44.8 (CH_3); 132.4 (C_4, C_4'); 151.9 (C_5, C_5'); EA ($C_6H_6N_8O_4$, 254.14): calcd, C: 28.35; H: 2.38; N: 44.09; Found, C: 28.40; H: 2.38; N: 43.71.

1,3′-Dimethyl-5,5′-dinitro-4,4′-bis(1,2,3-triazole) (**2c-14iii**): White solid. T_m: 165 °C, T_d: 237 °C; IR (cm^{-1}) n 3032, 2923, 2889, 2856, 1541, 1501, 1445, 1403, 1388, 1356, 1302, 1275, 1222, 1208, 1192, 1144, 1059, 1034, 992, 843, 835, 774, 746, 710, 688, 642, 590, 568, 459. 1H NMR (CD_3CN): 4.15 (3H, s, CH_3); 4.46 (3H, s, CH_3); 13C NMR (CD_3CN): 38.3 (N_2-CH_3); 40.5 (N_1'-CH_3); 125.4 (C_4'); 130.2

(C_4). The signals for C_5 and C_5' were not detected. EA ($C_6H_6N_8O_4$, 254.14): calcd, C: 28.35; H: 2.38; N: 44.09; Found, C: 28.59; H: 2.46; N: 45.53.

2,3′-Dimethyl-5,5′-dinitro-4,4′-bis(1,2,3-triazole) (**2c-14iv**): Colorless crystals. T_m: 160 °C, IR (cm^{-1}) ν 3040, 2967, 1548, 1532, 1489, 1437, 1388, 1351, 1307, 1274, 1238, 1214, 1054, 1012, 843, 776, 752, 725, 708, 654, 610, 470; ^1H NMR (CD_3CN): 4.12 (3H, s, CH_3); 4.41 (3H, s, CH_3); ^{13}C NMR (CD_3CN): 38.2 (N_2-CH_3); 45.1 (N_1'-CH_3); 125.0 (C_4'); 129.0 (C_4); 152.4 (C_5'); 152.9 (C_5). EA ($C_6H_6N_8O_4$, 254.14): calcd, C: 28.35; H: 2.38; N: 44.09; Found, C: 28.40; H: 2.38; N: 43.72.

A general procedure was followed to synthesize compounds **2c-15 (i-iii)**. Firstly, a solution of compound diammonium 4,4′-bis(5-nitro-1, 2, 3-2H-triazolate) (0.52 g, 2 mmol) in 30 mL of acetonitrile was prepared. Then, a freshly prepared O-p-toluenesulfonylhydroxylamine (THA) solution (0.94 g, 5 mmol) in 60 mL dichloromethane was added to the above solution. The resulting reaction mixture was stirred for a duration of 48 h at room temperature. After the completion of the reaction, the solvent was removed by evaporation, leaving behind a residue. This residue was then subjected to extraction with hot ethyl acetate (3 × 10 mL). The organic extracts obtained from the extraction were evaporated to obtain the desired product. To purify the obtained compounds **2c-15 (i-iii)**, silica chromatography was employed. In this purification process, a mixture of 5:3 ethyl acetate-hexane was used as the eluent. By passing the mixture through the silica gel column, the desired compounds were separated and collected. The final yield of compounds **2c-15 (i-iii)** was reported to be 0.21 g (41%). This general procedure provided a successful method for synthesizing compounds **2c-15 (i-iii)** using diammonium 4,4′-bis(5-nitro-1, 2, 3-2H-triazolate) and O-p-toluenesulfonylhydroxylamine as starting materials.

2,2′-Diamino-5,5′-dinitro-4,4′-bis(1,2,3-triazole) (**2c-15i**): Yellow solid. T_d: 235 °C; IR (cm^{-1}) n 3341, 3313, 3255, 1549, 1454, 1391, 1356, 1294, 1194, 1160, 1061, 953, 852, 831, 759, 725, 648, 551, 474. ^1H NMR (CD_3CN): 6.98 (4H, s, NH_2); ^{13}C NMR (CD_3CN): δ 130.3 (C_4, C_4'); 149.2 (C_5, C_5'); ^{15}N NMR (CD_3CN): δ − 26.72 (–NO_2), − 48.49 (N_1/N_1'), − 61.19 (N_3/N_3'), − 120.20 (N_2/N_2'), − 280.07 (–NH_2). EA ($C_4H_4N_{10}O_4$, 256.14): calcd, C: 18.76; H: 1.57; N: 54.68; Found, C: 19.18; H: 1.53; N: 53.67.

3,3′-Diamino-5,5′-dinitro-4,4′-bis(1,2,3-triazole) (**2c-15ii**): White solid. T_d: 198.2 °C; IR (cm^{-1}) n 3337, 3291, 3264, 1603, 1555, 1511, 1460, 1397, 1383, 1360, 1315, 1262, 1012, 939, 839, 771, 756, 461. ^1H NMR (CD_3CN): 6.42 (4H, s, NH_2); ^{13}C NMR (CD_3CN): δ 130.3 (C_4, C_4'); 151.1 (C_5, C_5'); ^{15}N NMR: δ − 17.05 (N_1/N_1'), − 27.29 (–NO_2), − 43.49 (N_2/N_2'), − 118.23 (N_3/N_3'), − 293.49 (–NH_2). EA ($C_4H_4N_{10}O_4$, 256.14): calcd, C: 18.76; H: 1.57; N: 54.68; Found, C: 18.99; H: 1.54; N: 52.92.

2,3′-Diamino-5,5′-dinitro-4,4′-bis(1,2,3-triazole) (**2c-15iii**): Yellow solid. T_d: 191 °C; IR (cm^{-1}) n 3339, 3273, 3196, 1612, 1549, 1511, 1485, 1448, 1389, 1354, 1311, 1254, 1165, 1042, 1011, 949, 845, 833, 756, 741, 645, 579, 401, 457. ^1H NMR (CD_3CN): 6.33 (2H, s, NH_2), 7.08 (2H, s, NH_2); ^{13}C NMR (CD_3CN): 124.0, 126.7, 149.5, 151.0; ^{15}N NMR (CD_3CN): δ − 17.37 (N_1), − 26.12 (–NO_2), − 27.99 (–NO_2), − 42.93 (N_6), − 46.99 (N_2), − 60.99 (N_8), − 118.03 (N_7), − 123.34 (N_3),

− 279.20 (–NH_2), − 299.03 (–NH_2). EA ($C_4H_4N_{10}O_4$, 256.14): calcd, C: 18.76; H: 1.57; N: 54.68; Found, C: 19.23; H: 1.62; N: 53.90.

4,4′-Bis(5-nitro-1,2,3-triazole) (**2c-16**): To prepare compound diammonium 4,4′-bis(5-nitro-1, 2, 3-2H-triazolate) (0.26 g, 1 mmol), it was dissolved in 10 mL of water. Then, potassium acetate (1.32 g, 13 mmol) was added to the solution. The temperature of the reaction mixture was maintained between 40 to 45 °C. Next, Oxone (4.37 g, 14 mmol) was added gradually in small portions over a period of 30 min. During this addition, the pH was adjusted to 4–5 by adding potassium acetate. The resulting reaction mixture was stirred for 48 h. After the completion of the reaction, the mixture was cooled to room temperature. To acidify the solution, concentrated H_2SO_4 was added. The resulting bright yellow solution was then extracted with ethyl acetate (3 × 15 mL). The organic phases were combined, and the resulting solution was dried using anhydrous sodium sulfate. To obtain the salt form of compound **2c-16**, ammonia gas was bubbled into the ethyl acetate solution for 5 min. Afterward, the solvent was removed, leaving behind a residue. This residue, which appeared as a yellow solid, was further dried under vacuum conditions. The final yield of the yellow solid obtained was reported to be 0.10 g (34%). T_d: 251.0 °C; ^{1}H NMR: 7.15 (8H, br, NH_4^+); ^{13}C NMR (D_2O): 122.8; 126.2, 130.2, 136.6, 141.6.

4-Nitro-5-(5-amino-1,2,4-triazol-3-yl)-2H-1,2,3-triazole (**2c-17**): To synthesize methyl 4-nitro-2H-1,2,3-triazole-5-carboxylate (1.806 g, 10.50 mmol), a drop-wise addition method was employed. The compound was added drop-wise to a vigorously stirring mixture of aminoguanidine sulfate (10.3 g, 41.8 mmol) in freshly prepared NaOMe solution. The NaOMe solution was prepared using 0.990 g (43.04 mmol) of sodium in 30 mL of dry MeOH (methanol). This reaction was carried out at a temperature of 0 °C. After the addition, the resulting mixture was heated to reflux for a duration of 24 h. Following the reflux period, the mixture was cooled to room temperature. To extract the desired product, the mixture was carefully poured over ice-water (30 mL). The pH of the solution was adjusted to 5 by adding 37% hydrochloric acid. After allowing the mixture to sit for 20 min, a precipitate formed. This precipitate was separated from the liquid phase by filtration. The obtained solid was then washed with cold water and dried under ambient air conditions. As a result, 1.317 g (6.72 mmol, 64%) of pale **2c-17** was obtained as a brown solid. In summary, the synthesis of methyl 4-nitro-2H-1,2,3-triazole-5-carboxylate involved the addition of the compound to a mixture of aminoguanidine sulfate in NaOMe solution, followed by a reflux reaction and subsequent work-up steps. The final product, pale **2c-17**, was obtained as a brown solid with a yield of 64%. T_d: 263.32 °C. ^{1}H NMR (300 MHz, DMSO-d_6): d 6.22 (br) ppm; ^{13}C NMR (75 MHz, DMSO-d_6): d 157.26, 150.62, 147.73, 130.07 ppm. IR (KBr): n 3447, 3152, 1678, 1535, 1471, 1369, 1311, 1257, 1209, 1082, 967, 824, 730 cm^{-1}. Elemental analysis of $C_4H_4N_8O_2$ (196.13): calcd C 24.50, H 2.06, N 57.13%; found: C 24.41, H 2.10, N 57.25%.

4-Nitro-5-(5-nitroimino-1,2,4-triazol-3-yl)-2H-1,2,3-triazole (**2c-18**): To perform the reaction, **2c-17** (0.784 g, 4.00 mmol) was added in small portions to 100% nitric acid (10 mL) at a temperature of − 10 °C. The mixture was then kept at 0 °C for a duration of 2 h to allow the reaction to proceed. After the reaction time, the mixture was poured onto a bed of ice weighing 50 g. The resulting mixture was then subjected to extraction with ether using 10 mL portions, repeated four times. The organic phase was separated and dried using magnesium sulfate as a drying agent. The solvent was subsequently evaporated to obtain 0.877 g (3.64 mmol) of product **2c-18** as a yellowish solid. The overall yield of this reaction was reported to be 91%. In summary, the synthesis of compound **2c-18** involved the addition of **2c-17** to nitric acid, followed by extraction with ether and subsequent drying and solvent evaporation steps. The final product obtained was a yellowish solid. T_d: 168.45 °C. ^{1}H NMR (300 MHz, acetone-d6): d 8.29 (br), 5.94 (br) ppm; ^{13}C NMR (75 MHz, acetone-d6): d 153.91, 151.98, 141.57, 129.25 ppm. IR (KBr): n 3439, 2350, 2324, 1691, 1642, 1516, 1479, 1371, 1296, 1155, 1115, 977, 832, 656 cm^{-1}. Elemental analysis of $C_4H_3N_9O_4$ (241.13): calcd C 19.92, H 1.25, N 52.28%; found: C 20.04, H 1.19, N 52.32%.

5-Nitro-4-(5-nitro-1,2,4-triazol-3-yl)-2H-1,2,3-triazole (**2c-19**): compound **2c-17** (0.392 g, 2.00 mmol) was added to a mixture of 30% H_2O_2 (3.5 mL) and concentrated H_2SO_4 (6.8 mL). The addition was done under stirring and cooling conditions using an ice bath. After allowing the reaction to proceed for 24 h at room temperature, the reaction mixture was poured into cold water (35 mL). To extract the desired product, the aqueous solution was subjected to extraction with ethyl acetate (3 × 15 mL). The organic solution containing the product was then dried using anhydrous $MgSO_4$ as a drying agent. The solvent was subsequently removed under vacuum, resulting in the isolation of the product as a white solid. The obtained yield of the product was reported to be 0.348 g (1.54 mmol), which corresponds to a yield of 77%. To summarize, the synthesis of compound **2c-17** involved the addition of the compound to a mixture of 30% H_2O_2 and concentrated H_2SO_4, followed by a reaction at room temperature. The reaction mixture was then extracted with ethyl acetate, and the product was isolated as a white solid through drying and solvent evaporation. The overall yield achieved was 77%. T_d: 210.83 °C. ^{1}H NMR (300 MHz, DMSO-d_6): d 5.11 (br) ppm; ^{13}C NMR (75 MHz, DMSO-d_6): d 162.83, 150.61, 145.77, 128.19 ppm. IR (KBr): n 3306, 3197, 2360, 2341, 1715, 1700, 1557, 1541, 1523, 1386, 1230, 1186, 1055, 833, 777 cm^{-1}. Elemental analysis of $C_4H_2N_8O_4$ (226.11): calcd C 21.25, H 0.89, N 49.56%; found: C 21.31, H 0.88, N 49.64%.

General procedure for the synthesis of compounds (**2c-(20-21)**): To synthesize the desired product, the sodium salt (10 mmol) of either 4-chloro pyrazole, 5-aminotetrazole, 3,4-dinitropyrazole, 3-nitro-1,2,4-triazole, or 3-nitro-5-amino-1,2,4-triazole was dissolved in a small volume of methanol. Then, 1,3-dinitro-triazole (10 mmol) in methanol (50 mL) and sodium carbonate (1.06 g, 10 mmol) were added to the methanol solution. The resulting reaction mixture was stirred for 6 h at room temperature. After completion of the reaction, the solvent was removed under vacuum, leaving behind a residue. This residue was dissolved in water (50 mL), and the solution was acidified with dilute HCl. A white precipitate formed, which was collected and washed with water. The filtrate obtained from the precipitation step was subjected to extraction with ether (3 × 30 mL). The organic phase was then washed with brine and water before being dried using $MgSO_4$ as a drying agent. The solvent was subsequently removed by evaporation. As a result, some white solid was obtained. The two sets of solids were combined, and the pure product was obtained through recrystallization from a mixture of ethanol and water. In summary, the synthesis involved dissolving the sodium salt of the starting material in methanol, followed by addition of 1,3-dinitro-triazole and sodium carbonate. After the reaction, the resulting precipitate was collected, washed, and further processed to obtain the pure product through recrystallization.

5-Amino-1-(3-nitro-1H-1,2,4-triazol-5-yl)-3-nitro-1,2,4-triazole (**2c-20**): White solid. Td: 252 °C. ^{1}H NMR (300 MHz, DMSO-d_6): d 6.85 ppm (s, 2H); ^{13}C NMR (75 MHz, DMSO-d_6): d 164.3, 159.7, 156.8, 149.5 ppm. IR (KBr): n 3563, 3405, 3308, 3243, 2912, 2855, 1875, 1655, 1531, 1410, 1358, 1307, 1188, 1134, 1091, 1022, 973, 843, 767, 719, 700, 619 cm^{-1}. Elemental analysis for $C_4H_5N_9O_5$ (259.14): calcd C 18.54, H 1.94, N 48.65%. Found: C 19.00, H 2.03, N 47.54%.

3,5′-Dinitro-2′H-1,3′-bi(1,2,4-triazole) (**2c-21**): White solid. T_m: 232 °C, T_d: 250 °C. ^{1}H NMR (300 MHz, CD_3CN): d 9.17 ppm (s, 1H); ^{13}C NMR (75 MHz, CD3CN): d 164.7, 159.9, 149.2, 147.0 ppm. IR (KBr): n 3621, 3123, 3027, 2951, 1602, 1569, 1510, 1426, 1307, 1167, 1040, 963, 890, 834, 710, 644 cm^{-1}. Elemental analysis for $C_4H_2N_8O_4$ (226.11): calcd C 21.25, H 0.89, N 49.56%. Found: C 21.39, H 0.90, N 49.29%.

3,5′-Dinitro-1′H-[1,3′-bi(1,2,4-triazol)]-1′-amine (**2c-22**) and 3,5′-dinitro-1′H-[1,3′-bi(1,2,4-triazol)]-20-amine (**2c-23**): Firstly, the sodium salt of **2c-21** (0.99 g, 4 mmol) was dissolved in acetonitrile (20 mL). Then, freshly prepared O-p-toluenesulfonylhydroxylamine (THA) (1.87 g, 10 mmol) in dichloromethane (100 mL) was added to the reaction mixture. The resulting mixture was stirred for 24 h at room temperature. After the completion of the reaction, the mixture was filtered to remove any solid impurities. The filtrate was then evaporated, and the remaining residue was subjected to purification using silica chromatography. In the chromatographic purification step, ethyl acetate:n-hexane (1:2) was used as the eluent. This process allowed for the separation of compounds **2c-22** and **2c-23**. The yield of these compounds was reported to be 0.46 g, corresponding to a 48% yield.

2c-22: white solid. T_m: 180 °C, T_d: 242 °C. ^{1}H NMR (300 MHz, CD_3CN): d 9.25 ppm (s, 1H). 6.21 ppm (s, 2H); ^{13}C NMR (75 MHz, CD_3CN): d 164.6, 157.7, 149.4, 144.4 ppm. IR (KBr): n 3345, 3269, 3196, 3126, 2957, 2861, 1767, 1639, 1562, 1501, 1427, 1307, 1226, 1192, 1082, 988, 890, 835, 739, 698, 646, 579 cm^{-1}. Elemental analysis for $C_4H_3N_9O_4$ (241.12): calcd C 19.92, H 1.25, N 52.28%. Found: C 20.18, H 1.34, N 51.34%.

2c-23: white solid. T_m: 119 °C, T_d: 167 °C. 1H NMR (300 MHz, CD_3CN): d 9.05 ppm (s, 1H). 6.83 ppm (s, 2H); 13C NMR (75 MHz, CD_3CN): d 171.8, 164.5, 148.1, 146.6 ppm. IR (KBr): n 3427, 3386, 3324, 2158, 2944, 2876, 1695, 1621, 1523, 1465, 1356, 1237, 1168, 1075, 997, 913, 864, 784, 723, 661, 579 cm^{-1}. Elemental analysis for $C_4H_3N_9O_4$ (241.12): calcd C 19.92, H 1.25, N 52.28%. Found: C 20.21, H 1.25, N 51.53%.

4-Nitro-2-(1-nitro-1,2,4-triazole-3-yl)-1,2,3-triazole (**2c-24**): A solution of 69.5 g (0.643 mol) of N_2O_5 in 800 mL of CH_2Cl_2 was prepared, and to this solution, 30 g of 4-nitro-2-(1,2,4-triazole-3-yl)-1,2,3-triazole was added in small portions at a temperature of -5 °C. The addition was done gradually. After the addition, the reaction mixture was allowed to stir for 30 min. It was then poured onto ice and filtered. The resulting precipitate was washed with cold ethanol and subsequently dried under reduced pressure at a temperature of 50 °C. The yield obtained from this step was 33 g (88%). A second crop of the desired product, bitriazole **2c-25**, was obtained from the neutralized filtrate by performing an ether extraction. This yielded 1.9 g of bitriazole **2c-25**. For further purification, recrystallization of the product was carried out using methanol as the solvent. This process resulted in the formation of pale yellow needles. However, it should be noted that these needles were found to decompose at a temperature of 164 °C. Anal. calcd for $C_4H_2N_8O_4$: C, 21.25; H, 0.89; N, 49.56; 0, 28.30 (%). Found: C, 21.29; H, 0.93; N, 49.34; 0, 28.22 (%).

(E)-1,2-di(4H-1,2,4-triazol-4-yl)diazene (**2c-26**): To carry out the reaction, 10.56 g of SDCI (48 mmol) was dissolved in 80 mL of water, and then 15 mL of AcOH (acetic acid) was added to the solution. The resulting mixture was allowed to stir for 6 h. After 6 h, 4-amino-1,2,4-triazole (4 g, 48 mmol) was added to the reaction mixture. The entire mixture was then stirred at room temperature for an additional hour. Following the reaction time, a precipitate formed in the mixture. This precipitate was filtered off, and further purification was carried out through recrystallization from water. This recrystallization process resulted in the formation of compound **2c-26** with a yield of 90%. T_d, 300 °C; ^{1}H NMR (400 MHz, D_2O): 7.22 (s, 4H); ^{13}C NMR (400 MHz, D_2O): 138.57; TOF-EIMS: m/z (%): 164 (M^+, 30), 96 (40), 69 (100), 68 (5), 42 (35); Anal. calcd for $C_4H_4N_8$ (164): C, 29.27; H, 2.44; N, 68.29. Found: C, 29.27; H, 2.40; N, 68.40.

(E)-3,5-dichloro-4-((3-chloro-4H-1,2,4-triazol-4-yl)diazenyl)-4H-1,2,4-triazole (**2c-27**): To perform the reaction, 10.56 g of SDCI (48 mmol) was dissolved in 80 mL of water, and then 15 mL of AcOH (acetic acid) was added to the solution. The resulting mixture was stirred for 6 h. After 6 h, 4-amino-1,2,4-triazole (4 g, 48 mmol) was added to the reaction mixture. The entire mixture was then stirred at room temperature for an additional hour. By increasing the molar ratio of SDCI to 4-amino-1,2,4-triazole, a precipitate formed in the mixture. This precipitate was filtered off, and further purification was carried out through recrystallization from water. The recrystallization process resulted in the formation of compound **2c-27** with a yield of 90%. T_m, 217–220 °C; ^{1}H NMR (400 MHz, DMSO-d6): 9.68 (s, H); ^{13}C NMR (400 MHz, DMSO-d6): 141.3, 139.5, 136.8; EIMS: m/z (%): 266 (M^+, 6), 164 (2), 130 (32), 102 (30), 75 (55), 47 (100); Anal. cacld for $C_4HC_{13}N_8$ (266): C, 18.05; H, 0.38; N, 42.11. Found: C, 18.01; H, 0.31; N, 42.18.

3-Azido-4,4′-azo-1,2,4-triazole (**2c-28**): A solution of compound **2c-26** (0.5 g, 2.5 mmol) in DMF (15 mL) was prepared. Then, NaN_3 (0.25 g, 3.8 mmol) was added to the solution at room temperature (RT). The resulting mixture was stirred for 2 h. After completion of the reaction, the solvent was evaporated, leaving behind a residue. This residue was then dissolved in acetone and filtered to remove any impurities. The filtrate was concentrated, and further purification was achieved through recrystallization from a mixture of acetone and petroleum ether. The final product yield obtained from this process was reported to be 0.26 g, corresponding to a 50% yield. T_d: 190 °C (DSC, 10 °C min^{-1}); ^{1}H NMR (400 MHz, DMSO-d6): 9.17 (s, 1H), 9.29 ppm (s, 2H); ^{13}C NMR (400 MHz, DMSO-d6): 135.0, 138.0, 147.2 ppm; IR (KBr): 2921, 2172, 1553,1469, 1423, 1167, 1031 cm^{-1}; MS (ESI):m/z: 206 [M+H]$^+$.

3,3′-Di(azido)-4,4′-azo-1,2,4-triazole (**2c-29**): A solution of compound **2c-27** (0.25 g, 1.1 mmol) in DMF (15 mL) was prepared, and then NaN_3 (0.18 g, 2.75 mmol) was added to the solution at room temperature (RT). The resulting mixture was stirred for 4 h. After the completion of the reaction, the solvent was evaporated, leaving behind a residue. This residue was subsequently dissolved in acetone and filtered to remove any impurities. The filtrate was then concentrated, and further purification was achieved through recrystallization from acetone. Yield: 0.12 g, 44%; T_d:186 °C (DSC, 10 °C min^{-1}); ^1H NMR (400 MHz, DMSO-d_6): 9.90 ppm (s, 2H); ^{13}C NMR (400 MHz, DMSO-d_6): 135.1,153.8 ppm; IR (KBr): 3090, 2145, 1500, 1417, 1180, 1095 cm^{-1}; MS (ESI):m/z: 247 $[M+H]^+$.

3,3′,5-Tri(azido)-4,4′-azo-1,2,4-triazole (**2c-30**): A solution of compound 3,3′,5-tri(chlorine)-4,4′-azo-1,2,4-triazole (0.53 g, 2 mmol) in DMF (15 mL) was prepared, and then NaN_3 (0.45 g, 8 mmol) was added to the solution at room temperature (RT). The resulting mixture was stirred for 12 h. After the completion of the reaction, the solvent was evaporated, leaving behind a residue. This residue was subsequently dissolved in acetone and filtered to remove any impurities. The filtrate was then concentrated, and further purification was achieved through recrystallization from a mixture of acetone and petroleum ether, following the previously reported method by our group.

3,3′,5,5′-Tetra(azido)-4,4′-azo-1,2,4-triazole (**2c-31**): To initiate the reaction, a solution of compound 3,3′,5,5′-tetra(chlorine)-4,4′-azo-1,2,4-triazole (0.30 g, 1.0 mmol) in DM (15 mL) was prepared, and then NaN_3 (0.29 g, 4.5 mmol) was added to the solution at room temperature (RT). The resulting mixture was stirred for 2 h. Following the 2-h stirring period, the reaction mixture was poured into water and stirred for an additional 1 h. Afterward, the mixture was filtered to remove any solid residues. The resulting residue was then dissolved in CH_2Cl_2 and purified through recrystallization from a mixture of CH_2Cl_2 and petroleum ether. The final product yield obtained from this process was reported to be 0.13 g, corresponding to a 40% yield.; T_d: 136 °C (DSC, 10 °C min^{-1});^{13}C NMR (400 MHz, CD_3COCD_3): 154.3 ppm.

3-Amino-4,4′-azo-1,2,4-triazole (**2c-32**): To perform the reaction, Compound 3-diiminotriphenylphosphorane-4,4′-azo-1,2,4-triazole (1.75 g, 4 mmol) was added to a stirring solution of 1 M HCl (60 mL). The resulting mixture was then heated at 50 °C for 3 h, monitoring the progress of the reaction using thin-layer chromatography (TLC) until no starting material remained. After completion of the reaction, a precipitate formed in the solution. This precipitate was filtered and subsequently washed with $CHCl_3$/MeOH to obtain compound **2c-32**. Yield: 0.54 g, 75%; T_d: 248 °C (DSC, 10 °C min^{-1}); ^1H NMR (400 MHz, DMSO-d_6): 9.56 (s, 2H), 9.08 (s, 1H), 8.77 ppm (s, 2H); ^{13}C NMR (400 MHz, DMSO-d_6): 131.0, 138.3, 149.4 ppm; IR (KBr): 3039, 2960, 1703, 1503, 1470, 1361, 1222, 1175, 1062, 889, 617 cm^{-1}; MS (EI):m/z: 179 $[M]^+$; elemental analysis calcd (%) for $C_4H_5N_9$: C 26.82, H 2.81, N 70.37; found: C 26.80, H 2.81, N 70.39.

3,3′-Di(amino)-4,4′-azo-1,2,4-triazole (**2c-33**): To carry out the reaction, compound 3,3′-di(iminotriphenylphosphorane)-4,4′-azo-1,2,4-triazole (2.86 g, 4 mmol) was added to a stirring solution of 1 M HCl (80 mL). The resulting mixture was then heated at 50 °C for 3 h, monitoring the progress of the reaction using thin-layer chromatography (TLC) until no starting material remained. After completion of the reaction, a precipitate formed in the solution. This precipitate was filtered and subsequently washed with $CHCl_3$/MeOH to obtain compound **2c-33**. The yield of the final product, compound **2c-33**, was reported to be 0.59 g, corresponding to a 75% yield. T_d: 245 °C; ^{1}H NMR (400 MHz, DMSO-d_6): 8.55 (s, 4H), 9.11 ppm (s, 2H); ^{13}C NMR (400 MHz, DMSO-d_6): 131.2, 148.8 ppm; IR (KBr): 3292, 3209, 3144, 3123, 2957, 1685, 1642, 1531, 1462, 1328, 1224, 1117, 1062, 967 cm^{-1}. MS (EI):m/z: 194 $[M]^+$; elemental analysis calcd (%) for $C_4H_6N_{10}$: C 24.74, H 3.09, N 72.16; found: C 24.54, H 3.30, N 71.26.

3,3′,5-Tri(amino)-4,4′-azo-1,2,4-triazole (**2c-34**): Compound 3,3′,5-tri(iminotriphenylphosphorane)-4,4′-azo-1,2,4-triazole (3.96 g, 4 mmol) was added to a stirring solution of 1 M HCl (100 mL). The resulting mixture was then heated at 50 °C until no starting material remained, as confirmed by thin-layer chromatography (TLC). After the completion of the reaction, a precipitate formed in the solution. This precipitate was subsequently filtered and washed with $CHCl_3$/MeOH to obtain compound **2c-34**. Yield: 0.50 g, 60%; T_d: 246 °C; ^{13}C NMR (100 MHz, D_2O): 131.2, 145.9, 148.7 ppm; IR (KBr): 3318, 3249, 3167, 3129, 2963, 2796, 2723, 1515, 1456, 1365, 1265, 1140 cm^{-1}; MS (ESI):m/z: 210 $[M+H]^+$; elemental analysis calcd (%) for $C_4H_7N_{11}$: C 22.97, H 3.35, N 73.68; found: C 23.15, H 3.30, N 73.43.

3,3′,5,5′-Tetra(amino)-4,4′-azo-1,2,4-triazole (**2c-35**): Compound 3,3′,5,5′-tetra(iminotriphenylphosphorane)-4,4′-azo-1,2,4-triazole (5.06 g, 4 mmol) was added to a stirring solution of 1 M HCl (120 mL). The resulting mixture was then heated to 50 °C until no starting material remained, as confirmed by thin-layer chromatography (TLC). After completion of the reaction, a precipitate formed in the solution. This precipitate was subsequently filtered and washed with $CHCl_3$/MeOH to obtain compound **2c-35**. Yield: 0.49 g, 45%; T_d: 290 °C; ^{1}H NMR (400 MHz, DMSO-d6): 9.66 ppm (s); ^{13}C NMR (400 MHz, D_2O): 145.2 ppm; IR (KBr): 3462, 3295, 3265, 3174, 3112, 2943, 2658, 2742, 2695, 1515,1456, 1365, 1265, 1140 cm^{-1}; MS (ESI):m/z: 225 $[M+H]^+$; elemental analysis calcd (%) for $C_4H_8N_{12}$: C 21.43, H 3.57, N 75.00; found: C 21.21,H 3.40, N 75.18.

3,3′-Dione-4,4′-azo-1,2,4-triazole (**2c-37**): A solution of sodium dichloroisocyanurate (SDIC) (5.09 g, 23 mmol) in water (40 mL) was prepared, and then acetic acid (5 mL) was added to the solution with vigorous stirring at 30 °C for 1 h. Following this, 4-amino-1,2,4-triazole chloride (3.09 g, 23 mmol) was added to the solution. The reaction mixture was maintained at 25 °C for 2 h. After the reaction was complete, the mixture was filtered, resulting in a yellow filtrate. The water in the filtrate was evaporated, yielding a light-yellow solid. Yield: 1.22 g, 47%. T_d: 258.1 °C. IR (KBr): 3428, 3158, 3065, 2945, 1711, 1539, 1431, 1403, 1387, 1348, 1274, 1257, 1206, 987, 816, 723, 616, 602 cm^{-1}; ^{1}H NMR (DMSO-d6) d: 8.74, 3.45 ppm; ^{13}C NMR (DMSO-d6) d: 150.72, 127.83, 32.47 ppm; elemental analysis calcd (%) for $C_6H_8N_8O_2$ (224.18): C 32.15, H 3.60, N 49.98; found: C 32.38, H 3.62, N 50.17.

1,1′-Azobis-1,2,3-triazole (**2c-38**): 1-Amino-1,2,3-triazole (1.68 g, 20 mmol) was dissolved in 40 mL of CH_3CN. The resulting solution was then cooled to a temperature range of −2 to 0 °C. Meanwhile, a solution of sodium dichloroisocyanurate (SDCI, 4.44 g, 20 mmol) in 10 mL of water and 5.0 mL of CH_3COOH was prepared and added dropwise to the cooled solution. The reaction mixture was stirred at 0 °C for 30 min. Afterward, the solution was neutralized using $NaHCO_3$ and then filtered to remove any insoluble solids. Subsequently, the filtrate was concentrated, and the product was isolated as a small amount of yellow solid. This solid was further purified through recrystallization from acetone. Yield 78%. T_d: 193.8 °C. ^{1}H NMR (400 MHz, DMSO-d6, 25 °C): 9.17 (s, 2H), 8.21 ppm (s, 2H). ^{13}C NMR (100 MHz, DMSO-d6, 25 °C): 134.8, 118.0 ppm. IR (KBr): 3127 (=CH), 1624, 1482, 1319, 1224, 1171 cm^{-1}. FAB-MS: m/z 165.0 $[M+H]^+$. Anal. caclcd for $C_4H_4N_8$ (164): C, 29.27; H, 2.44; N, 68.29. Found: C, 29.38; H, 2.42; N, 68.20.

3,3′,5,5′-Tetra(azidomethyl)-4,4′-azo-1,2,4-triazole (**2c-39**): Hydrazine monohydrate (37.5 g, 0.75 mol) was added dropwise to a solution of 70% aqueous glycolic acid (54.3 g, 0.50 mol) at 0 °C. The resulting solution was then heated at 120 °C for 6 h. Following this initial heating period, the reflux condenser was replaced with a downward condenser, and the reaction mixture was further heated at 160 °C for 18 h. This allowed excess hydrazine and water to distill off. After cooling, the resulting yellowish crystalline solid was recrystallized from water to obtain analytically pure 4-amino-4H-1,2,4-triazole-3,5-dimethanol (21.1 g, 58%) as colorless flakes. T_m, 206–208 °C; elemental analysis calcd (%) for $C_4H_8N_4O$ (144.13): C 33.33, H 5.59, N 38.87; found: C 33.29, H 5.41, N 39.05; ^{1}H NMR (300 MHz, DMSO-d6): 4.54 (d, 4H), 5.37 (t, 2H), 5.81 ppm (s, 2H); ^{13}C NMR (75 MHz, DMSO-d6): 52.7, 154.4 ppm; IR (KBr): n 3344, 3220, 2877, 2687, 1639, 1518, 1479, 1461, 1440, 1379, 1351, 1305,1280, 1204, 1068, 1029, 976, 953, 875, 819, 774, 720, 498 cm^{-1}.

A suspension of compound 4-amino-4H-1,2,4-triazole-3,5-dimethanol (28.8 g, 0.20 mol) in thionyl chloride (80 mL) was prepared. The resulting mixture was stirred at room temperature for 8 h, during which an exothermic reaction occurred. After the reaction period, the excess thionyl chloride was removed under reduced pressure. The resulting yellowish solid was then dried under vacuum. To obtain analytically pure product, recrystallization from ethanol was performed. This process yielded 4-amino-3,5-bis(chloromethyl)-4H-1,2,4-triazole (33.4 g; 76%) as colorless needles. T_m, 139–141 °C; elemental analysis calcd (%) for $C_4H_7Cl_3N_4$ (217.48): C 22.09, H 3.24, N 25.76; found: C 22.24, H 3.24, N 25.83; ^{1}H NMR (300 MHz, DMSO-d_6): 4.92 ppm (s,4H); ^{13}C NMR (75 MHz, DMSO-d_6): 32.9, 152.5 ppm; IR (KBr): n 3414, 3253, 3145, 2970, 2642, 1637, 1617, 1576, 1417, 1328, 1264, 1165, 1105, 1034, 975, 937, 919, 888, 807, 759, 668, 646, 605, 568, 513 cm^{-1}.

Compound 4-amino-3,5-bis(chloromethyl)-4H-1,2,4-triazole (0.36 g, 2 mmol) was dissolved in 5 mL of DMSO. Sodium azide (0.39 g, 6 mmol) was then added to the solution at room temperature. The resulting mixture was stirred for 6 h and subsequently poured into a solution of ice water (20 mL). To extract the desired product, the resulting solution was subjected to three extractions using ethyl acetate (3 × 25 mL). The organic layers from each extraction were combined and dried over anhydrous magnesium sulfate. Evaporation of the solvent under ambient conditions yielded 4-amino-3,5-di(azidomethyl)-1,2,4-triazole as a white solid with a yield of 88%. T_m, 105 °C; T_d, 209 °C; ^{1}H NMR (300 MHz, DMSO-d_6): 6.07 (s, 2H), 4.59 ppm (s, 4H); ^{13}C NMR (75 MHz, DMSO-d_6): 151.3, 42.7 ppm; IR (KBr): n 3268 (m), 3015 (w), 2144 (s), 2101 (s), 164 0 (m), 1507 (m), 1476 (m), 1438 (m), 1351 (m), 1252 (s), 1190 (s), 1069 (m), 967 (m), 886 (m), 814 (m), 785 (w), 650 (s), 557 cm^{-1} (m); elemental analysis calcd (%) for $C_4H_6N_{10}$ (194.16): C 24.74, H 3.11, N 72.14; found: C 24.79, H 3.10, N 71.48.

To initiate the reaction, a solution of compound 4-amino-3,5-di(azidomethyl)-1,2,4-triazole (0.29 g, 2 mmol) in acetonitrile (10 mL) was cooled using an ice-salt bath. Concurrently, a mixture of sodium dichloroisocyanurate (SDCl, 0.44 g, 2 mmol) and AcOH (0.5 mL) in water (5 mL) was prepared. This mixture was then added dropwise to the cooled solution at 0 °C. After the addition, the reaction mixture was stirred for 30 min at 0 °C. Subsequently, it was neutralized to a pH of 7–8 using a sodium bicarbonate solution. The resulting precipitate was isolated through filtration, washed with cold water, and dried under vacuum to yield a white solid identified as 2c-39. The overall yield of the product was 54%. T_d, 136 °C; ^{1}H NMR (300 MHz, DMSO-d_6): d = 4.96 ppm (s, 8H); ^{13}C NMR (75 MHz, DMSO-d_6): 148.3, 44.5 ppm; IR (KBr): n 2926 (w), 2127 (s), 2102 (s), 1545 (m), 1414 (s), 1324 (s), 1275 (s), 1221 (m), 1101 (vw), 1024 (w), 1001 (w), 924 (vw), 810 (w), 646 (w), 558 cm^{-1} (w); elemental analysis calcd (%) for $C_8H_8N_{20}$ (384.28): C 25.00, H 2.10, N 72.90; found: C 25.03, H 2.14, N 72.14; impact sensitivity: IS < 2 J; friction sensitivity: FS < 5 N.

3,3′,5,5′-Tetra(nitroxymethyl)-4,4–azo-1,2,4-triazole (**2c-40**): A solution of compound 4-amino-4H-1,2,4-triazole-3,5-dimethanol (0.29 g, 2 mmol) in acetonitrile (10 mL) was cooled using an ice-salt bath. Simultaneously, a mixture of sodium dichloroisocyanurate (SDCl, 0.44 g, 2 mmol) and AcOH (0.5 mL) in water (5 mL) was prepared. This mixture was then added dropwise to the cooled solution at 0 °C. After the addition, the reaction mixture was stirred for 30 min at 0 °C. Following this, it was neutralized to a pH of 7–8 using a sodium bicarbonate solution. The resulting precipitate was isolated by filtration, washed with cold water, and dried under vacuum. The product obtained from this process was identified as 3,3′,5,5′-tetra(hydroxymethyl)-4,4′-azo-1,2,4-triazole, and it was obtained as a white solid with a yield of 70%. T_m, 88 °C; T_d, 214 °C; ^{1}H NMR (300 MHz, DMSO-d_6): 5.80 (s, 4H), 4.86 ppm (s, 8H); ^{13}C NMR (75 MHz, DMSO-d_6): 151.6, 54.6 ppm; IR (KBr): 3354 (w), 3301 (w), 3125 (w), 2925 (w), 2855 (w), 2769 (vw), 1741 (m), 1703 (w), 1603 (m), 1563 (m), 1541 (s), 1433 (s), 1368 (m), 1321 (s), 1278 (vs), 1229 (vs), 1102 (s), 1073 (vs), 1016 (s), 988 (m), 802 (s), 759 (m), 698 (m), 645 (m), 560 cm^{-1} (m); elemental analysis calcd (%) for $C_8H_{12}N_8O_4{\cdot}H_2O$ (302.25): C 31.79, H 4.67, N 37.07; found: C 31.85, H 4.59, N 37.10; impact sensitivity: IS > 40 J; friction sensitivity: FS > 360 N.

Fuming nitric acid (1.0 mL) was added dropwise to anhydrous acetic acid (2 mL) at 0 °C. Then, 3,3′,5,5′-tetra(hydroxymethyl)-4,4′-azo-1,2,4-triazole (0.28 g, 1.0 mmol) was slowly added to the mixture at 0 °C. The resulting reaction mixture was stirred for an additional 30 min and then poured into ice water. As a result of this reaction, white solids identified as **2c-40** were generated with a yield of 80%. T_d, 146 °C;

^{1}H NMR (300 MHz, DMSO-d_6): 6.09 ppm (s, 8H); ^{13}C NMR (75 MHz, DMSO-d_6): d = 146.4, 64.3 ppm; IR (KBr): n 2998 (w), 2966 (w), 1776 (vw), 1651 (vs), 1530 (m), 1415 (s), 1364 (w), 1279 (s), 1034 (m), 976 (m), 947 (m), 844 (s), 791 (w), 753 (w), 648 (w), 617 (w), 561 cm^{-1} (w); elemental analysis calcd (%) for $C_8H_8N_{12}O_{12}$ (464.22): C 20.70, H 1.74, N 36.21; found: C 21.02, H 1.78, N 34.71; impact sensitivity: IS = 6 J; friction sensitivity: FS = 80 N.

5,5′-Dinitro-3,3′-azo-1H-1,2,4-triazole (**DNAT**): To prepare the compound, 5-amino-3-nitro-1H-1,2,4-triazole (**ANTA**), a slow addition of 80 g (0.62 mol) of **ANTA** to 1.4 L of concentrated hydrochloric acid was performed with stirring. The resulting suspension was warmed to 45 °C to dissolve ANTA into the acid solution. A warm solution composed of 700 mL of water and 68.6 g of potassium permanganate (0.43 mol) was then gradually added over 10 min to the mixture maintained at approximately 50 °C. To control the temperature increase caused by the slightly exothermic reaction, ice was added as necessary. After stirring for 4 h and cooling to room temperature, the solution was refrigerated overnight and subsequently filtered cold using a fritted funnel. The crude product was dissolved in 1.5 L of boiling water and then filtered while hot to remove insoluble impurities. The solution was chilled overnight, and the resulting product was collected by filtration. The collected product was then dried for 8 h at 105 °C, yielding 47.6 g (60% yield based on pure **DNAT**) of anhydrous **DNAT**, which contained a small percentage of 5-chloro-5′-nitro-3,3′-azo-1H-1,2,4-triazole. To refine the crude product, 50 mL of ethanol was added, and the mixture was triturated with heating for 15 min before being cooled in an ice bath. The thick mixture was filtered to remove excess ethanol and later washed with chilled ethanol (approximately 200 mL). The resulting yellow cake was then dried in an oven at 105 °C for 8 h. The yield of pure **DNAT** obtained from this process was 42.3 g (54%). In the aforementioned work, the position of the ring proton was determined to be on the second nitrogen position adjacent to the carbon bearing the azo linkage. ^{13}C NMR (DMSO-d_6) 163.10, 162.80. Elemental Analysis, calculated for $C_4H_2N_{10}O_4$: C, 18.91; H, 0.79; N, 55.12. Found: C, 18.56; H, 0.99; N, 55.66.

5,5′-Diamino-3,3′-azotriazole (**DAAT**): A solution of 3,5-diamino-1,2,4-triazole (36.0 g, 0.36 mol) in 130 mL water was prepared and vigorously stirred at room temperature. Acetic anhydride (40.8 mL) was added dropwise to the solution under vigorous stirring. After stirring for 1 h, the resulting precipitate was filtered off, washed with water, and dried at room temperature. The product obtained from this process was identified as 1-acetyl-3,5-diamino-1H-1,2,4-triazole, which appeared as a colorless powder. The yield of the product was 48.3 g (0.34 mol), corresponding to a 95% yield of 1-acetyl-3,5-diamino-1H-1,2,4-triazole. ^{1}H NMR (DMSO-d_6, 25 °C): 7.35 (s, 2 H, NH_2), 5.64 (s, 2 H, NH_2), 2.33 (s, 3 H, CH_3). ^{13}C NMR (DMSO-d_6, 25 °C): 170.5 (C=O), 162.2, 157.0, 23.6 (CH_3). IR (ATR, 25 °C): 3414 (m), 3388 (vs), 3295 (m), 3127 (s), 1709 (s), 1640 (vs), 1568 (s), 1448 (m), 1393 (s), 1365 (vs), 1336 (s), 1178 (m), 1116 (m), 1066 (m), 1043 (m), 973 (m), 839 (w), 757 (w), 699 (w), 669 (w) cm^{-1}.

A mixture of 1-acetyl-3,5-diamino-1,2,4-triazole (10.0 g, 70.9 mmol) and decaline (100 mL) was heated to reflux without stirring at a temperature range of 187–190 °C for 6 h. The resulting solid was then filtered off, washed with petroleum ether

(100 mL) and diethyl ether (100 mL), and dried in ambient air. The product obtained from this process was identified as 3-acetylamino-5-amino-1H-1,2,4-triazole, which appeared as a colorless powder. The yield of the product was 9.6 g (68.0 mmol), corresponding to a 96% yield of 3-acetylamino-5-amino-1H-1,2,4-triazole. ^{1}H NMR (D_2O, NaOH, 25 °C): 2.03 (s, 3 H, CH_3). ^{13}C NMR ([D_2O, NaOH, 25 °C): 174.1 (C=O), 162.5 (C–NH_2), 154.1 (C–NHAc), 22.7 (CH_3). IR (ATR, 25 °C): 3426 (m), 3251 (s), 1682 (vs), 1597 (vs), 1581 (vs), 1451 (s), 1295 (m), 1269 (m), 1080 (m), 1026 (w), 1010 (w), 816 (w), 714 (m) cm^{-1}. Elemental analysis $C_4H_7N_5O$: calcd: C 34.04, H 5.00, N 49.62%; found: C 34.11, H 4.86, N 49.12%.

To prepare 5,5′-diamino-3,3′-azotriazole (**DAAT**), a solution of 3-acetylamino-5-amino-1H-1,2,4-triazole (2.0 g, 14.2 mmol) in sodium hydroxide (32%, 15 mL) was prepared. Over a period of 10 min, potassium permanganate (1.54 g, 9.7 mmol) was added to the solution at 0 °C. The mixture was allowed to warm to room temperature and then heated to reflux for 3 h after adding an additional 5 mL of sodium hydroxide. The manganese oxide generated during the reaction was removed by filtration. The resulting filtrate was acidified with concentrated hydrochloric acid until the pH reached 6. The resulting precipitate was filtered off, yielding 5,5′-diamino-3,3′-azotriazole (**DAAT**) as an orange solid. The yield of **DAAT** obtained from this process was 0.93 g (4.8 mmol), corresponding to a 68% yield. ^{13}C NMR (D_2O, NaOH): 170.2 (C–N=N), 165.3 (C–NH_2). IR (ATR, 25 °C): 3452 (s), 3351 (s), 2677 (m), 1624 (vs), 1490 (m), 1465 (m), 1413 (w), 1349 (m), 1142 (m), 1102 (m), 1051 (m), 880 (w), 799 (w), 759 (w), 708 (vw), 676 (vw) cm^{-1}.

5,5′-Dinitrimino-3,3′-azo-1H-1,2,4-triazole (**2c-41**): 5,5′-Diamino-3,3′-azotriazole (**DAAT**) (0.35 g, 1.8 mmol) was added to concentrated sulfuric acid (1.75 mL). Then, nitric acid (0.30 mL, 7.2 mmol) was added to the mixture at 0 °C. The resulting mixture was stirred at 0 °C for 30 min, allowed to warm to room temperature, and further stirred for 1 h. To precipitate the product, the mixture was poured onto ice (10 g). The resulting precipitate was then filtered off, washed with water, and dried at 60 °C. The dried product obtained from this process appeared as a yellow solid, referred to as compound **2c-41**. DSC (Onset, 5 °C min^{-1}): T_d = 209 °C. ^{13}C NMR (DMSO-d6, 25 °C): 159.7 (C–N=N), 153.8 (C–N–NO_2). ^{14}N NMR (DMSO-d6, 25 °C): − 19 (NO_2). IR (ATR, 25 °C): 3066 (s), 1695 (w), 1586 (vs), 1542 (m), 1522 (s), 1493 (m), 1436 (m), 1273 (s), 1238 (s), 1139 (m), 1075 (m), 1039 (m), 979 (m), 845 (w), 770 (w), 721 (w) cm^{-1}. Elemental analysis $C_4H_4N_{12}O_4$: calcd: C 16.91, H 1.42, N 59.15%; found: C 18.27, H 1.83, N 58.25%. Sensitivities (grain size: 100–500 μm): FS: 20 N, IS: 2 J, ESD: 0.1 J.

5,5′-Bis (trinitromethyl)-3,3′-azo-1,2,4-triazole Hemihydrate (**2c-42**): A mixture of sulfuric acid (98%, 6 mL) and nitric acid (100%, 5 mL) was prepared. To this mixture, [5-(5-carboxymethyl-1H-[1,2,4]triazole)-2H-[1,2,4]-triazol-3-yl]-acetic acid (1 g, 3.57 mmol) was added at 0 °C. The resulting mixture was stirred for 15 h at room temperature. After completion of the reaction, the reaction mixture was poured into ice water, causing a precipitate to form. The precipitate was then filtered out. The mother liquor was subjected to extraction with ethyl acetate, and the organic extract was dried over magnesium sulfate ($MgSO_4$). The solvent was subsequently evaporated to yield additional product. The combined solids obtained from the filtration and evaporation steps were identified as **2c-42**. The yield of the yellow solid **2c-42** was determined to be 55%. T_d, 150 °C (onset). IR (KBr): n 3467, 1630, 1617, 1593, 1442, 1342, 1282, 1194, 1097, 1052, 964, 842, 801, 758 cm^{-1}. ^{13}C NMR: 165.7, 148.8, 123.8. ^{15}N NMR: 31.62 ($(NO_2)_3$), – 33.0 (NdN), – 60.35 (N_2), – 131.79 (N_3), – 132.26 (N_1). Anal. calcd for $C_6H_3N_{14}O_{12.5}$, 471: C, 15.29; H, 0.64; N, 41.62. Found: C, 15.56; H, 0.32; N, 41.50.

1,1′-Dimethyl-5,5′-bis(trinitromethyl)-3,3′-azo-1,2,4-triazole (**2c-43**): A solution of **2c-42** (100 mg, 0.21 mmol) in diethyl ether was prepared. To this solution, a 2 M solution of trimethylsilyl diazomethane (0.5 mL) was added dropwise at 0 °C. The resulting reaction mixture was stirred for 30 min, then filtered to remove any solids. The precipitate obtained from the filtration step was washed with diethyl ether (5 mL). The product obtained from this process was identified as **2c-43**. The yield of the yellow solid **2c-43** was determined to be 76%; T_d, 150 °C (onset). IR (KBr): n 3441, 1625, 1593, 1474, 1443, 1346, 1290, 1220, 1108, 962, 844, 801, 754 cm^{-1}. ^1H NMR: 4.39. ^{13}C NMR: 159.7, 147.7, 122.8. Anal. calcd for $C_8H_6N_{14}O_{12}$, 490: C, 19.60; H, 1.23; N, 40.00. Found: C, 19.69; H, 1.13; N, 38.00.

1,1′-(Oxybis-(methylene))bis(5-nitro-3-(trinitromethyl)-1H-1,2,4-triazole) (**2c-44**) and 1,1′-(oxybis(methylene))bis(5-nitro-3-(trinitromethyl)-2H-1,2,4-triazole) (**2c-45**): 5-(Trinitromethyl)-3-nitro-triazole was added to pre-cooled 98% sulfuric acid (5 mL) at 0 °C. Then, paraformaldehyde (0.50 g) was added to the solution gradually over a period of 5 min. The reaction temperature was allowed to slowly increase to 25 °C. The resulting mixture was left at room temperature for 12 h and then poured into 100 g of crushed ice. The target compounds formed during the reaction could be detected using nuclear magnetic resonance (NMR), but unfortunately, they could not be separated as individual compounds.

1,1′-(Oxybis(methylene))bis(3-chloro-5-nitro-1H-1,2,4-triazole) (**2c-46**): A solution of potassium 3,5-dinitro-1,2,4-triazolate (1.14 g, 6.0 mmol) and bis(chloromethyl)ether (0.35 g, 3.04 mmol) in dimethylformamide (5 mL) was prepared. The reaction mixture was stirred for 72 h at room temperature. The resulting white precipitate that formed during the reaction was removed by filtration. The filtrate obtained from the filtration step was then concentrated. The concentrated solution resulted in the formation of an oily residue, which was subsequently diluted with water. This dilution led to the precipitation of a white solid. The precipitated solid was filtered and washed with dichloromethane (10 mL × 3) to yield the product **2c-46**. T_d: 280 °C. ^1H NMR: d 5.80 ppm (s, 4H). ^{13}C NMR: d 161.9, 145.6, 77.9 ppm. IR (KBr): n 3105, 3029, 2978, 2922, 2851, 1558, 1497, 1437, 1396, 1372, 1308, 1246, 1148, 1083, 1052, 1017, 936, 841, 758, 687, 648, 592, 566 cm^{-1}. Elemental analysis for $C_6H_4Cl_2N_8O_5$ (339.05): calcd C 21.25, H 1.19, N 33.05%. Found: C 21.37, H 1.20, N 32.93%. IS: 28 J. FS: 240 N.

1,1′-(Oxybis(methylene))bis(3-nitro-1H-1,2,4-triazole) (**2c-47**): In a solution of potassium 3-nitro-1,2,4-triazolate (1.52 g, 10 mmol) and bis(chloromethyl)ether (0.58 g, 5.04 mmol) in dimethylformamide (5 mL), the reaction mixture was stirred for 72 h at room temperature. Following the completion of the reaction, a white precipitate formed, which was subsequently removed by filtration. The resulting filtrate was then concentrated, resulting in the formation of an oily residue. This oily residue was further diluted with water, leading to the precipitation of a white solid. The obtained white solid was filtered and subsequently washed three times with dichloromethane (10 mL each time). This process yielded the desired product, **2c-47**, with a yield of 0.43 g (42%). White solid. T_m: 165 °C. T_d: 320 °C. ^{1}H NMR: d 9.03 ppm (s, 2H), 5.92 ppm (s, 4H). ^{13}C NMR: d 162.5, 147.9, 77.6 ppm. IR (KBr): n 3129, 3031, 1980, 2924, 853, 2718, 1552, 1509, 1473, 1429, 1412, 1366, 1308, 1250, 1233, 1190, 1101, 1032, 939, 882, 835, 762, 652 cm^{-1}. Elemental analysis for $C_6H_6N_8O_5$ (270.16): calcd C 26.67, H 2.24, N 41.48%. Found: C 26.77, H 2.32, N 40.49%. IS: 20 J. FS: 240 N.

1,1′-(Ethane-1,2-diyl)bis(3-nitro-1H-1,2,4-triazol-5-amine) (**2c-48**): A suspension of 3-nitro-1H-1,2,4-triazol-5-amine (**ANTA**, 1.29 g, 10 mmol) in 5 mL dimethylformamide (DMF) was prepared. To this suspension, tetraethylammonium bromide (0.21 g, 1 mmol) and potassium hydroxide (KOH) (0.56 g, 10 mmol) were added. The reaction mixture was then heated to 70 °C for 30 min. Subsequently, dibromoethane (1.13 g, 6 mmol) was slowly added dropwise to the reaction mixture. The resulting orange suspension was maintained at 70 °C overnight. After completion of the reaction, the reaction mixture was poured into cold water (20 mL) and stirred for an additional 30 min. The resulting precipitate was then filtered, washed with water, and dried in the air. This process afforded the desired product, **2c-48**, as a yellowish solid with a yield of 0.86 g (61%). ^{1}H NMR: δ7.05 (s, 4H); 4.39 (s, 4H); ^{13}C NMR (d6-DMSO): δ 159.6, 156.8, 45.6; IR (KBr pellet): υ 3445, 3408, 3295, 3165, 1629, 1559, 1513, 1452, 1412, 1311, 1260, 1088, 864, 729 cm^{-1}; elemental analysis (%) calcd for $C_6H_8N_{10}O_4$ (284.19): C, 25.36; H, 2.84; N, 49.29; found: C, 25.29; H, 3.05; N, 48.06.

N,N′-(1,1′-(ethane-1,2-diyl)bis(3-nitro-1H-1,2,4-triazole-5,1-diyl))dinitramide (**2c-49**): To a solution of compound **2c-48** (0.57 g, 2 mmol), small portions were added to 10 mL of ice-cold 100% nitric acid (HNO_3). The resulting mixture was stirred at 0 °C for 30 min, forming a clear solution. To remove the excess nitric acid, air was blown onto the mixture, causing it to solidify. The solidified residue was then washed with a small amount of cold ethanol (EtOH) and subsequently dried under vacuum. This process yielded **2c-49** as a white solid, with a weight of 0.65 g (86% yield). ^{1}H NMR: δ11.65 (s, 2H),4.63 (s, 4H); ^{13}C NMR (DMSO-d6): δ 160.2, 150.4, 47.5; ^{15}N NMR (DMSO-d6): δ − 26.50, − 26.54, − 92.47, − 147.97, − 175.86, − 183.61; IR (KBr pellet): υ 3134, 2986, 1628, 1570, 1540, 1500, 1451, 1411, 1381, 1305, 1242, 1117, 978, 912, 841, 797, 709 cm^{-1}; elemental analysis (%) calcd for $C_6H_6N_{12}O_8$ (374.19): C, 19.26; H, 1.62; N, 44.92; found: C, 18.64; H, 1.59; N, 44.13.

4-Methyl-1,1′-bi(1,2,3-triazole) (**2c-50**), 1,1′-bi(1,2,3-triazole) (**2c-51**) and 4-methyl-1-(4H-1,2,4-triazol-4-yl)-1H-1,2,3-triazole (**2c-52**): A solution of triethylamine (Et_3N) (0.22 g, 2.2 mmol) in 10 mL of methanol was prepared. This solution was then added to a suspension containing amino azoles (2.0 mmol) and dichlorotosylhydrazone (2.0 mmol). The resulting reaction mixture was stirred at 0 °C for a period of 2–12 h. After the desired reaction time, the precipitate formed was collected by filtration. It was then washed consecutively with methanol (CH_3OH), filtered again, and dried under vacuum at 40 °C. The product obtained from this reaction, N′-(1-((1H-1,2,3-triazol-1-yl)imino)propan-2-ylidene)-4-methylbenzenesulfonohydrazide, was obtained as a light yellow solid. The yield of this compound was determined to be 0.56 g when amino azoles (2.0 mmol) and dichlorotosylhydrazone were reacted for a duration of 12 h.

A suspension of intermediate compound N′-(1-((1H-1,2,3-triazol-1-yl)imino)propan-2-ylidene)-4-methylbenzenesulfonohydrazide (0.5 mmol) in methanol (4 mL) was prepared. Dropwise addition of a solution containing diisopropylethylamine (DiPEA) (0.23 g, 1.8 mmol) in 2 mL of methanol was performed at either 65 °C or room temperature. During the reaction, the mixture gradually turned clear, and stirring was continued for 72 h. Upon completion of the reaction, any volatile components were removed under reduced pressure. The resulting residue was then subjected to purification by column chromatography, utilizing a mobile phase of petroleum ether/ethyl acetate (1:1 ratio). The yields of the resulting compounds, **2c-(50-52)**, generally ranged from 75 to 95%.

2c-50: White solid (0.071 g). Yield: 95%. DSC: T_m 74.1 °C, T_d, onset = 160.8 °C, T_d, peak = 201.6 °C. ^{1}H NMR (300 MHz, DMSO-d6) ppm: 9.01 (d, J = 1.2 Hz, 1H), 8.73 (s, 1H), 8.15 (d, J = 0.9 Hz, 1H), 2.38 (s, 3H). 13C NMR (75 MHz, DMSO-d6) ppm: 142.7, 133.7, 126.6, 123.5, 10.65. HRMS (ESI-TOF) m/z: $[M+H]^+$ calcd for $C_5H_7N_6$, 151.0732; found, 151.0730. IR (KBr): 3162, 3146, 1550, 1475, 1227, 1004, 805, 623 cm^{-1}.

2c-51: White solid (0.051 g). Yield: 75%. DSC: T_m, 132.0 °C, T_d, onset = 155.2 °C, T_d, peak = 185.3 °C. ^{1}H NMR (500 MHz, DMSO-d6) ppm: 9.04 (d, 1H), 8.17 (d, 1H). ^{13}C NMR (75 MHz, DMSO-d6) ppm: 133.8, 126.5. HRMS (ESI-TOF) m/z: $[M+H]^+$ calcd for $C_4H_5N_6$, 137.0575; found, 137.0597. IR (KBr): 3137, 3117, 1480, 1405, 1279, 1013, 939, 805, 602 cm^{-1}.

2c-52: White solid (0.066 g). Yield: 88%. DSC: T_m, 120.4 °C, T_d, onset = 185.0 °C, T_d, peak = 223.8 °C. ^{1}H NMR (300 MHz, DMSO-d6) ppm 9.30 (s, 2H), 8.58 (s, 1H), 2.35 (s, 3H). ^{13}C NMR (75 MHz, DMSO-d6) ppm: 142.3, 142.1, 123.7, 10.6. HRMS (ESI-TOF) m/z: $[M+H]^+$ calcd for $C_5H_7N_6$, 151.0732; found, 151.0729. IR (KBr): 3182, 3121, 1566, 1520, 1075, 1020, 612 cm^{-1}.

4,4′,6,6′-Tetra(azido)hydrazo-1,3,5-triazine (**2d-1**): A suspension of 184.5 g (1 mol) of cyanuric chloride in 500 mL of tetrahydrofuran was vigorously stirred at a temperature range of − 15 °C to − 10 °C. Simultaneously, a solution of 75 g (1.5 mol) of hydrazine hydrate (100%) in 100 mL of water was prepared. Once the desired temperature was reached, the hydrazine hydrate solution was added to the cyanuric chloride suspension. The resulting mixture was stirred at 0 °C for an additional 15–30 min. To precipitate the product, 2 L of water were added to the reaction mixture. The formed precipitate was then collected by filtration, washed with water, and subsequently dried. This process yielded 154 g of 2,2′-4,4′-tetrachlorohydrazo-1,3,5-triazine, with a yield of 94%. Anal. calcd for $C_6H_2Cl_4N_8$: C, 21.97; H, 0.61; C, 43.24; N, 34.16. Found: C, 22.20; H, 0.71; Cl, 42.98; N, 33.88.

A solution containing 1.68 g (0.02 mol) of sodium bicarbonate in 50 mL of water was prepared. This solution was then added to a suspension of 3.28 g (0.01 mol) of 2,2′-4,4′-tetrachlorohydrazo-1,3,5-triazine in 100 mL of chloroform. To initiate the reaction, a slow stream of chlorine gas was introduced into the mixture at a temperature of 20–25 °C. The addition of chlorine gas continued until a homogeneous red-colored chloroform layer was formed. The organic phase, which contained the desired product, was separated from the aqueous layer. It was then washed with water, dried using sodium sulfate, and subjected to evaporation until dryness. The resulting residue obtained after evaporation was further treated by slurring it in ether. The slurry was then filtered to isolate the final product, 2,2′,4,4′-tetrachloro-azo-1,3,5-triazine, with a yield of 2.7 g (83%). Anal. calcd for $C_6Cl_4N_8$: C, 22.1 l; Cl, 43.50; N, 34.37. Found: C, 22.35; Cl, 43.38; N, 34.10.

2,2′,4,4′-Tetrachloro-azo-1,3,5-triazine underwent diazotization to yield **2d-1**. Elemental analysis (%) calcd for $C_6H_2N_{20}$: C 20.34, H 0.57, N 79.09; found: C 20.04, H 0.75, N 79.44. IR (cm^{-1}): 2172 (vs), 2129 (vs), 3221 (s), 3091 (s); 1541 (vs), 1352 (vs), 1252 (vs), 972 (vs), 806 (vs), ^{1}H NMR (300 MHz, DMSO-d_6): 10.59 ppm, ^{13}C NMR (75 MHz, DMSO-d_6): 168.3, 169.6, 170.2 ppm.

4,4′,6,6′-Tetra(azido)azo-1,3,5-triazine (**2d-2**): **2d-1** was suspended in a solution of water and chloroform (H_2O:$CHCl_3$, 1:2 ratio). The suspension was then subjected to oxidation using chlorine gas at room temperature, resulting in the formation of **2d-2**. Elemental analysis (%) calcd for C_6N_{20}: C 20.46, H 0.00, N 79.54; found: C 20.82, H 0.07, N 79.18, IR (cm^-1): n (N_3) 2208 (vs), 2155 (vs), 2132 (vs); n (triazine) 1549 (vs), 1521 (vs), 1435 (vs), 1161 (vs), 1011(vs), 823 (vs), ^{13}C NMR (75 MHz, DMSO-d_6, 25 °C): d = 173.7, 176.4 ppm.

(1E,1′E,1"E,1‴E)-2,2′,2",2‴-(diazene-1,2-diylbis(1,3,5-triazine-6,2,4-triyl))tetrakis(N′-nitrohydrazine-1-carboximidamide) (**2d-3**): To prepare N-methyl-N′-nitroguanidine, an alkaline solution was made by dissolving nitroguanidine (0.1 mol, 10.4 g) in 30 mL of water containing KOH (0.215 mol, 12 g). The solution was heated to 40 °C to ensure dissolution of the nitroguanidine. Methylamine hydrochloride (0.2 mol, 13.5 g) was then added with stirring at this temperature. The temperature was then raised to 59 °C and maintained within the range of 59–61 °C for a duration of 30 min. After heating, the resulting clear solution was cooled using an ice-water bath and brought down to a temperature of 5–6 °C. A white precipitate formed, which was filtered off, washed with cold water (30 mL), and dried in air. This process yielded N-methyl-N′-nitroguanidine as a white solid with a weight of 10.0 g (84.7% yield) and a melting point ranging from 151 to 154 °C. Subsequently, N-methyl-N′-nitroguanidine (0.085 mol, 10.0 g) was dissolved in water (100 mL) by adding 30 mL of 70% HNO_3. The resulting solution was cooled in an ice bath, and $NaNO_2$ (0.18 mol, 12.4 g) dissolved in water (20 mL) was added over a period of 5 min. The mixture was stirred for 30 min, after which the resulting precipitate was filtered and washed with cold water. This step yielded N-methyl-N-nitroso-N′-nitroguanidine as a yellow crystalline product with a weight of 8.75 g (T_m, 112–113 °C).

An aqueous solution (50 mL) of bis-(4,6-dihydrazino-[1,3,5]triazin-2-yl)-diazene (0.31 g, 1 mmol) was prepared. N-methyl-N-nitroso-N′-nitroguanidine (4.2 mmol, 0.62 g) was then added to the solution at room temperature. The mixture was stirred at room temperature for a duration of 18 h. Upon completion of the reaction, the resulting precipitate formed in the mixture was filtered off, washed with cold water (50 mL), and subsequently dried. This process yielded the desired product, **2d-3**, as a yellow powder. The yield obtained was 74%, with a weight of 0.49 g. $T_d = 201$ °C; IR (KBr): n = 3302, 1590, 1539, 1499, 1376, 1277, 1056, 812, 552, 474 cm^{-1}; 1H NMR: d = 9.80 (s, 1H, –NH–), 9.33 (s, 1H, –NH–), 8.56 (s, 1H, NH), 8.15 ppm (s, 1H, NH); ^{13}C NMR: d = 167.5, 161.4 ppm; elemental analysis (%) calcd for $C_{10}H_{16}N_{28}O_8$ (656.42): C 18.30, H 2.46, N 59.75; found: C18.61, H 3.03, N 56.40.

4,6-Diazido-*N*-(4,6-diazido-1,3,5-triazin-2-yl)-1,3,5-triazin-2-amine **(2d-4)**: To maintain an alkaline condition in the reaction mixture, a solution of sodium hydroxide weighing 0.8 g (0.02 mol) in 8 mL of water was added dropwise with stirring at a temperature range of 0–5 °C. This solution was added to a separate solution containing 50 mL of acetone in which 4.8 g (0.02 mol) of 2-amino-4,6-dichloro-s-triazine and 3.7 g (0.02 mol) of cyanuric chloride were already dissolved at the same temperature. After an additional stirring period of 1.5 h, the resulting mixture was neutralized using sodium carbonate. Subsequently, the mixture was poured into 500 mL of ice-water. As a result, a precipitate formed, namely 4,6-dichloro-N-(4,6-dichloro-1,3,5-triazin-2-yl)-1,3,5-triazin-2-amine. This precipitate was collected through filtration and subsequently dried over phosphorus pentoxide.

A solution of sodium azide weighing 3.9 g (60 mmol) in 8 mL of water was combined with a solution of compound 4,6-dichloro-N-(4,6-dichloro-1,3,5-triazin-2-yl)-1,3,5-triazin-2-amine weighing 3.13 g (10 mmol) in 100 mL of acetone. The resulting mixture was stirred for a duration of 2 h at room temperature. Afterward, the solvent was removed by evaporation under reduced pressure. The residue obtained was then treated with 100 mL of water. The solid product formed was filtered off, washed with water on the filter, and air-dried. Subsequently, the product underwent recrystallization from a 2:1 mixture of ethanol and acetone. The final yield of the product was 3.22 g, representing a 95% yield. The obtained crystals were colorless., T_m, 220–222 °C. IR spectra (microcrystals) cm^{-1}: 3300 (NH), 2199, 2167, 2146 and 2124 (N_3), 1612, 1590, 1545, 1510, 1449, 1347, 1313, 1242, 1217, 1197, 1162, 971, 800, 763, 735, 714, 704, 689. ^1H NMR spectrum, ppm: 11.86 (NH). ^{13}C NMR spectrum, ppm: 165.3 (C-2); 170.2 (C-4,6). ^{15}N NMR spectrum, ppm: – 138.6; – 145.9; – 165.4; – 165.6; – 170.6; – 263.0. Mass spectrum: 340 $[M+H]^+$, 339 $[M]^+$ (100), 313 $[M-CN]^+$ (5), 243 (13), 119 (9), 118 (9), 107 (4), 94 (11), 93 (46), 92 (14), 78 (71), 67 (54), 66 (27), 54 (21), 53 (23), 52 (12), 41 (5). Found, %: C 21.32; H 0.34; N 78.34. C_6HN_{19}. Calculated, %: C 21.25; H 0.30; N 78.46.

3,3′-Azobis(6-amino-1,2,4,5-tetrazine) (**2d-5**): Triaminoguanidine monohydrochloride (7.03 g, 0.05 mol) was dissolved in water (50 mL) to form a solution. To this solution, 2,4-pentanedione was added dropwise with stirring at a temperature of 25 °C. The mixture was allowed to stir at this temperature for 0.5 h, after which it was heated to 70 °C and maintained at that temperature for 4 h. During this heating period, solid precipitated from the solution. After completion of the reaction, the mixture was cooled and the resulting product was filtered off. The solid product was washed with water and subsequently dried. This process yielded 5.77 g (85%) of pure 3,6-bis(3,5-dimethylpyrazol-1-yl)-1,2-dihydro-1,2,4,5-tetrazine. T_m, 150 °C; ^1H NMR: 2.21(s, 6H), 2.47 (s, 6H), 5.95 (s, 2H), 8.09 (bs, 2H); ^{13}C NM: 613.4, 13.7, 109.8, 142.3, 145.8, 149.9; ^{15}N NNM: – 92.6, – 156.8, – 182.6, – 266.5.

In a mechanically stirred 5-L, 3-necked flask, 3,6-bis(3,5-dimethylpyrazol-1-yl)-1,2-dihydro-1,2,4,5-tetrazine weighing 270.0 g (1.0 mol) was dissolved in 2-propanol (2 L). Anhydrous hydrazine weighing 16.0 g (0.5 mol) was added all at once to the solution with efficient stirring. The resulting mixture was refluxed for a duration of 2 h and then allowed to cool and stir overnight. The orange-colored product formed was filtered off and washed with 2-propanol. This process yielded 143.0 g of pure 1,2-bis(6-(3,5-dimethyl-1H-pyrazol-1-yl)-1,2,4,5-tetrazin-3-yl)hydrazine. The mother liquor, which contained additional product, was left to stand for a few weeks and subsequently refiltered. This step provided an additional 7.3 g of product identical to the first crop. The total yield obtained from both crops was 79%. T_m, 215–21 °C. IR (KBr): 3194, 1575, 1490, 1415 cm^{-1}; ^1H NMR (270 MHz DMSO-d_6): 2.21 (s, 6H), 2.49 (s, 6H), 6.22 (s, 2H), 11.2 (brs, 2H); ^{13}C NMR (270 MHz DMSO-d_6): 12.3, 13.0, 109.5, 142.3, 151.3, 158.9, 162.7; elemental analysis calcd for $C_4H_{16}N_4$: C 44.21, H 4.24, N 51.55; found: C 43.95, H 4.40, N 51.39.

In a 5-L, 3-necked flask equipped with a mechanical stirrer, N-bromosuccinimide weighing 267.4 g (1.5 mol) was dissolved in acetonitrile (2 L). Compound 1,2-bis(6-(3,5-dimethyl-1H-pyrazol-1-yl)-1,2,4,5-tetrazin-3-yl)hydrazine weighing 143.0 g (0.38 mol) was then added to the flask portion-wise over a period of 15 min. The resulting mixture was allowed to stir for 1 h. The purple precipitated product formed was filtered off and washed with tert-butyl methyl ether. It was subsequently air-dried, resulting in a yield of 197.0 g of 1,2-bis(6-(4-bromo-3,5-dimethyl-1H-pyrazol-1-yl)-1,2,4,5-tetrazin-3-yl)diazene, representing a yield of 98%. T_m, 195–197 °C. IR (KBr): 1501, 1450 cm^{-1}; ^1H NMR (270 MHz, $CDCl_3$): 2.44 (s, 6H), 2.84 (s, 6H); ^{13}C NMR (270 MHz, $CDCl_3$): 13.0, 14.4, 104.4, 142.2, 155.3, 159.3, 167.9; elemental analysis calcd for $C_{14}H_{12}N_{14}Br_2$: C 31.36, H 2.26, N 36.57; found: C 30.99, H 2.21, N 36.51.

In a 4-L beaker, rapid bubbling of ammonia was carried out through DMSO (1 L) until 7.0 g (0.41 mol) were absorbed. To this solution, 1,2-bis(6-(4-bromo-3,5-dimethyl-1H-pyrazol-1-yl)-1,2,4,5-tetrazin-3-yl)diazene weighing 110.4 g (0.206 mol) was added with stirring within 1 min. The resulting reaction mixture was allowed to stir for 15 min before being diluted with 2-propanol (1 L). The red solid product formed was then filtered off, washed with 2-propanol, and air-dried. This process yielded 34.3 g of **2d-5**. To remove the DMSO solvate, this material was boiled in water (250 mL) for a few minutes. The resulting mixture was filtered, and the solid product obtained was washed with water and subsequently air-dried. This final step provided pure **2d-5**. T_m, 252 °C. IR (KBr): 3372, 3283, 3194, 1629, 1506 cm^{-1}; ^1H NMR (270 MHz, DMF-d_6): 8.93 (br s, 4H); ^{13}C NMR (270 MHz, DMF-d_6): 163.6, 167.9; elemental analysis calcd for $C_4H_4N_{12}$: C 21.82, H 1.83, N 76.35; found: C 21.46, H 2.11, N 67.70.

N-oxides of 3,3′-azobis(6-amino-1,2,4,5-tetrazine) (**2d-6**): A heterogeneous mixture of 90% hydrogen peroxide (14.4 mL, 0.5 mol) in methylene chloride (400 mL) was prepared. Trifluoroacetic anhydride (82 mL, 0.57 mol) was added to the mixture in small portions with stirring and cooling at 0 °C. The reaction was allowed to stir for 10 min before compound **2d-5** (10.0 g, 45 mmol) was added. The reaction temperature was maintained at 0 °C for 1 h, after which it was gradually raised to 15 °C over a period of 5 h. The reaction mixture was then kept at 15 °C for an additional 18 h. The reaction mixture was poured into ice water (1 L), followed by filtration. The resulting solid was washed extensively with water and air-dried. Elemental analysis of the oxidized material was conducted to determine the oxygen content, revealing that the mixture contained approximately 3.5 oxygen atoms per molecule. CHN, theory 17.40% C, 1.45% H, 60.89% N; found 17.60% C, 1.48% H, 60.10% N.

4,4′,5,5′-Tetranitro-1H,1′ H-2,2′-biimidazole-1,1′-diamine (**2e-1**): In a 200 mL round-bottomed flask, 4,4′,5,5′-tetranitro-1H,1′H-2,2′-biimidazole (**TNBI**) weighing 1.58 g (10 mmol) was dissolved in excess aqueous ammonia. The solvent was removed by blowing air over the liquid surface, and the remaining residue was dried under vacuum. To this ammonium salt, acetonitrile (100 mL) was added, followed by the freshly prepared O-tosylhydroxylamine in CH_2Cl_2 (100 mL). The resulting mixture was stirred at room temperature for 12 h. The solvent was then removed using rotary evaporation, and ethyl acetate (50 mL) was added to the residue. The white precipitate formed was filtered, and the ethyl acetate solvent was subsequently removed. The isolated product, 4,4′,5,5′-tetranitro-1H,1′H-2,2′-biimidazole-1,1′-diamine (**2e-1**), was obtained through chromatography using a hexane/ethyl acetate eluent (100:50). The final product obtained was a white solid weighing 1.82 g, representing a yield of 53%. T_m, 206 °C, T_d, 217 °C (onset); ^{1}H NMR (DMSO-d_6): 8.55 (s, 2H), 6.87 (s, 4H); ^{13}C NMR (DMSO-d_6): 137.0, 134.8, 131.6; IR (KBr): 3360, 3259, 1532, 1504, 1389, 1366, 1312, 1147, 849, 816 cm^{-1}.

N,N′-(4,4′,5,5′-Tetranitro-1H,1′H-[2,2′-biimidazole]-1,1′-diyl)dinitramide (**2e-2**): To a cold solution of 4,4′,5,5′-tetranitro-1H,1′H-[2,2′-biimidazole]-1,1′-diamine (**2e-1**) weighing 344 mg (1 mmol) in concentrated sulfuric acid solution (5 mL H_2SO_4), fuming nitric acid (2 mL) was added dropwise at − 15 °C. The resulting reaction mixture was stirred at − 15 °C for 2 h. The formed precipitate was then filtered, washed with trifluoroacetic acid, and dried under vacuum to obtain the product **2e-2**. The final product was a yellowish solid with a yield of 81%. ^{1}H NMR (DMSO-d_6): 8.67 (br, 2H); ^{13}C NMR (DMSO-d_6): 135.8, 132.4, 129.0 ppm; IR (KBr): 3628, 3200, 1651, 1545, 1460, 1377, 1341, 1315, 1167, 1053, 953, 852, 807, 758, 710, 699, 571, 513 cm^{-1}.

4,4′-dinitro-5,5′-diamino-1H,1′H-2,2′-bisimidazole (**2e-3**): To a 28% aqueous ammonia solution (15 mL) at room temperature, 4,4′,5,5′-tetranitro-1H,1′H-2,2′-biimidazole (**TNBI**) weighing 1.0 g (3.18 mmol) was added. The reaction mixture was then heated to 100 °C and maintained at this temperature for 12 h. After cooling, the resulting mixture was acidified with 18% HCl until reaching a pH of 1. The formed precipitate was collected by filtration, washed with water (100 mL), and air-dried. This process yielded an orange-yellow solid weighing 0.66 g, representing an 82% yield. T_d (onset): 300 °C. ^{1}H NMR (DMSO-d_6): 13.21 (s, 1H), 7.25 (s, 2H) ppm. ^{13}C NMR (DMSO-d_6): 145.2, 130.3, 128.9 ppm. IR (KBr): 3422, 3316, 1662, 1497, 1426, 1348, 1261, 1186, 913, 755, 717, 650 cm^{-1}.

4,4′-Dinitro-5,5′-diazo-2,2′-bisimidazole zwitterion (**2e-4**): Method a: At 0 °C, nitric acid (100%, 5 mL) was gradually added to trifluoroacetic anhydride (10 mL), and compound **2e-3** (0.60 g, 2.36 mmol) was introduced into the resulting solution.

The reaction mixture was then stirred for 1 h at 0 °C. Following the completion of the reaction, the formed precipitate (0.20 g, 31%) was collected by filtration and washed with trifluoroacetic acid (3 mL). The filtrate obtained from the filtration step was concentrated, leading to the isolation of additional product (40 mg, 6%).

Method b: Sodium nitrite (0.6 g, 8.6 mmol) was added to concentrated sulfuric acid (5 mL) at 5 °C. Then, compound **2e-3** (0.51 g, 2.0 mmol) was introduced into the solution, and the mixture was stirred at room temperature for 1 h. Afterward, the solution was poured into ice water, resulting in the precipitation of a yellow solid (0.38 g, 69%) from the water phase. The solid product was collected by filtration and dried in air. The filtrate obtained from the filtration step was subjected to extraction with ethyl acetate (30 mL × 2). The combined organic phases were then dried using anhydrous sodium sulfate and concentrated by air. This process yielded the final product (0.06 g, 11%) as a yellow solid. T_d: 199 °C (onset). ^{13}C NMR (DMSO-d_6): 156.6, 148.7, 103.1 ppm. ^{15}N NMR (DMSO-d_6): − 25.7, − 38.7, − 101.6, − 111.8, − 146.0 ppm. IR (KBr): 2237, 2201, 1550, 1460, 1433, 1367, 1339, 1319, 1266, 1189, 1112, 832, 750, 692, 653, 597 cm^{-1}.

4,4′-dinitro-1H,1′H-2,2′-biimidazole (**2e-5**): A solution of 2,2′-biimidazole (**2e-6**) weighing 3.35 g (0.025 mol) was prepared by dissolving it in 40 mL of acetic acid (AcOH) and 20 mL of acetic anhydride (Ac_2O) at a temperature of 80 °C. To this solution, 2.81 mL (0.065 mol) of 99% nitric acid (HNO_3) was added dropwise. The resulting mixture was stirred at temperatures ranging from 90 to 100 °C for a duration of 8 h. After completion of the reaction, the reaction mixture was cooled, and the formed solid was filtered and subjected to crystallization using dimethylformamide (DMF) as the solvent. This process resulted in the formation of the product, which weighed 2.2 g (39%) and exhibited a melting point of 360 °C.

2,2-Biimidazole (**2e-6**): In a solution of sodium bisulfate (18.3 g, 175.5 mmol) in 90 mL of water, 50 mL of ethanol was added. To this mixture, 12.77 g (38 mmol) of 40% glyoxal was introduced, and the resulting solution was stirred for 1.5 h at room temperature. The formed white solid product was then filtered and washed with ethanol and diethyl ether to remove impurities. Next, the solid was redissolved in a solution containing 138 mL of aqueous ammonia (25%) and 5 g of ammonium carbonate. The resulting mixture was refluxed for 4 h and subsequently allowed to cool down to room temperature. After filtration and washing of the obtained solid with ethanol and diethyl ether, a pale-yellow product was obtained with high purity, weighing 1.25 g (38%). T_m > 300 °C. ^{1}H NMR (DMSO-d_6), 7.06 (m, 4H), 12.55 (s, 2H).

4,4′-Dinitro-1H,1′H-2,2′-biimidazole-1,1′-diamine (**2e-7**): 4,4′-dinitro-1H,1′H-2,2′-biimidazole (1.54 g, 10 mmol) was dissolved in 30 mL of dimethylformamide (DMF). To this solution, potassium carbonate (2.07 g, 15 mmol) was added and stirred at a temperature of 50 °C for 2 h. In an ice bath, freshly prepared O-tosylhydroxylamine in dichloromethane (100 mL) was added to the reaction mixture, ensuring that it was kept below 10 °C. The resulting mixture was allowed to warm and stirred at room temperature for 24 h. The organic solvent was subsequently removed using rotary evaporation. Ethyl acetate (50 mL) was added to dissolve the aminoimidazole, and then the mixture was filtered, washed with brine, and dried over sodium sulfate (Na_2SO_4). The compound **2e-7** was isolated via chromatography using a hexane–ethyl acetate gradient (ranging from 100:100 to 10:100). This process yielded a yellow solid product weighing 1.07 g (42% yield). T_m, 285 °C, T_d, 308 °C (onset); ^1H NMR (DMSO-d_6): 8.57 (s, 2H), 6.87 (s, 4H); ^{13}C NMR (DMSO-d_6): 143.3, 133.4, 123.6; IR (KBr): 3330, 3266, 3127, 1541, 1498, 1399, 1344, 1298, 1001, 820 cm^{-1}.

1H,1′H-2,2′-Biimidazole-1,1′-diamine (**2e-8**). 1H,1′H-2,2′-biimidazole (134 mg, 1 mmol) was suspended in a solution of aqueous KOH (560 mg of KOH in 20 mL of distilled water) and stirred at a temperature of 90 °C for 0.5 h. The reaction mixture was then allowed to cool down to 40 °C before the addition of hydroxylamine-O-sulfonic acid (684 mg, 6 mmol) over a period of 20 min, ensuring that the temperature stayed below 50 °C. After stirring overnight at 50 °C, the mixture was extracted with ethyl ether (5–10 mL). The organic layers were combined, dried using sodium sulfate (Na_2SO_4), and concentrated under reduced pressure. The resulting residue was subjected to purification by chromatography using a hexane/ethyl acetate mixture (20:100). This process yielded a white solid product weighing 39 mg (24% yield). T_m, 154 °C, T_d, 292 °C (onset); ^1H NMR (DMSO-d_6): 7.24 (s, 2H), 6.96 (s, 2H), 6.87 (s, 4H); ^{13}CNMR (DMSO-d_6): 134.5, 125.0, 120.9 ppm; IR (KBr pellet): 3267, 3180, 3119, 1637, 1504, 1437, 1411, 1327, 1294, 1106, 1023, 751 cm^{-1}.

Diammonium-4,4′,5,5′-tetranitro-2,2′-biimidazolate (**2e-10**): A suspension of 4,4′5,5′-tetranitro-2,2′-biimidazole dihydrate (**TNBI**) weighing 10.00 g (28.56 mmol) was prepared in 250 mL of deionized water. Ammonium hydroxide (28–30% NH_3) measuring 10.0 mL (71.9 mmol, 2.5 equiv.) was added dropwise with stirring. The resulting pale-yellow suspension was stirred at a temperature of 20 °C for 24 h before being cooled to 5 °C. The formed yellow solid product was collected by vacuum filtration and washed with 50 mL of isopropyl alcohol. It was then air-dried for 1 h and further dried in a vacuum oven at 60 °C for 24 h. This process yielded 9.45 g (27.14 mmol) of the desired compound, **2e-10**, as yellow needle-like crystals with a high yield of 95%. T_d, 269 °C; ^1H NMR (DMSO-d_6): 3.58 (s, 3 H) ppm. ^{13}C NMR (DMSO-d_6): 140.19, 143.84 ppm. IR: 3210, 2927, 2721, 2486, 1475, 1441, 1387, 1365, 1304, 1242, 1114, 950, 854, 810, 754, 706 cm^{-1}.

1,1′-Dimethyl-4,5′-dinitro-2,2′-biimidazole (**2e-11**): A suspension of **2e-5** (2.24 g, 0.01 mol) in 100 mL of dimethylformamide (DMF) was prepared. To this suspension, a solution of 3% CH_2N_2 (0.04 mol) in diethyl ether (Et_2O) was added dropwise. The resulting mixture was allowed to stand at room temperature for 10 h. Afterward, the excess CH_2N_2 was decomposed using acetic acid (AcOH), and the solvent was removed under vacuum. The residue obtained after solvent removal was subjected to column chromatography. This process resulted in the isolation of 1.26 g (50%) of the desired product. T_m, 174–175 °C.

1,1′-Dimethyl-4,4′,5,5′-tetranitro-2,2′-biimidazole (**DMTNBI, 2e-12**): A solution of dimethyl sulfate (4 g, 31.7 mmol) was combined with bis ammonium 4,4′,5,5′-tetranitro-2,2′-bi-1H-imidazolate (2 g, 5.75 mmol) and heated at a temperature of 70 °C for 18 h without stirring. After cooling the reaction mixture to room temperature, water (20 mL) was added, resulting in a light orange mixture. The mixture was further stirred for 3 h. The resulting yellow solid product was then filtered and washed with water, yielding 1.47 g (74.8% yield). To obtain larger crystals, the product was recrystallized using ethyl acetate via slow evaporation over the course of one week. This process produced large orange crystals of the desired compound. ^{1}H NMR (DMSO-d$_6$) 4.07 ppm. ^{13}C NMR (DMSO-d$_6$): 35.87, 131.71, 134.11 139.18 ppm.

4-Chloro-3′,5′-dinitro-1′H-1,4′-bipyrazole (**2e-13**): A solution was prepared by combining 3,4,5-trinitropyrazole (158 mg, 1 mmol), 4-chloropyrazole (102 mg, 1 mmol), and NaOH (80 mg, 2 mmol) in 2 mL of water. The resulting mixture was stirred at a temperature of 160 °C in a sealed tube for 20 h. After the reaction, the mixture was cooled and acidified using 20% H_2SO_4 until reaching a pH of 1. The formed precipitate was filtered, washed with cold water, and subjected to recrystallization from a mixture of ethanol and water. This process yielded yellow crystals weighing 181 mg (62% yield). T_m, 236 °C (DSC, 10 °C min^{-1}). IR (neat): 3135, 2377, 2351, 2319, 1954, 1618, 1512, 1419, 1388, 1354, 1292, 974 cm^{-1}; ^{1}H NMR (400 MHz, DMSO-d$_6$): d 8.31 (s, CH), 7.89 (s, CH) ppm; ^{13}C NMR (400 MHz, DMSO-d$_6$): d 147.3, 140.4, 131.7, 112.6, 110.5 ppm; MS (ESI): m/z (%): 257 [M-H]$^-$.

4-Chloro-3,5,3′,5′-tetranitro-1′H-1,4′-bipyrazole (**2e-14**): Compound **2e-13** weighing 257 mg (1 mmol) was dissolved in sulfuric acid (3.2 mL), and then 100% HNO_3 (0.4 mL) was added dropwise to the reaction mixture while stirring in an ice-salt bath. The resulting mixture was heated to a temperature of 100 °C and stirred for 20 h. After cooling the reaction mixture to room temperature, it was poured into iced water (20 mL). The formed precipitate was filtered, washed with cold water, and yielded white solids weighing 225 mg (65%). IR (neat): 3260, 1628, 1568, 1539, 1486, 1427, 1321, 1275 cm^{-1}; ^{1}H NMR (400 MHz, DMSO-d$_6$): d 6.36 (s, NH) ppm; ^{13}C NMR (400 MHz, DMSO-d$_6$): d 150.8, 150.2, 143.2, 108.1, 107.4 ppm; MS (ESI): m/z (%): 347.1, 348.65 (3: 1) [M-H]$^-$.

4-(4′-Amino-3′,5′-dinitro-1′-pyrazol)-3,5-dinitro-1H-pyrazole (**2e-15**): A solution was prepared by combining **2e-14** (500 mg, 1.4 mmol) with 25% ammonia (5 mL). The resulting mixture was stirred at 80 °C in a sealed tube for 20 h. After cooling, the mixture was filtered, and the obtained solid product was subjected to recrystallization from water. This process yielded yellow crystals of ammonium 4-(4′-amino-3′,5′-dinitro-1′-pyrazol)-3,5-dinitropyrazole, weighing 350 mg (70% yield). IR (neat): 3479, 3333, 3269, 1636, 1596, 1573, 1431, 1353, 1310, 1270, 1170 cm^{-1}; ^{1}H NMR (400 MHz, DMSO-d$_6$): d 7.53 (s, 2H, NH_2), 7.06 (s, 4H, NH_4^+) ppm; ^{13}C NMR (400 MHz, DMSO-d$_6$): d 150.8, 143.0, 131.7, 130.4, 109.7 ppm; MS (ESI): m/z (%): 328 [M]$^-$; elemental analysis calcd (%) for $C_6H_6N_{10}O_8$ (346.17): C 20.82, H 1.75, N 40.46; found: C 20.97, H 1.72, N 40.12.

To a single-necked flask containing ammonium 4-(4′-amino-3′,5′-dinitro-1′-pyrazol)-3,5-dinitropyrazole (200 mg, 0.6 mmol), 20% H_2SO_4 was added with stirring until the solution reached pH 1. The reaction mixture was then heated to 40 °C and stirred for 2 h. After cooling in an ice bath for 1 h, the precipitate was filtered off, resulting in yellow solids of compound **2e-15** weighing 162 mg (85% yield). IR (neat): 3493, 3378, 3241, 1646, 1620, 1555, 1518, 1435, 1337, 1312, 1278, 1263, 1159 cm^{-1}; ^{1}H NMR (400 MHz, DMSO-d$_6$): 11.54 (br, s, 1H, NH), 7.54 (s, 2H, NH_2) ppm; ^{13}C NMR (400 MHz, DMSO-d$_6$): 150.3, 143.0, 131.6, 130.3, 109.6 ppm; MS

(ESI): m/z (%): 328[M-H]$^-$; elemental analysis calcd (%) for $C_6H_3N_9O_8$ (329.14): C 21.89, H 0.92, N 38.30; found: C 21.99, H 0.90, N 38.01.

4-(3′,4′,5′-Trinitro-1′-pyrazol)-3,5-dinitro-1H-pyrazole (**2e-16**): Compound **2e-15** (300 mg, 0.8 mmol) was dissolved in sulfuric acid (30 mL), and $Na_2WO_4{\cdot}2H_2O$ (105 mg, 0.3 mmol) was added under stirring in an ice-salt bath. Dropwise addition of 30% hydrogen peroxide solution (7 mL) followed, and the resulting reaction mixture was stirred at 0 °C for 0.5 h. The reaction mixture was then allowed to warm up to room temperature and stirred for an additional 24 h. It was then poured into iced water. The formed precipitate was filtered, washed with cold water, and subjected to recrystallization from a mixture of ethanol and water. This process yielded yellow crystals weighing 131 mg (40% yield). IR (neat): 3546, 3488, 1626, 1591, 1542, 1463, 1453, 1405, 1370, 1326, 1247, 1200, 1111, 1003, 905, 873, 836, 817, 792, 763, 690, 664, 587, 517, 469, 377 cm^{-1}; ^{1}H NMR (400 MHz, DMSO-d$_6$): 14.01 (s, 1H, NH) ppm; ^{13}C NMR (400 MHz, DMSO-d$_6$): 155.9, 150.9, 146.3, 109.9, 106.9 ppm; MS (ESI): m/z (%): 358 [M-H]$^-$; elemental analysis calcd (%) for $C_6HN_9O_{10}$ (359.13): C 20.07, H 0.28, N 35.10; found: C 20.24, H 0.26, N 34.83.

4-(4′-Amino-3′,5′-dinitro-1′-pyrazol)-1-amino-3,5-dinitropyrazole (**2e-17**): Freshly prepared N-[[(4-methylphenyl)sulfonyl]oxy]-ethanimidic acid ethyl ester weighing 64 mg (0.25 mmol) was added to perchloric acid (0.6 mL) at room temperature and stirred for 2 h. The resulting reaction mixture was then poured into an ice-water slurry (6 mL), and after the ice melted, the mixture was extracted with dichloromethane (7 × 1 mL). The combined dichloromethane extracts were dried with Na_2SO_4. One portion of the dried dichloromethane extract was added to a solution containing ammonium 4-(4′-amino-3′,5′-dinitro-1′-pyrazol)-3,5-dinitropyrazole (346 mg, 0.2 mmol) in acetonitrile (25 mL). The resulting solution was stirred for 24 h under ambient conditions, followed by evaporation to dryness. The residue was resuspended in ethyl acetate and the suspension was filtered. The filtrate was evaporated and purified using column chromatography with a mixture of acetate and petroleum ether (6:1) as the eluent. This process yielded a yellow solid weighing 196 mg (57%). IR (neat): 3492, 2277, 2385, 2955, 2922, 2851, 1642, 1603, 1572, 1532, 1509, 1468, 1417, 1399, 1364, 1306, 1271, 1218, 1168, 1097, 998, 922, 871, 855, 816, 761, 743, 733, 709, 658, 635, 604 cm^{-1}; ^{1}H NMR (400 MHz, DMSO-d$_6$): 8.17 (s, 2H, NH$_2$), 7.71 (s, 2H, NH$_2$) ppm; ^{13}C NMR (400 MHz, DMSO-d$_6$): 144.3, 142.3, 137.1, 131.7, 130.3, 109.8 ppm; MS (ESI): m/z (%): 343 [M]$^-$; elemental analysis calcd (%) for $C_6H_4N_{10}O_8$ (344.16): C 20.94, H 1.17, N 40.70; found: C 21.16, H 1.15, N 40.25.

4,4′-Dinitro-1*H*,1′*H*-3,3′-bipyrazole (**2e-18**): 8.0 g (0.06 mol) of the bipyrazole compound was added to 92% H_2SO_4 (80 mL). Dropwise addition of HNO_3 (34.6 mL) was then performed on the mixture. The reaction mixture was gradually heated to a temperature range of 80–90 °C and stirred for 9 h at this temperature. Afterward, the reaction mixture was cooled to 20 °C and poured into ice water (600 mL). The resulting precipitate was filtered off, washed with ice water, and air-dried. This process yielded a white powder weighing 12.5 g (93% yield). T_m > 360 °C (H_2O-EtOH, 1:1). IR spectrum, cm^{-1}: 3134 (s), 3033 (m), 2954 (m), 2907 (m), 2865 (m), 2823 (w), 1525 (w), 1394 (w), 1323 (w), 1197 (w), 1047 (m), 913 (w), 848 (w),

757 (s), 608 (w). ^{1}H NMR spectrum, 14.35; 8.98. ^{13}C NMR spectrum: 135.6; 133.8; 132.3. ^{14}N NMR spectrum, – 18.89 (NO_2). Mass spectrum: 224 $[M]^+$ (100). Found, %: C 32.54; H 2.06; N 37.10. $C_6H_4N_6O_4$. Calculated, %: C 32.15; H 1.80; N 37.50.

General procedure of 1,1′,4,4′-tetranitro-1H,1′H-3,3′-bipyrazole (**2e-19**), 1,1′-dinitro-1H,1′H-3,3′-bipyrazole (**2e-22**), 2,2′,5,5′-tetranitro-2H,2′H-3,3′-bipyrazole (**2e-25**): Method I. Bipyrazole, specifically either **2e-18** or **2e-23** (16 mmol), was added to CF_3CO_2H (16 mL) at 0 °C. This was followed by dropwise addition of HNO_3 (3.9 mL) and acetic anhydride (10.92 mL). The resulting reaction mixture was stirred for 4 h at a temperature range of 0–5 °C. Afterward, the reaction mixture was poured into ice water (150 mL). The obtained precipitate was filtered off, washed with ice water, and air-dried. The filtrate was then subjected to extraction with CH_2Cl_2 (3 × 25 mL), and the organic layer was dried using anhydrous $CaCl_2$. The solvent was subsequently removed by evaporation under reduced pressure, resulting in the second crop of the product. Finally, both crops of the product were combined.

Method II. To a suspension of bipyrazole, specifically either **2e-18** or **2e-23** (1.00 g, 3 mmol), in 1,2-dichloroethane (100 mL) at a temperature range of 20–25 °C, potassium nitrate (1.79 g, 18 mmol) was added. This was followed by dropwise addition of $(CF_3CO)_2O$ (14 mL). The resulting reaction mixture was further stirred at 20–25 °C for 24 h. Afterward, the reaction mixture was poured into water (75 mL). The organic layer was separated, and the aqueous phase was subjected to extraction with EtOAc (2 × 35 mL). The organic fractions were combined, and the solvent was removed by evaporation under reduced pressure.

1,1′,4,4′-Tetranitro-1H,1′H-3,3′-bipyrazole (**2e-19**): Yield 4.71 g (96%, method I), 4.41 g (90%, method II), white powder, T_d, 187–188 °C ($CHCl_3$), 192–194 °C. IR spectrum: 3160 (s), 3147 (s), 1627 (s), 1525 (m), 1394 (m), 1367 (s), 1302 (s), 1280 (s), 1264 (s), 1185 (s), 1092 (s), 1035 (s), 963 (m), 902 (w), 823 (s), 771 (s), 736 (m). ^{1}H NMR spectrum,: 9.95 (2H, s, H-5). ^{13}C NMR spectrum: 135.5 (C-3); 134.3 (br. s, C-4); 131.9 (d, J = 200.2, C-5). ^{14}N NMR spectrum: – 65.28 (N–NO_2); – 22.15 (C–NO_2). Found, %: C 23.13; H 0.89; N 36.06. $C_6H_2N_8O_8$. Calculated, %: C 22.94; H 0.64; N 35.67.

1,1′-Dinitro-1H,1′H-3,3′-bipyrazole (**2e-22**): Yield 3.32 g (93%, method I), white powder, T_m 184 °C. IR spectrum: 3162 (s), 3146 (s), 1629 (s), 1503 (m), 1384 (m), 1365 (s), 1303 (s), 1281 (s), 1263 (s), 1187 (s), 1091 (s), 1033 (s), 964 (m), 904 (w), 821 (s), 772 (s), 734 (m). ^{1}H NMR spectrum: 8.92 (2H, d, J = 2.3, H-5); 7.18 (2H, d, J = 2.3, H-4). ^{13}C NMR spectrum: 143.9 (C-3); 128.6 (C-5); 107.5 (C-4). Found, %: C 32.52; H 1.86; N 37.68. $C_6H_4N_6O_4$. Calculated, %: C 32.15; H 1.80; N 37.50.

2,2′,5,5′-Tetranitro-2H,2′H-3,3′-bipyrazole (**2e-25**): Yield 4.02 g (82%, method I), 3.73 g (76%, method II), white powder, T_m, 165–167 °C. IR spectrum: 3178 (s), 3154 (s), 2982 (m), 1611 (s), 1518 (m), 1401 (m), 1377 (s), 1363 (s), 1315 (s), 1287 (s), 1259 (s), 1237 (m), 1175 (s), 1103 (s), 1045 (s), 959 (m), 836 (s). ^{1}H NMR spectrum: 7.85 (2H, s, H-4). ^{13}C NMR spectrum: 150.8 (br. s, C-5); 128.0 (C-3); 110.0 (C-4). ^{14}N NMR spectrum: – 24.13 (CNO_2); – 61.06 (NNO_2). Found, %: C 23.18; H 0.81; N 35.95. $C_6H_2N_8O_8$. Calculated, %: C 22.94; H 0.64; N 35.67.

4,4′-Dinitro-1H,1′H-[3,3′-bipyrazole]-5,5′-dicarbonitrile (**2e-20**): A solution of bipyrazole **2e-19** weighing 1.29 g (4.1 mmol) in MeCN (30 mL) was slowly added to a solution of KCN weighing 1.6 g (24.6 mmol) in water (20 mL) at 20 °C. The resulting reaction mixture was stirred at the same temperature for 30 min. The precipitate obtained, K_2SO_4, was removed by filtration and discarded. The solvent from the organic layer was then evaporated under reduced pressure, leaving behind a residue. This residue was subjected to extraction with Et_2O (3 × 80 mL). The organic layer was dried using anhydrous Na_2SO_4, and the solvent was evaporated under reduced pressure. The resulting solid residue was air-dried, yielding a white powder weighing 1.09 g (96% yield). T_m, 269–270 °C (EtOH). IR spectrum: 3561 (m), 3453 (m), 3238 (w), 2268 (m), 1574 (m), 1529 (s), 1475 (s), 1414 (m), 1370 (s), 1355 (s), 1210 (w), 1125 (m), 1064 (w), 960 (m), 833 (m), 712 (w). ^{13}C NMR spectrum: 135.0 (C-3); 130.3 (C-5); 121.1 (C-4), 111.1 (CN). ^{14}N NMR spectrum: – 28.08 (NO_2). Found, m/z: 273.0128 $[M\text{-}H]^-$. $C_8HN_8O_4$. Calculated, m/z: 273.0126. Found, %: C 35.21; H 0.76; N 41.07. $C_8H_2N_8O_4$. Calculated, %: C 35.05; H 0.74; N 40.87.

5,5′-Diazido-4,4′-dinitro-1H,1′H-3,3′-bipyrazole (**2e-21**): To prepare the compound, a solution of NaN_3 weighing 3.65 g (56.1 mmol) in water (25 mL) was stirred at room temperature. Dropwise addition of a solution containing bipyrazole **2e-19** weighing 1.47 g (22.6 mmol) in EtOAc (25 mL) was then performed. The resulting reaction mixture was stirred at the same temperature for 30 min. Afterward, the reaction mixture was acidified to a pH of 1–2 using a 20% H_2SO_4 solution. The subsequent workup procedure was similar to that used for the preparation of compound **2e-20**. This process yielded white crystals weighing 1.37 g (96% yield), T_m 93–95 °C (EtOH). IR spectrum: 2874 (w), 2153 (s), 2130 (m), 1580 (s), 1521 (s), 1443 (s), 1376 (s), 1224 (s), 955 (m), 831 (m), 790 (w). Found, m/z: 305.0258 $[M\text{-}H]^-$. $C_8HN_{12}O_4$. Calculated, m/z: 305.0249. Found, %: C 24.00; H 0.53; N 53.87. $C_6H_2N_{12}O_4$. Calculated, %: C 23.54; H 0.66; N 54.90.

4,4′,5,5′-Tetranitro-2H,2′H-3,3′-bipyrazole (**2e-24**): Bipyrazole **2e-23** (6.87 g, 0.03 mol) was combined with concentrated H_2SO_4 (47 mL) and subsequently treated with HNO_3 (6.2 mL) in a dropwise manner. The resulting reaction mixture was then heated at 60 °C for 12 h. After cooling, the mixture was poured into ice water (240 mL). The precipitate obtained was filtered, washed with a small amount of ice water, and air-dried. The filtrate was subjected to extraction with Et_2O (3 × 50 mL), and the organic layer was dried using anhydrous $MgSO_4$. The solvent was removed under reduced pressure by evaporation, resulting in the second crop of the product. The yield of 4,4′,5,5′-Tetranitro-2H,2H′-3,3′-bipyrazole monohydrate was 9.67 g (95%), and it appeared as colorless crystals. Its melting point was measured to be 187 °C in the presence of H_2O, and it decomposed at 238 °C. By maintaining the monohydrate over P_2O_5 at 120 °C and reduced pressure for 10 h, anhydrous bipyrazole **2e-24** was obtained with a yield of 9.14 g (95%). IR spectrum: 3616 (w), 3250 (m), 2701 (w), 2666 (w), 1570 (s), 1554 (s), 1530 (s), 1482 (s), 1425 (w), 1409 (w), 1365 (s), 1335 (s), 1107 (w), 1039 (w), 950 (w), 851 (w), 822 (m), 808 (m), 755 (w), 697 (w). ^1H NMR spectrum, ($(CD_3)_2CO$): 11.30 (2H, br. s, NH). ^{13}C NMR spectrum: 149.8 (C-5); 132.2 (br. s, C-3); 125.1 (C-4). ^{14}N NMR spectrum: – 28.0 (CNO_2). Mass spectrum, m/z (Irel, %): 314 $[M]^+$ (100). Found, %: C 23.09; H 0.84; N 35.17. $C_6H_2N_8O_8$. Calculated, %: C 22.94; H 0.64; N 35.67.

4,4′-Dinitro-5,5′-diamino-2H,2′H-(3,3′-bipyrazole) (**2e-26**): Ph_3P (1.24 g, 4.73 mmol) was added to a solution of azide **2e-21** (0.48 g, 1.57 mmol) in pyridine (7 mL) with stirring at 20–25 °C. The reaction mixture was stirred for 1.5 h and then treated with a dropwise addition of 20% aqueous ammonia solution (5 mL), followed by maintaining it for 15 h. The reaction mixture was poured into 25 mL of water, and then it was washed with 3 portions of 50 mL of PhH (phenylhexane) and 3 portions of 50 mL of Et_2O (diethyl ether). The aqueous layer was acidified to pH 1–2 with concentrated HCl. After that, the resulting precipitate was filtered out to obtain the first crop of the product. Meanwhile, the filtrate was extracted with 3 portions of 50 mL of EtOAc (ethyl acetate). The organic layer was then dried using anhydrous Na_2SO_4. The solvent was evaporated under reduced pressure, resulting in the second crop of the product. Both precipitates were combined and subjected to purification through silica gel flash column chromatography using $CHCl_3$-MeOH (chloroform–methanol) as the eluent (in a ratio of 3:1). This process yielded a light-yellow powder with a mass of 0.24 g, corresponding to a yield of 61%, $T_m > 360$ °C. IR spectrum: 3426 (m), 3145 (m), 1662 (s), 1634 (m), 1449 (s), 1366 (s), 1201 (s), 1133 (m), 949 (w), 830 (m), 770 (m), 739 (w). ^{1}H NMR spectrum: 12.49 (2H, br. s, NH pyrazole); 7.31 (2H, s, NH_2). ^{13}C NMR spectrum: 147.7 (C-3); 138.9 (C-5); 116.8 (C-4). ^{14}N NMR spectrum: – 18.10 (NO_2). Found, m/z: 255.0585 $[M+H]^+$. $C_6H_7N_8O_4$. Calculated, m/z: 255.0585.

N,N′,4,4′-Tetranitro-1H,1′H-[3,3′-bipyrazole]-5,5′-diamine (**2e-27**): A suspension of compound **2e-26** (0.64 g, 2.5 mmol) in 5 mL of CF_3CO_2H was stirred at 0 °C. To this mixture, HNO_3 (3.3 mL, 0.08 mol) was added dropwise. The resulting solution was then treated with Ac_2O (1 mL, 0.01 mol) and stirred for 2 h at a temperature range of 0–2 °C. The resulting precipitate was filtered out, washed with water, and air-dried to obtain a white powder. The yield of the product was 0.58 g, corresponding to a percentage yield of 67%. T_m, 172 °C. IR spectrum: 3152 (m), 3132 (w), 2940 (w); 1560 (s), 1440 (m), 1376 (s), 1354 (s), 1344 (s), 1296 (m), 1128 (m), 1018 (w), 960 (m), 904 (w), 834 (m). ^{13}C NMR spectrum: 137.2 (C-3); 135.2 (br. s, C-5); 117.1 (C-4). ^{14}N NMR spectrum: – 26.23 (CNO_2); – 40.25 ($NHNO_2$); – 206.74 ($NHNO_2$). Found, %: C 21.12; H 1.23; N 40.99. $C_6H_4N_{10}O_8$. Calculated, %: C 20.94; H 1.17; N 40.70.

4,4′-dinitro-5,5′-diazo-3,3′-bispyrazole zwitterion (**2e-28**): A suspension of compound **2e-26** (0.33 g, 1.29 mmol) was prepared in 10 mL of 20% H_2SO_4 and cooled in an ice bath. In a separate container, $NaNO_2$ (0.46 g, 3.56 mmol) was dissolved in 3 mL of water. The $NaNO_2$ solution was then added dropwise to the cooled suspension at 0 °C, and the resulting mixture was stirred for 1 h. The precipitate formed during this process was collected by filtration, washed with 10 mL of ice water, and subsequently dried in air. This procedure yielded product **2e-28** as a light-yellow solid, with a mass of 0.28 g and a yield of 80%. T_d (onset): 150 °C. ^{13}C NMR (DMSO-d_6): δ 147.9, 137.1, 117.0 ppm. ^{15}N NMR (DMSO-d_6): δ – 5.1, – 10.1, – 25.3, – 40.5, – 150.7 ppm. IR (KBr): $\upsilon = 2247$, 2221, 1598, 1510, 1482, 1452, 1353, 1277, 1224, 1137, 1127, 1078, 1033, 939, 830, 824, 763, 756, 579 cm^{-1}. Elemental analysis for $C_6N_{10}O_4$ (276.13): calcd C 26.10, H 0.00, N 50.73%. Found: C 26.26, H 0.27, N 49.96%.

General procedure for diazo-bridged nitroazoles (**2e-(29-35)**).

A solution of N-amino nitroazoles (1 mmol) in 5 mL of acetonitrile was cooled in an ice-salt bath. Meanwhile, a mixture of sodium dichloroisocyanurate (SDCl, 0.5 mmol), AcOH (0.2 mL), and 10 mL of water was prepared. This mixture was then added dropwise to the cooled solution at − 10 °C. After the addition, the reaction mixture was stirred for 30 min at − 10 °C. Subsequently, the reaction was neutralized to pH 7–8 using a sodium bicarbonate solution. The resulting precipitate was isolated by filtration, washed with cold water, and dried under vacuum to obtain the desired product. However, due to its poor solubility, it was not possible to record the ^{13}C NMR spectra for compound **2e-31**.

1,2-Bis(3,5-dinitro-1 H-pyrazol-1-yl)diazene (**2e-29**): White solid (105 mg, 62%); ^{1}H NMR (300 MHz, CD_3CN): d = 8.06 ppm (s, 2H); ^{13}C NMR (75 MHz, CD_3CN): d = 154.7, 147.0, 106.5 ppm; IR (KBr pellet): 3152, 2886, 1573, 1531, 1453, 1389, 1333, 1217, 1132, 1086, 986, 867, 739 cm^{-1}; elemental analysis (%) calcd for $C_6H_2N_{10}O_8$ (342.14): C 21.06, H 0.59, N 40.94; found: C 21.01, H 0.60, N 40.55.

1,2-Bis(3,4-dinitro-1 H-pyrazol-1-yl)diazene (**2e-30**): Yellow solid (87 mg, 51%); ^{1}H NMR (300 MHz, CD_3CN): d = 9.31 ppm (s, 2H); ^{13}C NMR (125 MHz, acetone-d_6): d = 149.1, 132.8, 130.1 ppm; IR (KBr pellet): 3138, 1563, 1528, 1405, 1364, 1331, 1233, 1135, 975, 860, 804, 750 cm^{-1}; elemental analysis (%) calcd for $C_6H_2N_{10}O_8$ (342.14): C 21.06, H 0.59, N 40.94; found: C 21.09, H 0.52, N 40.39.

N,N′-(1,1′-(Diazene-1,2-diyl)bis(3,5-dinitro-1 H-pyrazole-4,1-diyl))bis(N-methylnitramide) (**2e-31**): Yellow solid (120 mg, 41%); ^{1}H NMR (300 MHz, CD_3CN): d = 3.76 ppm (s, 3H); IR (KBr pellet): 1605, 1547, 1445, 1329, 1294, 1210, 1067, 892, 824, 790 cm^{-1}; elemental analysis (%) calcd for $C_8H_6N_{14}O_{12}$ (490.22): C 19.60, H 1.23, N 40.00; found: C 19.81, H 1.24, N 39.05.

1,2-Bis(2,4-dinitro-1 H-imidazol-1-yl)diazene (**2e-32**): White solid (92 mg, 54%); ^{1}H NMR (300 MHz, CD_3CN): d = 8.80 ppm (s, 2H); ^{13}C NMR (125 MHz, acetone-d_6): d 144.1, 141,4, 114.5 ppm; IR (KBr pellet): 3147, 1649, 1571, 1502, 150.2, 1360, 1317, 1246, 1125, 833 cm^{-1}; elemental analysis (%) calcd for $C_6H_2N_{10}O_8$ (342.14): C 21.06, H 0.59, N 40.94; found: C 21.37; H 0.58, N 40.32.

1,2-Bis(4,5-dinitro-1 H-imidazol-1-yl)diazene (**2e-33**): White solid (96 mg, 56%); ^{1}H NMR (300 MHz, CD_3CN): d = 8.48 ppm (s, 2H); ^{13}C NMR (125 MHz, CD_3CN): d = 139.7, 130.4, 129.4 ppm; IR (KBr pellet): 3138, 2924, 1691, 1544, 1468, 1359, 1318, 1259, 1206, 1132, 922, 846, 810 cm^{-1}; elemental analysis (%) calcd for $C_6H_2N_{10}O_8$ (342.14): C 21.06, H 0.59, N 40.94; found: C 21.28, H 0.53, N 40.52.

1,2-Bis(3-chloro-4-nitro-1 H-pyrazol-1-yl)diazene (**2e-34**): White solid (107 mg, 69%); An alternative method starting from 4,5-dinitro-1H-imidazol-1-amine was also investigated. In this method, 4,5-dinitro-1H-imidazol-1-amine (1 mmol) was dissolved in 5 mL of acetonitrile and cooled in an ice bath. A mixture containing sodium dichloroisocyanurate (1 mmol), AcOH (0.5 mL), and 10 mL of water was prepared separately. This mixture was then added dropwise to the cooled solution at a temperature of 0–5 °C. After the addition, the reaction mixture was stirred for 12 h at room temperature. Subsequently, the reaction was neutralized to pH 7–8 using a sodium bicarbonate solution. The resulting precipitate was isolated by filtration, washed with cold water, and dried under vacuum. This process yielded 120 mg of compound **2e-34**, corresponding to a 75% yield. ^{1}H NMR (300 MHz, CD_3CN): d = 8.51 ppm (s, 2H); ^{13}C NMR (125 MHz, acetone-d_6): d = 128.3, 120.7 ppm; IR (KBr pellet): 3129, 1553, 1499, 1477, 1358, 1306, 1257, 1194, 1099, 1011, 828 cm^{-1}; elemental analysis (%) calcd for $C_6H_2Cl_2N_8O_4$ (321.04): C 22.45, H, 0.63, N 34.90; found: C 22.41, H 0.71, N 34.95.

1,2-Bis(4-chloro-3,5-dinitro-1 H-pyrazol-1-yl)diazene (**2e-35**): White solid (72 mg, 35%); ^{13}C NMR (125 MHz, CD_3CN): d = 150.7, 143.2, 111.2 ppm; IR (KBr pellet): 1569, 1523, 1426, 1322, 1221, 1144, 1059, 888 cm^{-1}; elemental analysis (%) calcd for $C_6Cl_2N_{10}O_8$ (411.03): C 17.53, H 0.00, N 34.08; found: C 17.97, H 0.15, N 32.97.

1,2-Bis(5-azido-4-nitro-1 H-imidazol-1-yl)diazene (**2e-36**): To the suspension of **2e-35** (161 mg, 0.5 mmol) in a mixture of 20 mL of CH_3CN and 2 mL of H_2O, NaN_3 (260 mg, 4 mmol) was added. The resulting reaction mixture was stirred for 24 h at room temperature. After completion of the reaction, the solvent was evaporated under reduced pressure, and the residue was washed with water. This process yielded compound **2e-36**. White solid (113 mg, 68%); ^{1}H NMR (300 MHz, CD_3CN): d = 8.19 ppm (s, 2H); IR (KBr pellet): 3131, 2162, 1561, 1488, 1387, 1273, 1248, 1188, 1098, 832 cm^{-1}; elemental analysis (%) calcd for $C_6H_2N_{14}O_4$ (334.17): C 21.57, H 0.60, N 58.68; found: C 21.92, H 0.57, N 54.33. To ensure safety and prevent potential explosions, a very small amount of the sample, specifically less than 0.3 mg, was used. However, this limited quantity resulted in a significant relative error in the nitrogen value.

Ammonium 4-amino-3,5-dinitropyrazolate monohydrate (**N**): In the reaction, pyrazole (6.8 g, 0.1 mol) and N-chlorosuccinimide (NCS) (13.4 g, 0.1 mol) were added to CH_2Cl_2 (100 mL). The resulting mixture was stirred at room temperature overnight, leading to the formation of a clear solution. The solid product obtained was filtered, and the filter cake was washed with CH_2Cl_2 (10 mL). The combined filtrate was then subjected to washing with water (3 × 15 mL) and drying over Na_2SO_4. After removing the solvent, the crude product, 4-chloropyrazole, was dissolved in concentrated sulfuric acid (150 mL). To this solution, ice-cold 100% HNO_3 (20 mL) was added slowly, keeping the reaction temperature below 20 °C. The reaction mixture was stirred at room temperature for 2 h and then refluxed for 6 h. The resulting solution was cooled to room temperature, poured into crushed ice (800 g), and extracted with EtOAc (2 × 200 mL). The organic layers were combined, washed with water (50 mL), and concentrated to yield 4-chloro-3,5-dinitro-1H-pyrazole without the need for further purification. The obtained 4-chloro-3,5-dinitro-1H-pyrazole was then dissolved in aqueous ammonia (25%, 100 mL), and the reaction mixture was kept in a stainless steel vessel at 150 °C overnight. Afterwards, the reaction was cooled to 0–5 °C for 2 h and filtered to obtain the orange-colored product, referred to as compound **N**. The yield of compound **N** from these three steps was 11.9 g, corresponding to a 57% yield. ^{1}H NMR (DMSO-d_6): 7.25 (s, 4H); 6.65 ppm (s, 2H); ^{13}C NMR (DMSO-d_6): 143.6, 131.9 ppm; elemental analysis (%) calcd for

$C_3H_6N_6O_4 \cdot H_2O$ (208.13): C 17.31, H 3.87, N 40.38; found: C 17.14, H 3.74, N 40.11.

1,1′-(Ethane-1,2-diyl)bis(3,5-dinitro-1H-pyrazol-4-amine) (**2e-37**): To a solution of **N** (10 mmol, 2.08 g) and tetraethylammonium bromide (TEAB, 0.21 g, 1 mmol) in DMF (10 mL), dibromoethane (0.94 g, 5 mmol) was added dropwise. The reaction mixture was then heated to 70 °C and maintained at this temperature for 12 h. After cooling, the mixture was poured into 20 mL of cold water. The resulting mixture was filtered, washed with water, and dried in open air to obtain the yellow solid product, referred to as **2e-37**. The yield of the product was 1.48 g, corresponding to an 80% yield. ^{1}H NMR (DMSO-d_6): 7.33 (s, 4H); 5.09 ppm (s, 4H); ^{13}C NMR (DMSO-d_6): 140.2, 131.8, 130.4, 53.2 ppm; IR (KBr pellet): 3482, 3375, 1645, 1478, 1442, 1407, 1308, 1187, 1071, 888, 828, 781 cm^{-1}; elemental analysis (%) calcd for $C_8H_8N_{10}O_8$ (372.21): C 25.81, H 2.17, N 37.63; found: C 25.99, H 2.19, N 36.55.

N,N′-[1,1′-(Ethane-1,2-diyl)bis(3,5-dinitro-1H-pyrazole-4,1-diyl)]dinitramide (**2e-38**): To a solution of compound **2e-37** (1.12 g, 3 mmol), small portions were added to 10 mL of 100% nitric acid at a temperature below 10 °C. The reaction mixture was kept at 5 °C for 10 min, and then the nitric acid was removed by blowing air. The resulting residue was dried under vacuum, leading to the formation of compound **2e-38**. This product was obtained as a white solid with a yield of 1.29 g, corresponding to a 96% yield; ^{1}H NMR (DMSO-d_6): 7.38 (s, 2H), 5.25 ppm (s, 4H); ^{13}C NMR (DMSO-d_6): 146.4, 140.1, 110.5 ppm; IR (KBr pellet): 3314, 2149, 1670, 1613, 1508, 1451, 1371, 1274, 1074, 926, 792, 750, 704 cm^{-1}; elemental analysis (%) calcd for $C_8H_6N_{12}O_{12}$ (462.21): C 20.79, H 1.31, N 36.36; found: C 21.10, H 1.25, N 36.15.

1,1′-(Ethane-1,2-diyl)bis(3-nitro-4-diazo-1H-pyrazol-5-olate) (**2e-39**): In the reaction, $NaNO_2$ (207 mg, 3 mmol) was added in small portions to an ice-cold solution of **2e-37** (372 mg, 1 mmol) in 10 mL of 90% H_2SO_4. The reaction mixture was then warmed up to room temperature and allowed to react for 2 h. Afterward, the mixture was poured into crushed ice (50 g). The resulting yellow precipitate was collected by filtration, washed with water, and dried under vacuum. This process yielded compound **2e-39** as a yellow solid. The mass of the product obtained was 301 mg, corresponding to a yield of 90%; ^{1}H NMR (DMSO-d_6): 4.20 ppm (s, 4H); ^{13}C NMR (DMSO-d_6): 161.3, 13.3, 44.0 ppm; IR (KBr pellet): 2149, 1701. 1507, 1451, 1369, 1317, 1273, 1074, 924, 792, 784, 702 cm^{-1}; elemental analysis (%) calcd for $C_8H_4N_{10}O_6$ (336.18): C 28.58, H 1.20, N 41.66; found: C 28.48, H 1.26, N 40.90.

1,2-Bis(3,4,5-trinitro-1H-pyrazol-1-yl)ethane (**2e-40**): To a suspension of compound **2e-37** (1.12 g, 3 mmol) in concentrated sulfuric acid (15 mL), stirring was performed at 40 °C until a transparent solution was obtained. The resulting yellow solution was then cooled using an ice bath. Hydrogen peroxide (50%, 8 mL) was added dropwise, while maintaining the reaction temperature at 25–30 °C during the addition. After the addition, the reaction mixture was held at 25 °C for 6 h. Subsequently, the final mixture was poured into ice-water (100 g). The resulting white precipitate was collected by filtration, washed with water, and dried under vacuum. This process yielded compound **2e-40** as a white solid. The mass of the product obtained was 1.01 g, corresponding to a yield of 78%. ^{1}H NMR (CD_3CN): 5.31 ppm (s, 4H); ^{13}C NMR (CD_3CN): 144.5, 140.0, 124.8, 53.6 ppm; ^{15}N NMR (CD_3CN): − 31.93, − 33.45, − 36.40, − 81.06, − 184.58 ppm; IR (KBr pellet): 1729, 1591, 1561, 1541, 1481, 1398, 1340, 1314, 1192, 1072, 909, 846, 797, 769 cm^{-1}; elemental analysis (%) calcd for $C_8H_4N_{10}O_{12}$ (432.18): C 22.23, H 0.93, N 32.41; found: C 21.90, H 1.10, N 30.86.

1,2-Bis(5-azido-4,5-dinitro-1H-pyrazol-1-yl)ethane (**2e-41**): To prepare **2e-41**, a white solid, the following steps were followed: Sodium azide (520 mg, 8 mmol) was added in one portion to a solution of **2e-40** (864 mg, 2 mmol) in MeOH (5 mL). The resulting suspension was stirred at room temperature overnight. Afterward, the solvent was removed by rotary evaporation, and 15 mL of water was added. The resulting mixture was filtered, washed with water, and left to dry in air. This process yielded 755 mg (89%) of the desired product. ^{1}H NMR (DMSO-d_6): 4.56 ppm (s, 4H); ^{13}C NMR (DMSO-d_6): 147.3, 139.0, 117.7, 47.4 ppm; ^{15}N NMR (DMSO-d_6): − 26.07, − 27.21, − 99.71, − 145.00, − 152.02, − 193.96, − 294.85 ppm; IR (KBr pellet): 2162, 1550, 1497, 1450, 1383, 1339, 1271, 1130, 1053, 833, 804, 758 cm^{-1}; elemental analysis (%) calcd for $C_8H_4N_{14}O_8$ (424.21): C 22.65, H 0.95, N 46.23; found: C 22.69, H 0.99, N 45.66.

1,2-Bis(4-chloro-3,5-dinitro-1H-pyrazol-1-yl)ethane (**2e-42**): To synthesize 1,2-bis(4-chloro-1H-pyrazol-1-yl)ethane, the following steps were carried out: 4-chloro-1H-pyrazole (2.05 g, 20 mmol), finely ground KOH (1.12 g, 20 mmol), and tetraethylammonium bromide (TEAB, 0.42 g, 2 mmol) were combined and stirred at 40 °C until a viscous mixture formed. Dibromoethane (1.88 g, 10 mmol) was then added dropwise, and the reaction was maintained at 50 °C for 2 h. The resulting residue was recrystallized in a mixture of ethanol and water (1:1). This procedure yielded colorless crystals weighing 1.35 g (59% yield).; ^{1}H NMR ($CDCl_3$): 7.45 (s, 2H), 7.02 (s, 2H), 4.48 ppm (s, 4H); ^{13}C NMR ($CDCl_3$): 139.1, 128.4, 110.4, 52.2 ppm; IR (KBr pellet): 3117, 3038, 2967, 1628, 1526, 1383, 1335, 1312, 1256, 1200, 1136, 984, 970, 844, 816, 714, 650, 608 cm^{-1}; elemental analysis (%) calcd for $C_8H_8Cl_2N_4$ (231.08): C 41.58, H 3.49, N 24.25; found: C 41.38, H 3.35, N 24.15.

To obtain **2e-42**, a white solid, the following steps were performed: A solution of 1,2-bis(4-chloro-1H-pyrazol-1-yl)ethane (2.30 g, 10 mmol) in concentrated sulfuric acid (30 mL) was prepared, and then 100% nitric acid (5 mL) was added to this mixture. The resulting reaction mixture was refluxed for 6 h. After cooling, the mixture was poured into crushed ice (150 g). The formed precipitate was filtered, washed with water, and dried in air. This process yielded 3.36 g (82%) of the desired product, which was obtained as a white solid; ^{1}H NMR (DMSO-d_6): 5.24 (s, 4H); ^{13}C NMR (DMSO-d_6): 148.5, 143.1, 106.0, 53.0 ppm; IR (KBr pellet): 1567, 1506, 1437, 1331, 1055, 867, 622 cm^{-1}; elemental analysis (%) calcd for $C_8H_4Cl_2N_8O_8$ (411.07): C 23.37, H 0.98, N 27.26; found: C 23.37, H 0.94, N 27.03.

1,2-Bis(4-azido-3,5-dinitro-1H-pyrazol-1-yl)ethane (**2e-43**): To prepare **2e-42**, a white solid, the following steps were carried out: Sodium azide (3 mmol, 195 mg) was added to a solution of **2e-42** (1 mmol, 411 mg) in DMSO (3 mL). The resulting mixture was stirred at room temperature for 6 h. Afterward, the mixture was poured into water (15 mL). The formed precipitate was collected, washed with water, and dried under vacuum. This process yielded 369 mg (87%) of the desired product, which was obtained as a white solid. ^{1}H NMR (DMSO-d_6): 5.17 ppm (s, 4H); ^{13}C NMR (DMSO-d_6): 145.0, 138.1, 117.1 ppm; ^{15}N NMR (DMSO-d_6): – 26.43, – 31.68, – 79.28 (t, J = 1.5 Hz), – 147.24, – 149.16, – 185.60, – 302.69 ppm; IR (KBr pellet): 2140, 1583, 1516, 1446, 1327, 1261, 1069,1832, 791 cm^{-1}; elemental analysis (%) calcd for $C_8H_4N_{14}O_8$ (462.21): C 22.65, H 0.95, N 46.23; found: C 22.87, H 0.98, N 45.96.

1,1′-(Ethane-1,2-diyl)bis(3,4-dinitro-1H-pyrazol-5-amine) (**2e-44**): To obtain compound **2e-44**, the following steps were undertaken: Dibromoethane (0.94 g, 5 mmol) was added dropwise to a solution of 5-amino-3,4-dinitropyrazole (10 mmol, 1.73 g), KOH (0.56 g, 10 mmol), and tetraethylammonium bromide (TEAB, 0.21 g, 1 mmol) in DMF (10 mL). The reaction mixture was then maintained at 80 °C for 12 h. After cooling, the mixture was poured into water (40 mL) and extracted twice with ethyl acetate (EtOAc) (2 × 30 mL). The combined organic portions were concentrated under rotary evaporation, resulting in a residue. This residue was subsequently recrystallized in a mixture of ethanol and water (1:1), yielding compound **2e-44** weighing 1.15 g with a 62% yield. ^{1}H NMR (DMSO-d_6): d = 8.02 (s, 4 H), 4.39 ppm (s, 4H); ^{13}C NMR (DMSO-d_6): d = 147.8, 147.2, 108.3, 45.9 ppm; ^{15}N NMR (DMSO-d_6): – 23.30, – 24.11, – 104.65, – 213.30, 311.70 ppm (t, J = 91.8 Hz); IR (KBr pellet): 3448, 3346, 3315, 3204, 1640, 1551, 1522, 1461, 1358, 1327, 1248, 1057, 830, 785 cm^{-1}; elemental analysis (%) calcd for $C_8H_8N_{10}O_8$ (372.21): C 25.81, H 2.17, N 37.63; found: C 25.84, H 2.14, N 37.28.

N,N′-[1,1′-(Ethane-1,2-diyl)bis(3,4-dinitro-1H-pyrazole-5,1-diyl)]dinitramide (**2e-45**): To obtain the nitroamino product **2e-45**, the following steps were performed: Compound **2e-44** (0.74 g, 2 mmol) was added in small portions to ice-cold 100% HNO_3 (10 mL). The reaction mixture was then held at 0–5 °C for an additional 30 min. Nitric acid was removed by blowing air, and the resulting residue was dried under vacuum. This process yielded the nitroamino product **2e-45** weighing 0.88 g with a 95% yield. ^{1}H NMR (DMSO-d_6): 9.88 (s, 2H), 4.38 ppm (s, 4H); ^{13}C NMR (DMSO-d_6): 146.2, 145.7, 116.2, 47.0 ppm; IR (KBr pellet): 3297, 1626, 1566, 1523, 1495, 1364, 1337, 1317, 1232, 1156, 986, 945, 868, 817, 747 cm^{-1}; elemental analysis (%) calcd for $C_8H_6N_{12}O_{12}$ (462.21): C 20.79, H 1.31, N 36.36; found: C 20.56, H 1.32, N 35.48.

Sodium 4-amino-3,5-dinitropyrazolate monohydrate (**M**): To obtain the product **M**, the following steps were carried out: 4-Chloro-3,5-dinitropyrazole (0.14 mol, 27.00 g) was dissolved in concentrated ammonia solution (250 mL), resulting in a suspension. The suspension was then transferred to a steel autoclave and heated in an oven at 170 °C for 10 h. After that, the autoclave was cooled to room temperature, and the reaction mixture was acidified by adding concentrated hydrochloric acid until the pH reached 1. The cooled solution was extracted three times with ethyl acetate (3 × 100 mL). The organic phase was dried using magnesium sulfate and concentrated to dryness using a rotary evaporator. The resulting yellowish solid was dissolved in hot water (150 mL) and adjusted to pH 9 using a sodium hydroxide solution. The reaction mixture was cooled to 5 °C while vigorously stirring, and ethanol (500 mL) was slowly added. The precipitated solid was filtered and washed with ethanol and diethyl ether. The obtained product **M** appeared as an orange microcrystalline powder. The final yield of the product was 24.20 g (81.1%). ^{1}H NMR (DMSO-d$_6$, 25 °C, ppm) δ: 6.66 (s, 2H, NH_2), ^{13}C NMR (400 MHz, DMSO-d$_6$, 25 °C, ppm) δ: 144.1 (2C; C-3, C-5), 132.4 (1C; C-4).

Bis(4-amino-3,5-dinitropyrazolyl)methane (**2e-46**): To obtain **2e-46**, the following steps were performed: Sodium 3,5-dinitro-4-aminopyrazolate monohydrate (4.29 g, 22 mmol) was suspended in N,N-dimethylformamide (15 mL), and diiodomethane (2.68 g, 805 mL, 10 mmol) was added to the suspension. The mixture was stirred at 90 °C for 16 h and then poured into water (100 mL). A small amount of sodium thiosulfate solution was added to reduce the precipitated iodine until the suspension turned into a clean yellow color. The resulting precipitate was filtered, washed with a small amount of water, and dried in air, yielding **2e-46** as a yellow solid weighing 3.18 g (89%). DSC (5 °C min^{-1}): 310 °C (T_d); ^{1}H NMR (400 MHz, DMSO-d$_6$): 7.29, 7.16 ppm; ^{13}C NMR (400 MHz, DMSO-d$_6$): 142.1, 131.7, 130.9, 67.2 ppm; ^{15}N NMR (40.6 MHz, DMSO-d$_6$): – 23.8 (s, NO_2), – 29.3 (s, NO_2), – 73.7 (t, JN, H = 2.7 Hz, –CH_2), – 192.8, – 314.1 ppm (t, JN, H = 93.7 Hz, NH_2); elemental analysis: calcd (%) for $C_7H_6N_{10}O_8$ (258.19): C 23.47, H 1.69, N 39.10; found: C 23.73, H 1.81, N 38.90; BAM drop hammer: 11 J (<100 mm); friction tester: > 360 N (<100 mm); ESD: > 1 J (<100 mm).

Bis(3,4,5-trinitropyrazolyl)methane (**2e-47**): To obtain **2e-47**, the following steps were carried out: A solution of **2e-46** (1.00 g, 2.80 mmol) in concentrated H_2SO_4 (5 mL) was added dropwise to a mixture of 50% H_2O_2 (7.5 mL) and H_2SO_4 (25 mL) at 0 °C. The reaction mixture was stirred at 0 °C for 3 h and then left overnight at room temperature. Afterward, the reaction mixture was poured onto ice (100 g), resulting in the formation of a precipitate. Finally, the precipitate was isolated by filtration, washed with water, and dried in air, yielding **2e-47** as a product weighing 1.00 g (86% yield). DSC (5 °C min^{-1}): 205 °C (T_d); ^{1}H NMR (acetone-d$_6$, 25 °C): 7.84 ppm; ^{13}C NMR (acetone-d$_6$, 25 °C): 144.2, 138.8, 123.8, 67.3 ppm; ^{14}N NMR (acetone-d$_6$, 25 °C): – 29.1 (NO_2), – 31.0 (NO_2), – 33.5 (NO_2) ppm; elemental analysis: calcd (%) for $C_7H_2N_{10}O_{12}$ (418.15): C 20.11, H 0.48, N 33.50; found: C 19.91, H 0.82, N 32.55; BAM drop hammer: 4 J (<100 mm); friction tester: 112N (<100 mm); ESD: 0.6 J (<100 mm).

Bis(4-diazo-5-nitro-3-oxopyrazolyl)methane (**2e-48**): (method A): To obtain **2e-48**, the following steps were performed: **2e-46** (3 mmol, 1.075 g) was dissolved in sulfuric acid (15 mL) and cooled to 0 °C. A solution of sodium nitrite (9 mmol, 0.621 g) in water (2 mL) was then added dropwise to the reaction mixture. The resulting mixture was stirred at 0 °C for 2 h and then for an additional 4 h at room temperature. Afterwards, the mixture was poured onto ice, resulting in the formation of a yellow precipitate. The precipitate was filtered and washed with a small amount of cold water. The water phase was extracted three times with ethyl acetate (3 × 25 mL), dried over magnesium sulfate, and then evaporated under reduced pressure. This process yielded **2e-48** weighing 0.84 g with an 87% yield.

Method B: A solution of **2e-46** (3 mmol, 1.075 g) in 10 mL of sulfuric acid was cooled to 0 °C and added dropwise to 10 mL of fuming nitric acid at the same temperature. The reaction mixture was stirred at 0 °C for 4 h and then poured onto crushed ice. Once the ice melted, the water phase was extracted three times using 50 mL portions of ethyl acetate. The collected organic phase was dried over magnesium sulfate. Throughout the process, nitrogen gas was bubbled through the solution until a crude product was obtained. This crude product was subsequently recrystallized from ethyl acetate, resulting in a yield of 65% (0.63 g). DSC (5 °C min^{-1}): 194 °C (T_m, onset), 226 °C (T_d, onset); ^{1}H NMR (400 MHz, DMSO-d_6): 5.88 ppm (s, 2H, CH_2); ^{13}C NMR (101 MHz, DMSO-d_6): 161.3, 65.6, 53.8 ppm; ^{14}N NMR (28.9 MHz, DMSO-d_6): − 24.9 (NO_2), − 139.9 ppm (N+/N); ^{15}N NMR (40.6 MHz, DMSO-d_6): − 16.0 (N+/N), − 28.7 (NO_2), − 84.1 (–CH_2), − 139.6 (N+/N), − 193.0 ppm (Npyr); elemental analysis: calcd (%) for $C_7H_2N_{10}O_6$ (322.15): C 26.10, H 0.63, N 43.48; found: C 26.03, H 0.83, N 43.33; BAM drop hammer: 1.5 J (<100 mm); friction tester: 40 N (<100 mm); ESD: 0.10 J (<100 mm).

1,1′-(Oxybis(methylene))bis(4-chloro-3,5-dinitro-1H-pyrazole) (**2e-49**): To obtain **2e-49**, the following steps were performed: 4-Chloro-3,5-dinitro-1Hpyrazole (1.10 g, 5.71 mmol) was added to pre-cooled 98% sulfuric acid (5 mL) at 0 °C. Paraformaldehyde (0.50 g) was then added to the solution portion-wise over a span of 5 min. The reaction temperature was allowed to slowly rise to 25 °C. The mixture was kept at room temperature for 12 h and was subsequently poured into 100 g of crushed ice. After filtration and washing the solid with water, the residue was dried under vacuum conditions, resulting in the formation of **2e-49** as a white solid weighing 0.44 g (56% yield). T_m: 133 °C. T_d: 284 °C. ^{1}H NMR: d 6.11 ppm (s, 4H). ^{13}C NMR: d 150.8, 143.6, 109.3, 83.4 ppm. IR (KBr): n 1578, 1512, 1447, 1422, 1364, 1332, 1263, 1147, 1098, 1051, 1016, 945, 878, 774, 631, 565 cm^{-1}. Elemental analysis for $C_8H_4Cl_2N_8O_9 \cdot H_2O$ (445.09): calcd C 21.59, H 1.36, N 25.18%. Found: C 21.72, H 1.03, N 25.67%. IS: 30 J. FS: > 360 N.

1,1′-(Oxybis(methylene))bis(3,4-dinitro-1H-pyrazole) (**2e-50**): To obtain **2e-50**, the following steps were performed: 3,4-Dinitropyrazole (1.58 g, 10 mmol) was added to pre-cooled 98% sulfuric acid (5 mL) at 0 °C. Paraformaldehyde (0.50 g) was then added to the solution portion-wise over a span of 5 min. The reaction temperature was allowed to slowly rise to 25 °C. The mixture was kept at room temperature for 12 h and was subsequently poured into 100 g of crushed ice. Afterward, the resulting mixture was dried under vacuum conditions, yielding **2e-50** as a white solid weighing 1.2 g (68% yield). T_m: 121 °C. T_d: 298 °C. ^{1}H NMR: d 8.66 ppm (s, 2H), 5.72 ppm (s, 4H). ^{13}C NMR: d 149.2, 134.7, 128.6, 82.4 ppm. IR (KBr): n 3141, 2986, 2899, 1548, 1518, 1461, 1350, 1314, 1256, 1230, 1138, 1090, 1011, 930, 865, 806, 754, 634, 588, 513 cm^{-1}. Elemental analysis for $C_8H_6N_8O_9$ (358.18): calcd C 26.83, H 1.69, N 31.28%. Found: C 26.47, H 1.64, N 29.80%. IS: > 40 J. FS: > 360 N.

1,1′-(Oxybis(methylene))bis(3,5-dinitro-1H-pyrazole) (**2e-51**): To obtain **2e-51**, the following steps were performed: 3,5-Dinitropyrazole (1.58 g, 10 mmol) was added to pre-cooled 98% sulfuric acid (5 mL) at 0 °C. Paraformaldehyde (0.50 g) was then added to the solution portion-wise over a span of 5 min. The reaction temperature was allowed to slowly rise to 25 °C. The mixture was kept at room temperature for 12 h and was subsequently poured into 100 g of crushed ice. Afterward, the resulting mixture was dried under vacuum conditions, resulting in the formation of **2e-51** as a white solid weighing 0.59 g (66% yield). T_m: 118 °C. T_d: 299 °C. ^{1}HNMR: d

7.78 ppm (s, 2H), 6.13 ppm (s, 4H). ^{13}C NMR (300 MHz, d_6-DMSO): d 147.7, 135.1, 126.7, 81.2 ppm. IR (KBr): n 1636, 1565, 1470, 1346, 1297, 1140, 1086, 1045, 996, 951, 843, 785, 739 cm^{-1}. Elemental analysis for $C_8H_6N_8O_9$ (358.18): calcd C 26.83, H 1.69, N 31.28%. Found: C 27.05, H 1.67, N 30.70%. IS: 10 J. FS: 240 N.

1,1′-(Oxybis(methylene))bis(4-amino-3,5-dinitro-1H-pyrazole) (**2e-52**): To obtain **2e-52**, the following steps were performed: A solution of ammonium 4-amino-3,5-dinitropyrazolate (0.99 g, 5.0 mmol) and bis(chloromethyl)ether (0.28 g, 2.52 mmol) in dimethyl formamide (5 mL) was heated to 80 °C for 8 h. The resulting solution was then poured into 50 g of ice-water. The precipitate that formed was collected and washed with water and CH_2Cl_2 (10 mL × 3). Finally, the precipitate was dried in air, resulting in the formation of **2e-52** as a yellow solid weighing 0.81 g (70% yield). T_m: 172 °C. T_d: 203 °C. ^{1}HNMR: d 6.58 ppm (s, 4H), 6.02 ppm (s, 4H). ^{13}C NMR: d 142.9, 132.2, 131.6, 82.3 ppm. IR (KBr): n 3489, 3374, 1640, 1573, 1516, 1479, 1316, 1221, 1085, 1009, 939, 884, 829, 789, 689, 664, 612 cm^{-1}. Elemental analysis for $C_8H_8N_{10}O_9$ (388.21): calcd C 24.75, H 2.08, N 36.08%. Found: C 24.86, H 2.13, N 35.47%. IS: 25 J. FS: > 360 N.

1,1′-(Oxybis(methylene))bis(4,5-dinitro-1H-imidazole) (**2e-53**): To obtain **2e-53**, the following steps were performed: A suspension of 4,5-dinitro-1H-imidazolate (1.18 g, 6 mmol) in acetone (15 mL) was prepared, and then bis(chloromethyl) ether (0.30 g, 2.61 mmol) and sodium bromide (0.52 g, 5.03 mmol) were added to the suspension. The reaction mixture was stirred for 72 h at room temperature. The resulting white precipitate was removed by filtration, and the filtrate was then concentrated. The oily residue obtained was diluted with water, which led to the formation of a white solid. Finally, the white solid was dried in air, resulting in the production of **2e-53** weighing 0.70 g (65% yield) as a pale-yellow solid. T_m: 204 °C. T_d: 270 °C. ^{1}H NMR (300 MHz, d_6-DMSO): d 8.21 ppm (s, 2H), 5.67 ppm (s, 4H). ^{13}C NMR (300 MHz, d_6-DMSO): d 142.0, 136.5, 119.1, 73.2 ppm. IR (KBr): n 3105, 2968, 1734, 1663, 1541, 1495, 1401, 1366, 1342, 1278, 1227, 1136, 1084, 1009, 936, 865, 827, 766, 652, 593, 517 cm^{-1}. Elemental analysis for $C_8H_6N_8O_9$ (358.18): calcd C 26.83, H 1.69, N 31.28%. Found: C 27.14, H 1.75, N 30.51%. IS: > 40 J. FS: > 360 N.

3-Nitramino-4-tetrazolefurazan anion (**ATF$^-$**): To obtain 4-amino-3-(5-tetrazole) furazan (**HATF**), the following steps were performed: 4-amino-3-(5-tetrazole) furazan (HATF, 4.4 g, 28.7 mmol) was added in small portions to a mixture of 98% H_2SO_4 (25 mL) and fuming nitric acid (95%, 25 mL), which had been cooled using an ice-water bath at temperatures below 5 °C. The reaction mixture was then allowed to warm up to room temperature and stirred for 30 min before being poured into 500 mL of ice-water. The resulting solution was extracted with ethyl acetate, and the combined organic phase was washed with brine to remove any remaining acids. The organic phase was dried with $MgSO_4$ for 30 min before being filtered. Slow bubbling of ammonia gas into the filtrate resulted in the formation of a yellow precipitate. After filtration and washing with water and ethanol, yellow crystals weighing 5.4 g were obtained with an 81% yield.^{1}H NMR (DMSO-d_6, 300 MHz): d 7.43 (s, 8H) ppm; ^{13}C NMR (DMSO-d_6, 100 MHz): d 144.8, 150.2, 156.9 ppm. IR (KBr): 3185, 2812, 1697, 1592, 1513, 1437, 1371, 1292, 1155, 1045, 1012, 991, 914, 873,

824, 772, 742, 727, 585, 540, 488, 457 cm^{-1}. Elemental analysis for $C_3H_8N_{10}O_3$ (232.16): calculated: C: 15.52 H: 3.47 N: 60.33; found: C: 15.69 H: 3.63 N: 60.25%.

4-Amino-furazan-3-yl-tetrazol-1-ol (**2f-1**): Firstly, 4-amino-1,2,5-oxadiazole-3-carbohydroximoyl chloride (1.3 g, 8 mmol) was suspended in 15 mL of ethanol. At 0 °C, a solution of sodium azide (2.1 g, 32 mmol) in a minimal amount of water was added. The suspension was then stirred for 24 h, followed by the addition of 50 mL of water. The resulting precipitate, 4-amino-1,2,5-oxadiazole-3-carbohydroximoylazide, was removed by filtration, washed with water, and dried by blowing air. Without further purification, the 4-amino-1,2,5-oxadiazole-3-carbohydroximoylazide was suspended in 50 mL of diethyl ether. Next, gaseous HCl was bubbled through the reaction mixture at 0 °C with stirring for 2 h. At the same temperature, 20 mL of ethanol and 20 mL of concentrated hydrochloric acid were added dropwise. The reaction mixture was then stirred at room temperature for 8 h. All the solvent was removed under vacuum, leaving behind a solid residue. This residue was dissolved in 40 mL of water and extracted three times with 30 mL portions of ethyl acetate. The combined ethyl acetate extracts were washed with 30 mL of brine, dried over $MgSO_4$, and concentrated under vacuum to yield the white product. The final yield of **2f-1** was determined to be 1.1 g (80%). T_m (onset): 189 °C; T_d (onset):197 °C; ^1H NMR (DMSO-d_6, 300 MHz) 12.53 (b, 1H, OH), 6.48 (b, 2H, NH_2) ppm; ^{13}C NMR (DMSO-d_6, 300 MHz) 155.65, 137.64, 134.21 ppm; IR (KBr): 3442, 3339, 1628, 1439, 1283, 1134, 1063, 988, 872 cm^{-1}; Anal. calcd for $C_3H_3N_7O_2$ (169.03): C 21.31, H 1.79, N 57.98; found C 21.39, H 1.86, N 57.28.

4-Amino-3-(5-tetrazole) furoxan (**2f-3**): Initially, 4-amino-3-cyanofuroxan (1.000 g, 7.9 mmol), NaN_3 (0.619 g, 9.5 mmol), and $ZnBr_2$ (1.786 g, 7.9 mmol) were dissolved in 10 mL of water in a 25 mL round-bottomed flask. The reaction mixture was then stirred at room temperature for 3 h. Next, the mixture was acidified to pH 1–2 by adding 2% HCl. The resulting solution was filtered, and the filtrate was then concentrated to obtain a solid product. The crude product was further purified by recrystallization from ethanol (EtOH). This process yielded white crystals weighing 1.341 g, corresponding to an 83% yield. ^1H NMR (DMSO-d_6, 400 MHz): 6.78 ppm (br s, 2H; NH_2); ^{13}C NMR (DMSO-d_6, 100 MHz): 155.8, 144.9, 101.9 ppm; IR (neat): 3516, 3438, 3315, 1650, 1614, 1567, 1496, 1422, 1389, 1233, 1113, 1080, 1034, 1012, 950, 861, 762, 657 cm^{-1}; MS (ESI): m/z: 168.3.

4-Nitro-3-(5-tetrazole)furoxan (**2f-4**): Firstly, $Na_2WO_4{\cdot}2H_2O$ (1.980 g, 6.0 mmol) and 50% hydrogen peroxide solution (18.000 g, 264.6 mmol) were added to a three-necked round-bottomed flask fitted with a dropping funnel in an ice-salt bath at − 15 °C. The mixture was stirred. Then, sulfuric acid (14 mL) was added dropwise to the reaction mixture while stirring at 5 °C. The reaction mixture was allowed to warm up to room temperature, and then **2f-3** (1.000 g, 5.9 mmol) was added. The resulting mixture was stirred at room temperature for an additional 3 h. Afterward, the mixture was poured into iced water. The resulting solution was extracted with ethyl acetate (4 × 50 mL), and the combined organic phases were dried over $MgSO_4$. The solvent was evaporated under vacuum, and the crude product was obtained. Finally, the crude product was purified by column chromatography using a CH_2Cl_2/CH_3OH (10:1) solvent system. This process yielded a white solid known as **2f-4**, weighing

0.522 g, with a 45% yield. ^{1}H NMR (DMSO-d_6, 400 MHz): 9.28 ppm (br s, 1H; NH); ^{13}C NMR (DMSO-d_6, 100 MHz): 157.8, 145.5, 103.4 ppm; IR (neat): 3138, 1644, 1589, 1562, 1502, 1445, 1335, 1311, 1240, 1264, 1171, 1125, 1078, 1036, 981, 839, 791, 706, 660, 466, 435 cm^{-1}; MS (ESI): m/z: 198.0; elemental analysis calcd (%) for $C_3HN_7O_4$ (199.08): C 18.10, H 0.51, N 49.25; found: C 18.26, H 0.51, N 49.18.

Compound **a**: Initially, thionyl chloride was added to a cold solution of amimofurazan-3-acetic acid in MeOH (30 mL) at room temperature. The resulting mixture was left to stir overnight. The solvent was then evaporated, resulting in the formation of an ester solid. Subsequently, 98% hydrazine hydrate was added dropwise to a solution of the ester compound in 20 mL of ethanol (EtOH) at 0 °C while stirring. The reaction mixture was stirred for 1 h at room temperature. Following this, the precipitate formed was filtered, washed with cold water, and dried in air to obtain compound **a**.

Compound **b**: Firstly, a solution of N-methyl-N-nitroso-N′-nitroguanidine (0.48 g, 3.3 mmol) in 20 mL of methanol was added to a suspension of compound **a** (0.47 g, 3.0 mmol) in 30 mL of water under stirring. The resulting mixture was then heated at its boiling point for 2 h and subsequently cooled to room temperature. The precipitate formed was filtered off, washed with water, and dried in air, resulting in the formation of a white powder known as compound **b**. Found: C 24.67, H 3.25, N 45.80%.

Compound **c**: Initially, a solution of N-methyl-N-nitroso-N′-nitroguanidine (0.44 g, 3.0 mmol) in 20 mL of methanol was added to a suspension of compound **b** (0.43 g, 3.0 mmol) in 30 mL of water with stirring. The resulting mixture was heated at its boiling point for 2 h and then cooled to room temperature. Following cooling, the precipitate that formed was filtered off, washed with water, and dried in air, resulting in the formation of a pure white powder known as compound **c**.

3-Amino-4-(5-nitramino-1,2,4-triazol-3-yl) methylfurazan (**2f-5**): Method 1: Initially, a solution of potassium hydroxide (0.29 g, 5.2 mmol) in 15 mL of water was added to a suspension of compound **b** (0.49 g, 2.0 mmol) in 20 mL of water. The mixture was then heated to its boiling point and maintained at that temperature for 2 h. Afterward, the mixture was cooled to 50 °C and acidified to pH = 3 using 68% nitric acid. The resulting solution was left to cool overnight. Following the cooling period, the precipitate formed was filtered off, washed with cold water, and dried in air. This process yielded 0.45 g (1.7 mmol, 85%) of compound **2f-5** as a white powder.

Method 2. Firstly, a solution of N-methyl-N-nitroso-N′-nitroguanidine (0.53 g, 3.6 mmol) in 20 mL of methanol was added to a suspension of compound **a** (0.24 g, 1.5 mmol) in 30 mL of water under stirring. The mixture was then heated at its boiling point for 2 h and cooled to room temperature. Next, a solution of potassium hydroxide (0.28 g, 5.0 mmol) in 15 mL of water was added to the mixture. The resulting mixture was stirred at the boiling point for 2 h, followed by cooling to 50 °C. Subsequently, the mixture was acidified to pH = 3 using 68% nitric acid. The precipitate that formed was filtered off, washed with cold water, and dried in air. This process afforded compound **2f-5** weighing 0.48 g (1.8 mmol) with a yield of 90%. ^{1}H NMR (300 MHz, DMSO-d_6): 6.21 (br), 4.17 (s, 2H) ppm; ^{13}C NMR (75 MHz,

DMSO-d_6): 156.14, 152.98, 143.43, 19.98 ppm. IR (KBr): 3439, 3333, 2940, 1642, 1608, 1571, 1501, 1450, 1338, 1262, 1202, 995, 868, 578 cm^{-1}. Elemental analysis for $C_5H_6N_8O_3$ (226.16): calcd C 26.55, H 2.67, N 49.55%; Found: C 26.63, H 2.62, N 49.48%.

3-Methyl-4-(5-nitramino-1,2,4-triazol-3-yl) furoxan (**2f-6**): Method 1: Initially, a solution of potassium hydroxide (0.29 g, 5.2 mmol) in 15 mL of water was added to a suspension of compound **b** (0.49 g, 2.0 mmol) in 20 mL of water. The mixture was then heated at its boiling point for 2 h. Afterward, the mixture was cooled to 50 °C and acidified to pH = 3 using 68% nitric acid. The resulting solution was left to cool overnight. Following the cooling period, the precipitate formed was filtered off, washed with cold water, and dried in air. This process yielded 0.34 g (1.5 mmol, 75%) of compound **2f-6** as a yellowish powder.

Method 2. Firstly, a solution of N-methyl-N-nitroso-N′-nitroguanidine (0.53 g, 3.6 mmol) in 20 mL of methanol was added to a suspension of compound **b** (0.24 g, 1.5 mmol) in 30 mL of water under stirring. The mixture was then heated at its boiling point for 2 h and cooled to room temperature. Next, a solution of potassium hydroxide (0.28 g, 5.0 mmol) in 15 mL of water was added to the mixture. The resulting mixture was stirred at the boiling point for 2 h, followed by cooling to 50 °C. Subsequently, the mixture was acidified to pH = 3 using 68% nitric acid. The precipitate that formed was filtered off, washed with cold water, and dried in air. This process yielded compound **2f-6** weighing 0.39 g (1.7 mmol) with an 85% yield. ^{1}H NMR (300 MHz, DMSO-d_6): 4.66 (br), 2.38 (s, 3H) ppm; ^{13}C NMR (75 MHz, DMSO-d_6): 168.68, 156.64, 142.83, 52.33, 27.91 ppm. IR (KBr): 3470, 1619, 1578, 1521, 1461, 1319, 1276, 946, 852, 780, 635, 494 cm^{-1}. Elemental analysis for $C_5H_5N_7O_4$ (227.14): calcd C 26.44, H 2.22, N 43.17%; Found: C 26.37, H 2.15, N 43.27%

3-Amino-4-(5-nitramino-1,2,4-triazol-3-yl) furazan monohydrate (**2f-7**): Method 1. Initially, a solution of potassium hydroxide (0.29 g, 5.2 mmol) in 15 mL of water was added to a suspension of compound **c** (0.52 g, 2.0 mmol) in 20 mL of water. The mixture was then heated at its boiling point for 3 h. Afterward, the mixture was cooled to 50 °C and acidified to pH = 3 using 68% nitric acid. The resulting solution was left to cool overnight. Following the cooling period, the precipitate formed was filtered off, washed with cold water, and dried in air. This process yielded 0.35 g (1.6 mmol, 82%) of compound **2f-7** as a white powder.

Method 2: Initially, a solution of N-methyl-N-nitroso-N′-nitroguanidine (0.53 g, 3.6 mmol) in 20 mL of methanol was added to a suspension of compound **b** (0.21 g, 1.5 mmol) in 30 mL of water under stirring. The mixture was then heated at its boiling point for 2 h and cooled to room temperature. Next, a solution of potassium hydroxide (0.28 g, 5.0 mmol) in 15 mL of water was added to the mixture. The resulting mixture was stirred at the boiling point for 2 h, followed by cooling to 50 °C. Subsequently, the mixture was acidified to pH = 3 using 68% nitric acid. The precipitate that formed was filtered off, washed with cold water, and dried in air. This process yielded compound **2f-7** weighing 0.36 g (1.7 mmol), with an 82% yield. T_m: 158.83 °C. T_d: 195.50 °C. ^{1}H NMR (300 MHz, DMSO-d_6): 3.48 (br) ppm; ^{13}C NMR (75 MHz, DMSO-d_6): 155.86, 155.30, 145.98, 138.73 ppm. IR (KBr): 3422, 3329, 1683, 1648, 1618, 1574, 1510, 1444, 1404, 1309, 1241, 1160, 1013, 964, 578,

549 cm^{-1}. Elemental analysis for $C_4H_6N_8O_4$ (230.14): calcd C 20.88, H 2.63, N 48.69%; Found: C 20.39, H 2.74, N 46.65%.

3-Nitramino-4-(5-nitramino-1,2,4-triazol-3-yl) furazan (**2f-8**): Firstly, compound **2f-7** (0.64 g, 3.0 mmol) was added in small portions to 5 mL of 100% nitric acid at -5 °C. After the addition, the solution was stirred at 0–5 °C for 40 min. Then, the mixture was poured onto 50 g of ice and extracted with ether (10 mL × 3). The organic phase was dried using magnesium sulfate and the solvent was evaporated. This process resulted in the formation of 0.68 g (2.6 mmol) of compound **2f-8**, with a yield of 87%. T_d: 128.67 °C. ^{1}H NMR (300 MHz, DMSO-d_6): 4.92 (br) ppm. ^{13}C NMR (75 MHz, DMSO-d_6): 163.28, 160.21, 143.44, 141.46 ppm. IR (KBr): 3396, 1783, 1620, 1587, 1516, 1461, 1384, 1314, 1155, 1085, 1029, 995, 972, 861, 744 cm^{-1}. Elemental analysis for $C_4H_3N_9O_5$ (257.13): calcd C 18.69, H 1.18, N 49.03%.

3-Nitro-4-(5-nitramino-1,2,4-triazol-3-yl) furazan (**2f-9**): Initially, sodium tungstate dihydrate (2.20 g, 6.7 mmol) was added to a three-necked round-bottomed flask along with 30% hydrogen peroxide solution (30.00 g, 0.44 mol). The flask was placed in an ice-salt bath at − 10 °C and equipped with a stirrer and dropping funnel. Sulfuric acid (30.00 g, 0.31 mol) was then added drop-wise to the reaction flask under stirring at − 5 °C. The reaction mixture was allowed to warm to room temperature, and then compound **2f-7** (1.06 g, 5.0 mmol) was slowly added. After complete addition of **2f-7**, the reaction mixture was stirred at room temperature for another 24 h. Next, the mixture was extracted with dichloromethane (50 mL × 4). The combined organic extracts were washed with ice-water and dried using magnesium sulfate. The resulting yellow precipitate was concentrated on a rotary evaporator, leaving behind 0.88 g (3.6 mmol) of solid compound **2f-9** with a yield of 73%. T_d: 163.21 °C. ^{1}H NMR (300 MHz, DMSO-d_6): 5.62 (br) ppm. ^{13}C NMR (75 MHz, acetone-d_6): δ 159.57, 155.87, 143.66, 138.21 ppm. IR (KBr): 3111, 1707, 1570, 1426, 1379, 1321, 1283, 1181, 1132, 990, 890, 848, 828, 725, 653, 494 cm^{-1}. Elemental analysis for $C_4H_2N_8O_5$ (242.11): calcd C 19.84, H 0.83, N 46.28%; Found: C 19.73, H 0.79, N 46.81%.

3-Nitramino-4-(3-(trinitromethyl)-1,2,4-triazol-5-yl) furazan (**2f-10**): Initially, ethyl 3-ethoxy-3-iminopropionate (10.0 g, 62.8 mmol) was dissolved in methanol (150 mL). Then, 3-carbohydrazide-4-aminofurazan (9.0 g, 62.8 mmol) was added to the solution. The mixture was refluxed for 6 h. Afterward, the reaction mixture was cooled to room temperature, and the product was collected by filtration. It was then dried in air, resulting in the formation of ethyl 3-(2-(4-amino-1,2,5-oxadiazole-3-carbonyl)hydrazinyl)-3-iminopropanoate as a yellow solid. The yield of the product was 13.7 g (53.5 mmol), corresponding to an 85% yield. ^{1}H NMR (300 MHz, DMSO) δ 11.77 (s), 8.48 (s), 8.14 (s), 6.29 (s), 4.09 (s), 3.81 (s), 1.19 (s). ^{13}C NMR (126 MHz, DMSO) δ 167.35, 157.56, 155.17, 150.38, 145.24, 61.05, 35.68, 13.96. IR: ν = 2974, 2325, 1728, 1682, 1620, 1544.31, 1511, 1393, 1319, 1197, 1005, 867, 845, 644 cm^{-1}. Elemental analysis (%) calcd for $C_8H_{12}N_6O_4$ (256.09): C, 37.50; H, 4.72; N, 32.80; found: 37.98; H, 4.26; N, 32.04.

Firstly, compound ethyl 3-(2-(4-amino-1,2,5-oxadiazole-3-carbonyl) hydrazinyl)-3-iminopropanoate (5.0 g, 19.5 mmol) was added to a solution of sodium hydroxide (1.0 g, 25.0 mmol) in water (200 mL). The reaction mixture was heated at reflux for 2 h. Afterward, the mixture was cooled to room temperature and acidified with concentrated hydrochloric acid to reach a pH of 1. This led to the formation of a precipitate. The resulting precipitate was collected by filtration, washed with water, and dried in air. This process yielded 2-(5-(4-amino-1,2,5-oxadiazol-3-yl)-1,2,4-triazol-3-yl) acetic acid weighing 2.9 g (13.8 mmol), with a 71% yield. The product obtained was in the form of a white powder. ^{1}H NMR (500 MHz, DMSO) δ 6.43 (s), 3.95 (s). ^{13}C NMR (126 MHz, DMSO) δ 169.48, 155.28, 152.12, 151.09, 139.46, 32.49. IR: ν = 3474, 3331, 2988, 1702, 1670, 1623, 1577, 1527, 1422, 1331, 1216, 1153, 1012, 896, 694, 663 cm^{-1}. Elemental analysis (%) calcd for $C_6H_6N_6O_3$ (210.05): C, 34.29; H, 2.88; N, 39.99; found: C, 34.59; H, 3.02; N, 40.04.

Method A: Thionyl chloride (1.14 g, 9.6 mmol) was slowly added to a solution of compound 2 (1.0 g, 4.8 mmol) in ethanol at 0 °C. The mixture was then heated to room temperature and allowed to react for 3 h. After the reaction period, the solution was removed, resulting in the formation of ethyl 2-(5-(3-amino-1,2,5-oxadiazol-4-yl)-1,2,4-triazol-3-yl) acetate. The product weighed 1.0 g (4.2 mmol), corresponding to an 89% yield. It was obtained as a white powder.

Method B: Firstly, compound 3-(2-(4-amino-1,2,5-oxadiazole-3-carbonyl) hydrazinyl)-3-iminopropanoate (3.0 g, 11.7 mmol) was added to glacial acetic acid (30 mL). The solution was then refluxed for 6 h. After the reflux period, the solvent was removed, resulting in the formation of ethyl 2-(5-(3-amino-1,2,5-oxadiazol-4-yl)-1,2,4-triazol-3-yl) acetate. The product weighed 2.6 g (10.9 mmol), corresponding to a yield of 93%. It was obtained as a white powder.^{1}H NMR (500 MHz, DMSO) δ 6.43 (s), 4.14 (q), 4.05 (s), 1.19 (t). ^{13}C NMR (126 MHz, DMSO) δ 167.95, 155.25, 151.58, 151.19, 139.39, 61.15, 32.20, 13.97. IR: 3507, 3404, 2699, 1729, 1701.92, 1624, 1596, 1562, 1464, 1394, 1368, 1321, 1216, 1172, 1062, 1023, 981, 934, 861, 809.43, 766, 733, 668 cm^{-1}. Elemental analysis (%) calcd for $C_8H_{10}N_6O_3$ (238.08): C, 40.34; H, 4.23; N, 35.28; found: C, 40.44; H, 4.53; N, 35.10.

Compound 2-(5-(4-amino-1,2,5-oxadiazol-3-yl)-1,2,4-triazol-3-yl) acetic acid (0.50 g, 2.38 mmol) was added in small portions to a mixture of 98% H_2SO_4 (2 mL) and 100% nitric acid (1 mL) at 0 °C. The addition was done gradually. After the addition, the reaction mixture was warmed to 25 °C and stirred for 3 h. During this time, a precipitate formed. The resulting precipitate was collected by filtration and washed with trifluoroacetic acid. It was then dried in air, resulting in the formation of **2f-10** as a white powder. The yield of the product was 0.68 g (1.98 mmol), corresponding to an 83% yield. ^{1}H NMR (500 MHz, Acetone) 9.37 (br). ^{13}C NMR (126 MHz, Acetone) 150.40, 149.27, 146.27, 142.42, 123.35. IR: 3292, 2876, 1597, 1456, 1386, 1313, 1187, 1082, 1052, 989, 844, 801, 750 cm^{-1}. Elemental analysis (%) calcd for $C_5H_2N_{10}O_9$ (346.00): C, 17.35; H, 0.58; N, 40.47; found: C, 17.65; H, 0.62; N, 40.13.

3-Nitramino-4-(3-(trinitromethylene)-1, 2, 4-triazol-5-yl) furazan (**2f-12**): Method A: Compound ethyl 2-(5-(3-amino-1,2,5-oxadiazol-4-yl)-1,2,4-triazol-3-yl) acetate (1.00 g, 4.20 mmol) was added in small portions to a mixture of 98% H_2SO_4 (3.0 mL) and 100% nitric acid (1.5 mL). The reaction mixture was then warmed to

25 °C and stirred for 3 h. After the reaction period, the mixture was poured into ice water (50 mL) and extracted with diethyl ether (4 × 25 mL). The combined organic phases were dried over anhydrous Na_2SO_4 and filtered to remove any impurities. The concentrated ether solution was dissolved in ethanol, and two equivalents of aqueous ammonia, hydrazine monohydrate, and hydroxylamine were added successively. The resulting mixture was filtered, and the precipitate was collected. Finally, the collected precipitate was washed with ethanol to remove any residual impurities.

Method B: A suspension of compound **2f-10** (0.50 g, 1.44 mmol) in 20 mL of ethanol was prepared. Then, two equivalents of aqueous ammonia, hydrazine monohydrate, and hydroxylamine were respectively added to the suspension. The resulting mixture was stirred at room temperature for 2 h. During this time, a precipitate formed. The precipitate was collected by filtration and washed with ethanol to remove any impurities. To obtain a pure product, the solid was recrystallized in a mixture of ethanol and water. As a result of these steps, the desired product was obtained as salts of 3-nitramino-4-(3-(dinitromethanidyl)-1,2,4-triazol-5-yl) furazanate with a yield of 75–78%.

Firstly, salt (0.50 g, 1.49 mmol) was added to 20% HCl (10 mL) at room temperature. The mixture was then stirred for 3 h. After the stirring period, the resulting solution was extracted with ethyl acetate. This extraction process helped isolate the desired product. The ethyl acetate extract was then dried in air, resulting in the formation of **2f-12** as a white powder. The yield of the product was 0.37 g (1.24 mmol), corresponding to an 83% yield. ^{1}H NMR (300 MHz, Acetone) 8.60 (s), 7.98 (s). ^{13}C NMR (126 MHz, Acetone) 151.98, 149.18, 146.33, 146.27, 142.72, 129.63. IR: 3272, 1581, 1523, 1457, 1318, 1214, 1139, 1057, 983, 874, 812, 752, 706 cm^{-1}. Elemental analysis (%) calcd for $C_5H_3N_9O_7$ (301.02): C, 19.94; H, 1.00; N, 41.86; found: C, 19.98; H, 0.92; N, 41.66.

5-(5-Amino-1H-1,2,4-triazole-3-yl)-1H-tetrazole (**2f-13**): Thionyl chloride (18.7 mL, 258 mmol) was added dropwise to a suspension of 5-amino-1H-1,2,4-triazole-3-carboxylic acid (25.0 g, 195 mmol) in ethanol (260 mL) at 0 °C. The addition was done gradually. The mixture was then refluxed for 2 h. After the reflux period, the solvent was evaporated under vacuum to remove excess solvent. To the remaining highly viscous residue, a saturated solution of sodium acetate (200 mL) was added. This resulted in the formation of a precipitate. The precipitate was collected by filtration and washed with cold water. The obtained colorless solid was dissolved in concentrated aqueous ammonia (150 mL) and heated to 70 °C for 2 h. This step allowed for the conversion of the compound. After heating, the mixture was cooled to room temperature and then acidified with concentrated acetic acid to pH 4. This caused a precipitate to form. The resulting precipitate was collected by filtration, resulting in the yield of 5-amino-1H-1,2,4-triazole-3-carboxamide as a colorless solid. The final product weighed 17.5 g (136 mmol). ^{1}H NMR (DMSO-d_6): 12.5 (s), 7.42 (s), 6.03 ppm (s 2); ^{13}C NMR (DMSO-d_6): 161.9 (C=O), 158.4 (C-NH_2), 154.1 ppm; elemental anal. calcd (%) for $C_3H_5N_5O$: C 28.35, H 3.96, N 55.10; found: C 28.64, H 3.76, N 55.03.

Firstly, 5-amino-1H-1,2,4-triazole-3-carboxamide (5.0 g, 40 mmol) and phosphorus pentoxide (30 g, 105 mmol) were suspended in dry acetonitrile (1.0 L). The flask containing the suspension was then sealed. The suspension was stirred for 72 h at room temperature to allow for the reaction to occur. After the stirring period, the acetonitrile solvent was evaporated under vacuum. This step removed the solvent and left behind a solid residue. The remaining solid was dissolved in water (1.0 L), resulting in a clear solution. The clear solution was then extracted with ethyl acetate (3 × 200 mL) to separate the desired product. The combined organic layers were dried over magnesium sulfate to remove any residual moisture. Finally, the solvent was removed under vacuum, resulting in the formation of 5-amino-3-cyano-1H-1,2,4-triazole as a colorless powder. The final product weighed 3.1 g (28 mmol), corresponding to a yield of 70%. ^{1}H NMR (DMSO-d_6): 12.9 (s), 6.49 ppm (s); ^{13}C NMR (DMSO-d_6): 157.9 (C–NH_2), 136.2 (C–CN), 114.2 ppm (CN); elemental analysis calcd (%) for $C_2H_3N_5$: C 33.03, H 2.77, N 64.20; found: C 33.11, H 2.82, N 62.47.

Firstly, 3-amino-5-cyano-1H-1,2,4-triazole (2.61 g, 23.9 mmol), sodium azide (1.71 g, 26.3 mmol), and zinc chloride (4.0 g, 29.3 mmol) were suspended in water (120 mL). The reaction mixture was then refluxed for 16 h. Refluxing involves heating the mixture to its boiling point and maintaining it at that temperature. After the reflux period, the mixture was cooled to room temperature. To avoid the precipitation of zinc hydroxide, 2 M HCl (20 mL) was added to the mixture. The resulting precipitate was collected by filtration, washed with water to remove any impurities, and dried in air. As a result of these steps, 5-(5-amino-1H-1,2,4-triazole-3-yl)-1H-tetrazole (**2f-13**) was obtained as a colorless solid. The final product weighed 3.41 g (22.4 mmol), corresponding to a yield of 94%. ^{1}H NMR (DMSO-d_6): 6.50 ppm (s, –NH_2); ^{13}C NMR (DMSO-d_6): 158.2 (C-NH_2), 150.2, 148.2 ppm; IR: 3381 (m), 3327 (m), 3214 (w), 1645 (vs), 1583 (s), 1524 (m), 1444 (w), 1386 (m), 1346 (w), 1290 (s), 1210 (w), 1152 (w), 1108 (m), 1102 (m), 1068 (w), 1035 (w), 988 (w), 849 (w), 738 (w), 695 (m), 661 cm^{-1} (w); elemental analysis calcd (%) for $C_3H_4N_8$: C 23.69, H 2.65, N 73.66; found: C 24.01, H 2.66, N 71.95; MS: m/z (DEI+) = 152.1 [M+]; sensitivities (grain size < 100 mm): friction: 360 N, impact: 40 J; DSC (onset, 5 °C min^{-1}): T_d, 347 °C.

5-(3-Nitro-1,2,4–1H-triazol-5-yl)-2H-tetrazole (**2f-14**): Initially, a suspension of **2f-13** (0.5 g, 3.3 mmol) in 20% sulfuric acid (6.0 mL) was prepared. The resulting suspension was then added dropwise to a solution of sodium nitrite (30 equiv, 6.8 g, 98 mmol) in water (10 mL). The addition was performed at a temperature of 40 °C. After the complete addition, the mixture was stirred at 50 °C for 1 h to allow for the reaction to occur. Following the reaction period, the mixture was cooled down to room temperature. Sulfuric acid (20%) was added to the mixture until no evolution of NO_2 gas could be observed. Next, the reaction mixture was extracted with ethyl acetate (EtOAc) to separate the desired product. The organic layer was dried over magnesium sulfate ($MgSO_4$) to remove any remaining traces of water. Finally, the solvent was evaporated, resulting in the formation of **2f-14** as a colorless solid. The final product weighed 0.49 g (2.7 mmol), corresponding to a yield of 82%. ^{1}H NMR (DMSO-d_6): 7.32 ppm (s, H-triazole/tetrazole); ^{13}C NMR (DMSO-d_6): 163.0 (CNO_2), 148.3 (C-triazole), 145.4 ppm (C-tetrazole); ^{14}N NMR (DMSO-d_6): – 26 ppm; ^{15}N NMR (DMSO-d_6): – 19.1 (N_6), – 29.0 (N_4), – 89.4 (N_5), – 90.9

(N_2), – 141.9 (N_3), – 159.5 ppm (N_1); IR: 3135 (m), 3019 (w), 2970 (w), 2790 (m), 2730 (w), 2646 (w), 1754 (w), 1630 (w), 1555 (s), 1506 (w), 1486 (m), 1477 (m), 1460 (w), 1420 (m), 1384 (w), 1365 (s), 1338 (s), 1312 (vs), 1241 (m), 1174 (s), 1129 (s), 1075 (m), 1034 (m), 1034 (m), 1023 (m), 971 (s), 899 (s), 842 (vs), 774 (m), 767 (m), 708 (s), 692 (m), 644 (s), 622 (w), 606 cm^{-1} (w); Raman (200 mW): 1631 (100), 1608 (4), 1557 (6), 1504 (25), 1477 (15), 1446 (4), 1420 (42), 1366 (18), 1340 (5), 1315 (9), 1244 (4), 1212 (3), 1187 (17), 1175 (9), 1128 (26), 1078 (10), 1035 (2), 1019 (20), 970 (5), 844 (2), 776 (2), 768 (8), 468 (5), 468 (5), 395 (10), 334 (9), 261 (2), 247 (2), 236 (4) cm^{-1}; elemental analysis calcd (%) for $C_3H_2N_8O_2$: C 19.79, H 1.11, N 61.53; found: C 20.12, H 1.11, N 60.95; MS: m/z (DEI+) = 182.1 [M+]; sensitivities (grain size 100–500 mm): friction: 288 N, impact: 25 J, ESD: 0.85 J; DSC (onset, 5 °C min^{-1}): T_d, 211 °C.

5-(5-Nitrimino-1,4H-1,2,4-triazol-3-yl)-1H-tetrazole (**2f-15**): Initially, nitric acid (100%, 1.0 mL) was added slowly to a solution of **2f-13** (0.5 g, 3.3 mmol) in concentrated sulfuric acid (6.0 mL). The addition was done at a temperature of 0 °C. The mixture was then allowed to warm up to room temperature and stirred for 1 h to facilitate the reaction. After the stirring period, the clear solution was poured onto ice. This caused precipitation of the desired product. The precipitate was collected by filtration to separate it from the liquid phase. As a result, **2f-15** was obtained as a colorless solid. The final product weighed 0.4 g (2.4 mmol), corresponding to a yield of 65%. ^{1}H NMR(DMSO-d_6): d = 7.81 ppm (s, Htriazole/tetrazole); ^{13}C NMR (DMSO-d_6): 153.0, 148.4, 141.8 ppm; ^{14}N NMR (DMSO-d_6): – 23 ppm; IR: 3309 (m), 1681 (vs), 1650 (m), 1628 (m), 1376 (s), 1257 (m), 1194 (m), 1121 (m), 1091 (s), 1054 (m), 1035 (m), 972 (vs), 958 (s), 910 (s), 756 (m), 745 (m), 723 cm^{-1} (m); elemental analysis calcd (%) for $C_3H_3N_9O_3$: C 18.28, H 1.53, N 63.95; found: C 18.64, H 1.67, N 63.26; MS: m/z (DCI+) = 198.1 [M+]; sensitivities (grain size < 100 mm): friction: 20 N, impact: < 1 J, ESD: 0.05 J; DSC (onset, 5 °C min^{-1}): T_d, 215 °C.

5-(5-Azido-1H-1,2,4-triazol-3-yl) tetrazole (**2f-16**): Initially, a solution of sodium nitrite (0.72 g, 10 mmol) in water (6.0 mL) was prepared. The prepared sodium nitrite solution was then added dropwise to a suspension of **2f-13** (1.1 g, 7.0 mmol) in sulfuric acid (25 wt%, 30 mL). The addition was done at a temperature of 0 °C. After the complete addition, the mixture was allowed to warm up to room temperature and subsequently stirred at 40 °C for 30 min to facilitate the reaction. Following the stirring period, the mixture was cooled down to room temperature. A solution of sodium azide (5 equiv, 2.3 g, 35 mmol) in water (10 mL) was then added dropwise to the suspension. The resulting suspension was stirred overnight at room temperature to remove the excess sodium azide. Subsequently, the suspension was extracted with ethyl acetate (EtOAc) using 3 portions of 30 mL each. The organic layers were combined and the solvent was evaporated under vacuum. As a result, **2f-16** was obtained as a colorless solid. The final product weighed 0.8 g (4.5 mmol), corresponding to a yield of 64%. ^{1}H NMR (DMSO-d_6): 11.83 ppm (s, Htriazole/tetrazole); ^{13}C NMR (DMSO-d_6): 155.5, 148.5, 146.3 ppm; ^{14}N NMR (DMSO-d_6): – 143 (N_8) ppm; ^{15}N NMR (DMSO-d_6): – 11.5 (N_5), – 97.0 (N_4), – 143.2 (N_3), – 145.2 (N_7), – 152.0 (N_8), – 294.9 ppm (N_6); IR: 3141 (m), 3055 (w), 2929 (w),

2862 (w), 2381 (vw), 2233 (vw), 2148 (vs), 1754 (w), 1625 (m), 1559 (s), 1544 (s), 1493 (m), 1484 (s), 1412 (m), 1392 (s), 1322 (w), 1265 (s), 1194 (s), 1140 (w), 1081 (m), 1054 (m), 1043 (s), 981 (s), 981 (s), 827 (w), 819 (w), 813 (m), 811 (m), 800 (m), 790 (m), 784 (m), 775 (m), 773 (m), 769 (m), 764 (m), 759 (m), 755 (m), 753 (m), 743 (m), 738 (m), 736 (m), 727 (s), 710 (w), 696 cm^{-1} (s); Raman (200 mW): 2148 (7), 1627 (100), 1560 (8), 1485 (16), 1441 (4), 1415 (8), 1268 (8), 1243 (4), 1196 (3), 1143 (10), 1084 (2), 1062 (6), 1043 (13), 1012 (1), 981 (2), 801 (1), 766 (2), 559 (3), 514 (1), 412 (4), 403 (7), 358 (5), 295 (3), 295 (3) cm^{-1}; elemental analysis calcd for $C_3H_2N_{10}$: C 20.23, H 1.13, N 78.64; found: C 21.09, H 1.10, N 75.83. MS: m/z (DEI+) = 178.1 [M]+; sensitivities (grain size < 100 mm): friction: 20 N, impact: < 1 J, ESD: 0.05 J; DSC (onset, 5 °C min^{-1}): T_d, 164 °C.

3,5-Diamino-1-(1H-tetrazol-5-yl)-1H-1,2,4-triazole (**2f-17**): Initially, 10 mmol of cyanogen bromide was dissolved in 50 mL of dry acetonitrile at a temperature of 0 °C. To this solution, 40 mmol of dry sodium azide was added. The resulting mixture was then stirred in the temperature range of 0–5 °C for a duration of 4 h. Following the reaction period, the inorganic salt was filtered out from the mixture. It is crucial to promptly dissolve the collected salt in cold water after filtration. Subsequently, the resulting cyanogen azide solution was introduced into a separate solution containing either 5 mmol of secondary amine (or 2.5 mmol for diamine) or amino ester (or 4.5 mmol NaOH for ammonium hydrochloride). This step took place in 20 mL of water at 0 °C. The resulting mixture was then stirred overnight at room temperature (ambient temperature). Next, the solvent was evaporated by allowing it to naturally evaporate in air. In order to purify the product, it underwent washing steps using both water and acetonitrile as solvents. The final yield of the white solid product obtained was 0.75 g, corresponding to a yield of 63%. T_m, 248 °C. IR (KBr): 3442, 3347, 3143, 2659, 1684, 1632, 1582 cm^{-1}. ^1H NMR (300 MHz, DMSO-d_6): 7.3 (s, 2 H), 7.8 (br. s, 3 H) ppm. ^{13}C NMR (75.5 MHz, DMSO-d_6): 151.5, 154.6, 162.7 ppm. ^{15}N NMR (50.7 MHz, DMSO-d_6): – 326.5, – 315.3, – 221.8 (N_3), – 189.3 (N_5), – 160.4 (N_4), – 119.5 (N_1), – 6.5 (N_2) ppm. $C_3H_5N_9$ (167.13): calcd C 21.56, H 3.02, N 75.43; found C 21.29, H 3.11, N 73.43.

3-Nitro-1-(2H-tetrazol-5-yl)-1H-1,2,4-triazol-5-amine (**2f-18**): Initially, at 0 °C, 10 mmol of cyanogen bromide was dissolved in 50 mL of anhydrous acetonitrile. To this solution, 40 mmol of sodium azide was added. The resulting reaction mixture was stirred at a temperature range of 0–5 °C for a duration of 4 h. Afterward, the inorganic salt formed during the reaction was filtered off from the mixture. The resulting cyanogen azide solution was then added to a separate solution containing 5 mmol of **ANTA** (aromatic nitro compound) that had been neutralized. This step was carried out in 20 mL of water at 0 °C. The mixture was stirred for an additional 4 h at room temperature (ambient temperature). Next, the solvent was removed by allowing it to evaporate in air. To purify the product, it was washed with both water and acetonitrile as solvents. The final yield of the yellow solid product obtained was 78%. T_m, 246 °C. ^1H NMR (DMSO-d_6, 400 MHz): 7.96 (s, 1H), 6.31 (brs, 2H) ppm. ^{13}C NMR (DMSO-d_6, 400 MHz): 161.8, 156.8, 154.6 ppm. IR (neat): 3568, 3417, 3167, 1639, 1582, 1554, 1518, 1404, 1307, 1218, 1155, 1020, 990, 846, 766, 722,

612, 478 cm^{-1}. MS (ESI): 196 (M-H)$^-$. Anal. calcd for $C_3H_3N_9O_2$: C 18.28, H 1.53, N 63.95%; found C 18.28, H 1.54, N 63.93%.

3,4,5-triamino-1-tetrazolyl-1,2,4-triazole (**2f-20**): Initially, at 0 °C, cyanogen bromide (4.0 g, 38 mmol) was dissolved in 30 mL of dry acetonitrile. To this solution, sodium azide (11.0 g, 169 mmol) was added. The resulting mixture was stirred at a temperature range of 0–15 °C for a duration of 4 h. Following the reaction period, the inorganic salt formed during the reaction was removed by filtration. The resulting cyanogen azide solution was then added to a 30 mL water solution. This water solution was prepared by adding a NaOH solution (160 mg, 4 mmol) dropwise into a diaminoguanidinium chloride solution (500 mg, 4 mmol) at 0 °C. The mixture was stirred for an additional 4 h at room temperature (ambient temperature). Afterward, the precipitate was filtered and washed with water. It was then dried in air. The final yield of the light yellow solid product obtained is not specified. 267 mg (37%). IR (KBr): 3401, 3326, 3277, 3154, 1703, 1664, 1616, 1550, 1510, 1398, 1163, 1128, 1100, 1077, 1012, 908, 834, 762, 739, 696, 624, 486 cm^{-1}. $_1$H NMR (DMSO-d_6): 6.71 (br, 2H), 5.72 (br, 2H); ^{13}C NMR (DMSO-d_6): 154.28, 152.07, 148.27. Anal. calcd for $C_3H_6N_{10}$ (182.15): C 19.78, H 3.32, N 76.90; found: C 19.57, H 3.20, N 76.31.

Synthesis of 3-[(5-amino-1H-tetrazol-1-yl) methyl]-5-nitramino- 1,2,4-triazole (**2f-21**): Firstly, methyl 2-(5-amino-1H-tetrazol-1-yl)acetate (1a, 1.57 g, 10 mmol) was dissolved in 30 mL of ethanol. To this solution, hydrazine monohydrate (0.5 g, 10 mmol) was added. The resulting mixture was heated to 60 °C and stirred for 6 h. After cooling, a white precipitate formed. This precipitate was collected by filtration and washed with ethanol. The collected solid was then dried in air to obtain 2-(5-amino-1H-tetrazol-1-yl)acetohydrazide (1.34 g) as a white solid. The yield of the product was reported to be 85%. ^{1}H NMR: d 9.39 (s, 1H), 6.65 (s, 2H), 4.76 (s, 2H), 4.33 (s, 2H) ppm. ^{13}C NMR: d 164.5, 156.3, 45.7 ppm. IR (KBr): 3373, 3318, 3255, 3158, 2983, 1737, 1692, 1643, 1596, 1539, 1492, 1427, 1374, 1339, 1285, 1142, 1112, 1004, 937, 818, 750, 701, 628 cm^{-1}. Elemental analysis for $C_3H_7N_7O$ (157.13): calcd C 22.93, H 4.49, N 62.40%. Found: C 23.34, H 4.63, N 61.50%.

N-Methyl-N-nitroso-N′-nitroguanidine (1.62 g, 11 mmol) was added in small portions to a suspension of 2-(5-amino-1H-tetrazol-1-yl)acetohydrazide (1.57 g, 10 mmol) in 30 mL of water at room temperature. The reaction mixture was then heated to 60 °C and stirred for 24 h. After the reaction period, the resulting precipitate formed was collected by filtration, washed with cold water (50 mL), and dried in air. The final product obtained was 2-(2-(5-amino-1H-tetrazol-1-yl)acetyl)-1-[amino(nitroimino)methyl]hydrazine, which appeared as a white powder. The yield of the product was reported to be 75%, corresponding to 1.83 g of the final compound. T_d: 235 oC. ^{1}H NMR: d 10.39 (s, 1H), 9.75 (s, 1H), 8.74 (s, 1H), 8.26 (s, 1H), 6.68 (s, 2H), 4.97 (s, 2H) ppm. ^{13}C NMR: d 165.5, 161.0, 156.2, 45.8 ppm. IR (KBr): 3432, 3301, 3065, 2820, 1704, 1644, 1584, 1529, 1483, 1423, 1354, 1265, 1233, 1138, 1113, 1075, 1009, 980, 937, 820, 775, 749, 695, 611 cm^{-1}. Elemental analysis for $C_4H_8N_{10}O_3$ (244.17): calcd C 19.68, H 3.30, N 57.36%. Found: C 19.38, H 3.18, N 57.17%.

A solution of sodium hydroxide (0.2 g, 5 mmol) in 5 mL of water was added to a suspension of 2-(2-(5-amino-1H-tetrazol-1-yl)acetyl)-1-[amino(nitroimino)methyl]hydrazine (1.22 g, 5 mmol) in 20 mL of water. The reaction mixture was heated at reflux for 2 h. After the reflux, the reaction mixture was allowed to cool to room temperature and then acidified to pH = 2 using concentrated hydrochloric acid. The resulting precipitate formed was collected by filtration and washed with water. The collected solid was dried in air to obtain compound **2f-21** as a white solid. The yield of the product was reported to be 76%, corresponding to 0.85 g of compound **2f-21**. T_m: 148 °C. T_d: 204 °C. ^{1}H NMR: d 14.20 (br), 6.85 (s, 2H), 5.48 (s, 2H) ppm. ^{13}C NMR: d 156.0, 153.2, 146.2, 40.3 ppm. IR (KBr): 3383, 3350, 3280, 3226, 3069, 2995, 2934, 2781, 2718, 1660, 1599, 1562, 1498, 1471, 1450, 1217, 1132, 1094, 1073, 1026, 988, 914, 860, 780, 718, 648 cm^{-1}. Elemental analysis for $C_4H_6N_{10}O_2$ (226.16): calcd C 21.24, H 2.67, N 61.93%. Found: C 20.68, H 2.55, N 60.92%. IS: 15 J. FS: 240 N.

General procedure for the synthesis of compounds **2f-(28-30)**: Ammonium nitrate (1.66 g, 20 mmol) was added to a suspension of 3-nitro-1,2,4-triazole (1.13 g, 10 mmol) in 10 mL of absolute CH_2Cl_2 with stirring at room temperature. Trifluoroacetic anhydride (5 mL) in absolute CH_2Cl_2 (5 mL) was then added dropwise to the suspension obtained at the same temperature. The reaction mixture was stirred for 6 h at − 20 °C and poured into ice-water. The organic layer was separated, and the aqueous layer was extracted with CH_2Cl_2 three times using 10 mL portions each time. The organic phases were combined, washed with water, and dried over $MgSO_4$. Subsequently, the solvent was evaporated under vacuum. As a result, compound 1,3-dinitro-1,2,4-triazole was formed as a colorless oil. ^{1}H NMR (300 MHz, acetone-d_6): d 9.41 ppm (1H); ^{13}C NMR (75 MHz, acetone-d_6): d 147.0, 143.2 ppm.

The sodium salt (10 mmol) of either 4-chloro pyrazole, 5-aminotetrazole, 3,4-dinitopyrazole, 3-nitro-1,2,4-triazole, or 3-nitro-5-amino-1,2,4-triazole was dissolved in a small volume of methanol. Subsequently, 1,3-dinitro-triazole (10 mmol) in methanol (50 mL) and sodium carbonate (1.06 g, 10 mmol) were added to the methanol solution. The reaction mixture was stirred for 6 h at room temperature. The solvent was then removed under vacuum, leaving behind a residue. The residue was dissolved in water (50 mL), and the resulting solution was acidified by adding dilute hydrochloric acid (HCl). A white precipitate formed, which was collected and washed with water. The filtrate (remaining liquid) was extracted with ether three times using 30 mL portions each time. The organic phase obtained from the extraction was washed with brine and water, and then dried over magnesium sulfate ($MgSO_4$). The solvent was evaporated in air, resulting in the formation of some white solid. The two sets of white solids obtained earlier were combined, and the pure product was obtained by recrystallization from a mixture of ethanol and water.

5-Amino-1-(3-nitro-1H-1,2,4-triazol-5-yl)-1H-tetrazole monohydrate (**2f-28**): White solid. T_m: 121 °C. T_d: 239 °C. ^{1}H NMR (300 MHz, DMSO-d_6): d 6.48 ppm (s, 2H); ^{13}C NMR (75 MHz, DMSO-d_6): d 163.8, 155.7, 149.5 ppm. IR (KBr): 3413, 3305, 3257, 1868, 1649, 1640, 1565, 1509, 1407, 1378, 1310, 1193, 1094, 1058, 991,

858, 843, 762, 645, 540 cm^{-1}. Elemental analysis for $C_3H_5N_9O_3$ (215.13): calcd C 16.75, H 2.34, N 58.60%. Found: C 16.79, H 2.20, N 57.63%.

5-(4-Chloro-1H-pyrazol-1-yl)-3-nitro-1H-1,2,4-triazole (**2f-29**): White solid. T_m: 200.2 °C. ^{1}H NMR (300 MHz, DMSO-d_6): d 12.2 ppm (br), 8.82 ppm (1H), 8.16 ppm (1H); ^{13}C NMR (75 MHz, DMSO-d_6): d 160.5, 149.2, 142.6, 128.0, 113.4 ppm. IR (KBr): 3156, 3121, 2994, 2918, 2861, 2697, 1898, 1719, 1614, 1555, 1551, 1512, 1460, 1387, 1341, 1310, 1278, 1236, 1158, 1103, 999, 970, 943, 841, 755, 698, 636, 594 cm^{-1}. Elemental analysis for $C_5H_3ClN_6O_2$ (214.57): calcd C 27.99, H 1.41, N 39.17%. Found: C 28.03, H 1.36, N 38.92%.

5-(4,5-Dinitro-1H-pyrazol-1-yl)-3-nitro-1H-1,2,4-triazole monohydrate (**2f-30**): Colourless solid. T_m: 80 °C, T_d: 330 °C. ^{1}H NMR (300 MHz, CD_3CN): d 9.23 ppm (s, 1H); ^{13}C NMR (75 MHz, CD_3CN): d 164.3, 152.1, 149.5, 138.8, 118.9 ppm. IR (KBr): 3566, 3436, 3134, 1612, 1560, 1517, 1427, 1346, 1310, 1169, 1123, 1030, 1014, 934, 839, 808, 747, 710, 644, 581 cm^{-1}. Elemental analysis for $C_5H_4N_8O_7$ (288.13): calcd C 20.84, H 1.40, N 38.89%. Found: C 20.67, H 1.29, N 37.84%.

1-(3,4-Dinitro-1H-pyrazol-5-yl)-3-nitro-1H-1,2,4-triazole (**2f-31**): The sodium salt of 3-nitro-1,2,4-triazole was dissolved in a small amount of methanol. Then, 1,3,4-trinitropyrazole in methanol (50 mL) and sodium carbonate were added to the solution. The reaction mixture was stirred for 6 h at room temperature. Next, the solvent was removed under vacuum. The resulting sodium salt was purified by column chromatography. The purified sodium salt was then dissolved in 10 mL of water, and the pH of the solution was adjusted to 1 by adding a 10% HCl solution. The acidic solution was extracted with ether three times using 30 mL portions each time. The organic phase obtained from the extraction was dried over $MgSO_4$. The ether phase was concentrated to 20 mL, and ammonium solution was added. As a result, the ammonium salt precipitated as a light-yellow solid. The ammonium salt was then dissolved in acetone (30 mL), and the pH was adjusted to 1 by adding several drops of 36% HCl. After stirring at room temperature for 30 min, $MgSO_4$ was added, and the reaction mixture was stirred for another 30 min. The mixture was then filtered, and the filtrate was concentrated. Compound **2f-31** was obtained as a white solid. It should be noted that the compound is hygroscopic and can become sticky quickly.

1-(3,5-Dinitro-1H-pyrazol-4-yl)-3-nitro-1H-1,2,4-triazol-5-amine (**2f-32**): **ANTA** (2.58 g, 20 mmol), **TNP** (2.03 mg, 10 mmol), and KOH (1.4 g, 25 mmol) were dissolved in 50 mL of water to form a solution. The solution was stirred at 160 °C in a sealed bottle for 24 h. After the reaction period, the mixture was allowed to cool, resulting in the formation of a precipitate. The precipitate was filtered off, washed with cold water, and then subjected to recrystallization from H_2O. As a result, yellow crystals of the potassium salt (K^+) of compound **2f-32** were obtained. The yield of the product was reported to be 1.52 g, corresponding to a 47% yield. IR (neat): 3577, 3392, 3293, 3227, 3164, 1643, 1573, 1496, 1422, 1371, 1320, 1146, 1009, 844, 657 cm^{-1}; ^{1}H NMR (500 MHz, DMSO-d_6): 7.26 (s, 2H) ppm; ^{13}C NMR (100 MHz, DMSO-d_6): 161.0, 158.2, 151.3, 106.2 ppm; MS (ESI): m/z (%): 284 $[M-K]^-$; elemental analysis calcd (%) for $C_5H_2KN_9O_6$ (323.23): C 18.58, H 0.62, N 39.00; found: C 18.15, H 0.80, N 38.72.

A solution of $AgNO_3$ (1.70 g, 10 mmol) in 20 mL of water was added dropwise to a solution of the potassium salt (K^+) of compound **2f-32** (3.23 g, 10 mmol) in 20 mL of water under stirring. After one hour of stirring, the precipitate formed was removed by filtration. The collected precipitate was then dried in air, resulting in the formation of a brown solid, the silver salt (Ag^+) of compound **2f-32**. The yield of the product was reported to be 3.71 g, corresponding to a 95% yield; elemental analysis calcd (%) for $C_5H_2AgN_9O_6$ (392.00): C 15.32, H 0.51, N 32.16; found: C 15.15, H 0.63, N 31.81.

Ag^+ of **2f-32** (391 mg, 1 mmol) was added to a solution containing hydrochloric acid (2 mmol) in 20 mL of water. The resulting mixture was stirred at 80 °C for 5 h. After the reaction, the hot solution was filtered to remove any impurities. The filtrate was then evaporated under vacuum, leaving behind a residue. The residue was subjected to recrystallization from water, resulting in the formation of yellow crystals. The yield of the product was reported to be 214 mg, corresponding to a 75% yield. IR (neat): 3453, 3347, 1644, 1623, 1533, 1502, 1432, 1344, 1307, 1253, 1155, 1074, 996, 854, 831, 723, 670, 630, 572, 508 cm^{-1}; ^1H NMR (500 MHz, d_6-DMSO): 11.11 (s, 1H), 7.37 (s, 2H) ppm; ^{13}C NMR (100 MHz, DMSO-d_6): 161.3, 158.4, 149.7, 106.5 ppm; MS (ESI): m/z (%): 284 [M-H]-; elemental analysis calcd (%) for $C_5H_3N_9O_6$ (285.14): C 21.06, H 1.06, N 44.21; found: C 21.27, H 1.12, N 44.01.

N-(3,5-dinitro-1H-pyrazol-4-yl)-1H-tetrazol-5-Amine (**2f-33**): A solution containing 3,4,5-trinitropyrazole (203 mg, 1 mmol), **ATZ** (255 mg, 3 mmol), and KOH (224 mg, 4 mmol) in 10 mL of water was prepared. The solution was then stirred at 160 °C in a sealed tube for 30 h. After the reaction period, the mixture was cooled, and the pH was adjusted to 1 by adding 20% H_2SO_4. The resulting precipitate was filtered off, washed with cold water, and subjected to recrystallization from an ethanol–water solution. As a result, a deep yellow solid was obtained with a yield of 152 mg, corresponding to a 63% yield. $T_d = 279$ °C (DSC, 5 °C min^{-1}). ^1H NMR (500 MHz, DMSO-d_6, d): 9.24 (s, 2H, 2NH), 7.11 (t, 1H, NH) ppm; ^{13}C NMR (100 MHz, DMSO-d_6, d): 149.3, 115.52, 99.9 ppm; ^{15}N NMR (700 MHz, DMSO-d_6, d): – 16.50, – 48.52, – 148.92, – 323.40 ppm; IR (neat): 3301, 3211, 3084, 2760, 1624, 1472, 1327, 1079, 837, 556 cm^{-1}; anal. calcd for $C_4H_3N_90_4$: C 19.92, H 1.25, N 52.28; found: C 19.95, H 1.28, N 52.31; HRMS calcd for $C_4H_2N_9O_4^-$: 240.02352; found, 240.02350.

N-(1-(2-(4-Nitro-1H-pyrazol-1-yl) ethyl)-1H-tetrazol-5(4H)-ylidene)nitramide (**2f-34**): A total of 1-(2-(1H-Pyrazol-1-yl)ethyl)-1H-tetrazol-5-amine (0.50 g, 2.79 mmol) was added to 10 mL of 100% nitric acid in small portions at a temperature ranging from 0–2 °C. The resulting mixture was stirred at room temperature for 6 h. Afterward, the mixture was poured into 50 mL of cold water. A white precipitate formed and was collected by filtration. The collected precipitate was washed with water and then dried in ambient air. As a result, the yield of compound **2f-34** was measured to be 0.57 g, representing a 76% yield. T_m: 95 °C (onset). T_{dec} (5 °C min^{-1}): 136 °C (onset). Density: 1.65 g cm^{-3}. ^1H NMR: d 8.88 (s, 1H, PyrH), 8.23 ppm (s, 1H, PyrH), 6.95 (s, 1H, NH), 4.65 (m, 4H, CH_2). ^{13}C NMR: d 150.5, 136.0, 135.1, 131.5, 50.0, 46.4. ^{15}N NMR: d – 17.89, – 18.39, – 26.41, – 27.84

[t (J = 2.03 Hz)], – 72.52 [d (J = 12.17 Hz)], – 156.04, – 157.93, – 168.77, – 173.21. IR (ν, cm^{-1}): 3518, 3455, 3162, 3113, 3022, 1626, 1577, 1530, 1509,1480, 1447, 1409, 1315, 1256, 1142, 1035, 989, 964, 922, 887,818, 783, 754, 723, 598, 548 cm^{-1}. Elemental analysis for $C_6H_7N_9O_4$ (269.18): calcd C 26.77, H 2.62, N 46.83%. Found: C 27.38, H 2.66, N 46.28%. IS: 39 J. FS: > 360 N.

N-(1-(2-(4-chloro-3,5-dinitro-1H-pyrazol-1-yl)ethyl)-1H-tetrazol-5(4H)-ylidene) nitramide monohydrate (**2f-35**·H_2O): A solution of 1-(2-(4-chloro-1H-pyrazol-1-yl)ethyl)-1H-tetrazol-5-amine (0.43 g, 2 mmol) in concentrated sulfuric acid (10 mL) was prepared. To this solution, 2 mL of 100% nitric acid was added, and the reaction mixture was heated at 50 °C for 10 h. After cooling, the mixture was poured into crushed ice (100 g). The resulting precipitate was collected by filtration and washed with 10 mL of water. The collected precipitate was then dried in ambient air, yielding **2f-35**·H_2O as a white solid. The yield of the product was measured to be 0.47 g, corresponding to a 68% yield. T_m: 78 °C (onset). T_{dec} (5 °C min^{-1}): 130 °C (onset). Density: 1.75 g cm^{-3}. ^1H NMR: 5.06 [t (J = 5.85 Hz), 2H, CH_2], 4.75 [t (J = 5.85 Hz), 2H, CH_2]. ^{13}C NMR: d 150.8, 148.5, 143.2, 106.3, 52.7, 45.3. ^{15}N NMR: d – 18.48, – 25.94, – 28.13, – 28.22 [t (J = 2.53 Hz)], – 32.86, – 77.25 [t (J = 3.04 Hz)], – 155.70, – 159.12, – 174.32 (m), 181.17 (m). IR (ν, cm^{-1}): 3548, 3381, 1572, 1506, 1445, 1436, 1355, 1327, 1254, 1161, 1057, 1032, 982, 942, 867, 625, 447 cm^{-1}. Elemental analysis for $C_6H_7ClN_{10}O_7$ (366.63): calcd C 19.66, H 1.92, N 38.20%. Found: C 19.72, H 1.85, N 37.75%. IS: 24 J. FS: > 360 N.

1-(2-(4-Amino-3,5-dinitro-1H-pyrazol-1-yl) ethyl)-1H-tetrazol-5-amine (**2f-36**): Initially, cyanogen bromide (0.41 g, 3.85 mmol) was dissolved in 15 mL of dry acetonitrile at 0 °C. Then, sodium azide (1.25 g, 19.27 mmol) was added to the solution and the reaction mixture was stirred at 0–5 °C for 4 h. The resulting salt was removed by filtration, ensuring it was quickly dissolved in cold water. The obtained cyanogen azide solution was then added to a solution containing amino compound (0.50 g, 1.28 mmol) and sodium hydroxide (51 mg, 1.28 mmol) in 2 mL of water at 0 °C. After stirring for 24 h at room temperature, the solvent was evaporated by blowing air over the liquid surface. As a result, a yellow precipitate of 1-(2-(4-amino-3,5-dinitro-1H-pyrazol-1-yl)ethyl)-1H-tetrazol-5-amine was obtained. The product was further purified by washing with water (5 × 2 mL) and acetonitrile (5 mL). The yield of the product was reported as 0.22 g, corresponding to a 59% yield. T_m: 235 °C (onset). T_{dec} (5 °C min^{-1}): 257 °C (onset). Density: 1.67 g cm^{-3}. ^1H NMR: 7.31 [s, 2H, NH_2], 6.69 [s, 2H, NH_2], 4.94 [t (J = 5.25 Hz), 2H, CH_2], 4.63 [t (J = 5.21 Hz), 2H, CH_2]. ^{13}C NMR: d 155.7, 140.3, 131.8, 130.5, 53.2, 43.8. ^{15}N NMR: d 6.49, – 22.60, – 24.25 [t (J = 1.77 Hz)], – 28.23, – 71.43 [t (J = 3.04 Hz)], – 91.75, – 180.11, – 187.03, – 314.16 [t (J = 93.78 Hz)], – 336.80 [t (J = 87.19 Hz)]. IR (ν, cm^{-1}): 3475, 3456, 3399, 3360, 3335, 3137, 1632, 1590, 1508, 1471, 1429, 1311, 1116, 1082, 874, 830, 774, 666, 474. Elemental analysis for $C_6H_8N_{10}O_4$ (284.19): calcd C 25.36, H 2.84, N 49.29%. Found: C 25.63, H 2.73, N 48.49%. IS: > 40 J. FS: > 360 N.

1-(2-(4-Amino-3,5-dinitro-1H-pyrazol-1-yl) propyl)-1H-tetrazol-5-amine (**2f-37**): Initially, cyanogen bromide (0.79 g, 7.44 mmol) was dissolved in 25 mL of dry acetonitrile at 0 °C. Next, sodium azide (2.42 g, 37.20 mmol) was added to the solution, and the resulting mixture was stirred at 0–5 °C for 4 h. The formed salt was then removed by filtration, ensuring quick dissolution of the salt in cold water. The obtained cyanogen azide solution was added to a separate solution containing amino compound (1.00 g, 2.48 mmol) and sodium hydroxide (99 mg, 2.48 mmol) in 3 mL of water at 0 °C. After stirring for 24 h at room temperature, the solvent was evaporated by blowing air over the liquid surface. As a result, a yellow precipitate of 1-(2-(4-amino-3,5-dinitro-1H-pyrazol-1-yl)ethyl)-1H-tetrazol-5-amine was obtained. The product was further purified by washing with water (5 × 2 mL) and acetonitrile (5 mL). The yield of the final product was reported as 0.56 g, corresponding to a 76% yield. T_m: 190 °C (onset). T_{dec} (5 °C min^{-1}): 265 °C (onset). Density: 1.65 g cm^{-3} ^{1}H NMR: 7.27 [s, 2H, NH_2], 6.67 [s, 2H, NH_2], 4.63 [t (J = 6.92 Hz), 2H, CH_2], 4.17 [t (*J* = 6.87 Hz), 2H, CH_2], 2.32 (m, 2H, CH_2). ^{13}C NMR: d 155.4, 139.9, 131.8, 130.7, 52.1, 41.8, 27.7. ^{15}N NMR: d 4.96, -22.36, -24.13, -28.06, -72.47 [t (*J* = 3.04 Hz)], – 91.64, – 175.83, – 182.19, – 314.55 [t (*J* = 93.52 Hz)], – 336.74[t (J = 87.19 Hz)]. IR (ν, cm^{-1}): 3572, 3453, 3402, 3340, 3172, 1640, 1578, 1507, 1472, 1445, 1316, 1258, 1221, 1110, 1092, 1078, 1049, 1011, 1000, 968, 888, 831, 803, 790, 760, 742, 666. Elemental analysis for $C_7H_{10}N_{10}O_4$ (298.22): calcd C 28.19, H 3.38, N 46.97%. Found: C 28.05, H 3.41, N 45.67%. IS: > 40 J. FS: > 360 N.

1-(2-(3,4,5-Trinitro-1H-pyrazol-1-yl) ethyl)-1H-tetrazol-5-amine (**2f-38**): 1-(2-(4-Amino-3,5-dinitro-1H-pyrazol-1-yl)ethyl)-1H-tetrazol-5-amine (0.57 g, 2 mmol) was dissolved in 10 mL of concentrated sulfuric acid. The resulting solution was cooled to a temperature range of 0 to 5 °C using an ice bath. Dropwise addition of 30% hydrogen peroxide (5 mL) was carried out at a temperature of 15–20 °C. After the complete addition of hydrogen peroxide, the reaction mixture was stirred at room temperature for 24 h. The mixture was then poured into an ice-water mixture consisting of 80 mL of water. The resulting solution was extracted with ethyl acetate using two separate portions of 100 mL each. The organic layer was dried over sodium sulfate and concentrated. As a result, a pale-yellow solid of **2f-38** was obtained. The yield of the product was reported as 0.51 g, corresponding to an 82% yield. T_{dec} (5 °C min^{-1}):135 °C (onset). Density: 1.71 g cm^{-3}. ^{1}H NMR: 6.71 [s, 2H, NH_2], 5.07 [t (*J* = 5.28 Hz), 2H, CH_2], 4.73 [t (*J* = 5.28 Hz), 2H, CH_2]. ^{13}C NMR (300 MHz, CD_3CN, ppm): d 156.8, 144.1, 139.7, 124.6, 54.5, 44.8. ^{15}N NMR (500 MHz, CD_3CN, ppm): d 3.55, -23.53, – 31.53, – 33.04, – 36.01, – 80.51, – 94.31, – 178.89, – 182.15, – 338.54. IR (n, cm^{-1}): 3344, 3162, 2362, 2338, 1726, 1701, 1653, 1587, 1542, 1482, 1450, 1428, 1341, 1310, 1275, 1211, 1112, 1080, 1043, 997, 939, 883, 843, 802, 779, 745, 669, 658. Elemental analysis for $C_6H_6N_{10}O_6$ (314.18): calcd C 22.94, H 1.93, N 44.58%. Found: C 23.77, H 2.02, N 43.03%. IS: 30 J. FS: 240 N.

1-(2-(3,4,5-Trinitro-1H-pyrazol-1-yl) ethyl)-1H-tetrazol-5-amine (**2f-39**): First, **2f-37** (1.00 g, 3.35 mmol) was dissolved in 15 mL of concentrated sulfuric acid. The solution was then cooled to a temperature range of 0 to 5 °C using an ice bath. Next, hydrogen peroxide (30%, 8 mL) was added dropwise at a temperature of 15–20 °C. After the complete addition of hydrogen peroxide, the reaction mixture was stirred at room temperature for 24 h. The resulting mixture was poured into an ice-water mixture consisting of 120 mL of water. The solution was extracted with ethyl acetate using two separate portions of 150 mL each. The organic layer was dried over sodium sulfate and then concentrated. As a result, a pale yellow solid of **2f-39** was obtained. The yield of the product was reported as 0.89 g, corresponding to an 81% yield. T_m: 60 °C (onset). T_{dec} (5 °C min^{-1}): 134 °C (onset). Density: 1.67 g cm^{-3}. ^{1}H NMR: 6.71 [s, 2H, NH_2], 4.73 [t (J = 4.20 Hz), 2H, CH_2], 4.23 [t (J = 4.05 Hz), 2H, CH_2], 2.44 (m, 2H, CH_2). ^{13}C NMR (500 MHz, CD_3CN, ppm): d 156.2, 144.1, 139.7, 124.7, 54.0, 43.5, 28.2. IR (ν, cm^{-1}): 3395, 3181, 2256, 1697, 1644, 1586, 1541, 1481, 1454, 1434, 1365, 1338, 1244, 1199, 1122, 1070, 899, 845, 825, 806, 779, 743, 667, 472. Elemental analysis for $C_7H_8N_{10}O_6$ (328.20): calcd C 25.62, H 2.46, N 42.68%. Found: C 26.28, H 2.39, N 40.30%. IS: > 40 J. FS: > 360 N.

1-(2-(4-Chloro-3,5-dinitro-1H-pyrazol-1-yl) ethyl)-1H-tetrazol-5-amine (**2f-40**): A solution was prepared by dissolving ammonium 4-chloro-3,5-dinitropyrazolate (1.09 g, 5.21 mmol) and tetraethylammonium bromide (TEAB, 0.11 g, 0.52 mmol) in 5 mL of DMF. To this solution, compound **2f-37** (1.00 g, 5.21 mmol) was added, and the reaction mixture was maintained at 80 °C for 16 h. After cooling, the mixture was poured into 10 mL of cold water. The resulting precipitate was collected by filtration, washed with 5 mL of water, and then dried in the air. As a result, compound **2f-40** was obtained as an off-white solid. The yield of the product was reported as 1.05 g, corresponding to a 66% yield. T_m: 169 °C (onset). T_{dec} (5 °C min^{-1}): 221 °C (onset). Density: 1.71 g cm^{-3}. ^{1}H NMR (DMSO-d_6, ppm): 6.68 (s, 2H, NH_2), 5.03 [t (J = 5.26 Hz), 2H, CH_2], 4.68 ppm [t (J = 5.29 Hz), 2H, CH_2]. ^{13}C NMR (DMSO-d_6, ppm): d 155.7, 148.6, 142.8, 106.4, 54.0, 43.7 ppm. ^{15}N NMR (DMSO-d_6, ppm): 6.97, – 24.36 [t (J = 1.52 Hz)], – 27.38, – 32.31, – 75.54 [t (J = 3.04 Hz)], – 91.35, – 179.42, – 179.88, – 336.52 [t (J = 87.19 Hz)]. IR (ν, cm^{-1}): 3335, 3165, 1656, 1578, 1553, 1508, 1482, 1440, 1364, 1328, 1272, 1217, 1124, 1095, 1060, 993, 942, 866, 739, 662, 625. Elemental analysis for $C_6H_6ClN_9O_4$ (303.62): calcd C 23.74, H 1.99, N 41.52%. Found: C 23.75, H 2.02, N 40.99%. IS: > 40 J. FS: > 360 N.

1-(2-(4-Azido-3,5-dinitro-1H-pyrazol-1-yl) ethyl)-1H-tetrazol-5- amine (**2f-41**): A solution of sodium azide (195 mg, 3.00 mmol) was added to a solution of **2f-40** (304 mg, 1.00 mmol) in 3 mL of DMSO. The reaction mixture was then stirred at room temperature for 8 h. Afterward, the mixture was poured into 15 mL of water. The resulting precipitate formed was collected by filtration, washed with water, and dried under vacuum. As a result, compound **2f-41** was obtained as a pale-yellow solid. The yield of the product was reported as 262 mg, corresponding to an 85% yield. T_{dec} (5 °C min^{-1}): 159 °C (onset). Density: 1.70 g cm^{-3}. ^{1}H NMR (DMSO-d_6, ppm): 6.68 (s, 2H, NH_2), 5.00 [t (J = 5.30 Hz), 2H, CH_2], 4.66 ppm [t (J = 5.29 Hz), 2H, CH_2]. ^{13}C NMR (DMSO-d_6, ppm): d 155.6, 144.9, 137.7, 117.2, 53.4, 43.6. ^{15}N NMR (DMSO-d_6, ppm): 6.93, – 24.23, – 25.77, – 31.07, – 77.46, – 91.31, – 146.71, – 148.27, – 179.90, – 183.64, – 301.80, – 336.47 [t (J = 87.41 Hz)]. IR (ν, cm^{-1}): 3335, 3163, 2194, 2151, 2136, 1654, 1587, 1538, 1516, 1484, 1451, 1440, 1374, 1361, 1326, 1263, 1117, 1070, 943, 876, 761, 745, 664. Elemental analysis for

$C_6H_6N_{12}O_4$ (310.19): calcd C 23.23, H 1.95, N 54.19%. Found: C 23.32, H 2.04, N 53.50%. IS: 10 J. FS: 120 N.

1-(2-(3-Nitro-1H-pyrazol-1-yl) ethyl)-1H-tetrazol-5-amine (**2f-42**): First, 3-nitropyrazole (0.70 g, 6.19 mmol), finely ground KOH (1.04 g, 18.57 mmol), and tetraethylammonium bromide (TEAB, 0.13 g, 6.19 mmol) were mixed in 3 mL of DMF, resulting in a viscous mixture. A solution of 1-(2-bromoethyl)-1H-tetrazol-5-amine (1.19 g, 6.19 mmol) in 3 mL of DMF was added dropwise to the mixture. The reaction mixture was then maintained at 80 °C for 24 h. After cooling, the mixture was poured into 15 mL of cold water. Ethyl acetate was used for extraction, with two separate portions of 50 mL each. The organic layer was dried over sodium sulfate and then concentrated. As a result, an off-white solid residue was obtained. This residue was recrystallized in ethyl acetate/acetonitrile (1:1) to yield the white solid compound **2f-42**. The yield of the product was reported as 0.78 g, corresponding to a 56% yield. T_m: 165 oC (onset). T_{dec} (5 °C min^{-1}): 254 oC (onset). Density: 1.53 g cm^{-3}. ^{1}H NMR (DMSO-d_6, ppm): d 7.88 [d (J = 2.52 Hz), 1H, PyrH], d 7.01 [d (J = 2.50 Hz), 1H, PyrH], 6.71 (s, 2H, NH_2), 4.62 (m, 4H, CH_2). ^{13}C NMR (DMSO-d_6, ppm): d 155.5, 155.3, 134.6, 102.9, 51.1, 44.1. ^{15}N NMR (DMSO-d_6, ppm): 6.05, – 19.71, – 23.89, – 77.87 [t (J = 2.28 Hz)], – 91.78, – 169.11, – 179.47, – 336.27 [t (J = 87.44 Hz)]. IR (ν, cm^{-1}): 3323, 3143, 1652, 1597, 1548, 1536, 1501, 1456, 1442, 1378, 1361, 1301, 1214, 1203, 1182, 1134, 1113, 1061, 1041, 1000, 947, 922, 823, 786, 780, 753, 743, 691, 654. Elemental analysis for $C_6H_8N_8O_2$ (224.18): calcd C 32.15, H 3.60, N 49.98%. Found: C 32.02, H 3.56, N 49.28%. IS: > 40 J. FS: > 360 N.

1-(2-(3,4-Dinitro-1H-pyrazol-1-yl) ethyl)-1H-tetrazol-5-amine (**2f-43**). A solution was prepared by combining ammonium 3,4-dinitropyrazolate (1.20 g, 6.85 mmol) and tetraethylammonium bromide (TEAB, 0.14 g, 0.69 mmol) in 6 mL of DMF. To this solution, 1-(2-bromoethyl)-1H-tetrazol-5-amine (1.32 g, 6.85 mmol) was added, and the reaction mixture was maintained at 80 °C for 16 h. After cooling, the mixture was poured into 12 mL of cold water. The resulting precipitate was collected by filtration, washed with 5 mL of water, and then dried in the air. As a result, compound **2f-43** was obtained as a white solid. The yield of the product was reported as 1.05 g, corresponding to a 68% yield. T_m: 152 °C (onset). T_{dec} (5 °C min^{-1}): 232 °C (onset). Density: 1.59 g cm^{-3}. ^{1}H NMR (DMSO-d_6, ppm): d 9.09 (s, 1H, PyrH), 6.74 (s, 2H, NH_2), 4.66 ppm (m, 4H, CH_2). ^{13}C{^{1}H} NMR (DMSO-d_6, ppm): d 155.5, 147.1, 134.6, 126.0, 52.0, 43.6 ppm. ^{15}N NMR (DMSO-d_6, ppm): 6.62, – 24.40 [t (J = 1.36 Hz)], – 25.69, – 25.88, – 80.19 [t (J = 2.35 Hz)], – 91.66, – 176.16, – 179.88, – 336.15 [t (J = 87.41 Hz)]. IR (ν, cm^{-1}): 3329, 3148, 2926, 1660, 1598, 1552, 1468, 1439, 1369, 1349, 1312, 1273, 1165, 1137, 1113, 1037, 1013, 986, 930, 861, 813, 749. Elemental analysis for $C_6H_7N_9O_4$ (269.18): calcd C 26.77, H 2.62, N 46.83%. Found: C 26.84, H 2.49, N 45.12%. IS: > 40 J. FS: > 360 N.

(Z)-N-(1-(2-(3,4-Dinitro-1H-pyrazol-1-yl) ethyl)-1H-tetrazol-5- (4H)-ylidene)nitramide (**2f-44**): Method A. A solution of **2f-42** (0.22 g, 1.00 mmol) in 5 mL of concentrated sulfuric acid was prepared. To this solution, 100% nitric acid (1 mL) was added, and the reaction mixture was stirred at room temperature for 8 h.

Afterward, the mixture was poured into 30 mL of cold water. The resulting white precipitate formed was collected by filtration, washed with 5 mL of water, and then dried in the air. As a result, compound **2f-44** was obtained. The yield of the product was reported as 0.28 g, corresponding to an 89% yield. T_m: 90 °C (onset). T_{dec} (5 °C min^{-1}): 157 °C (onset). Density: 1.69 g cm^{-3}.

Method B. In small portions, **2f-43** (0.27 g, 1.00 mmol) was added to 4 mL of 100% nitric acid at a temperature range of 0–2 °C. The resulting mixture was stirred at room temperature for 6 h. Afterward, the mixture was poured into 30 mL of cold water. The white precipitate formed was collected by filtration, washed with water, and dried in the air. As a result, compound **2f-44** was obtained. The yield of the product was reported as 0.23 g, corresponding to a 93% yield. ^{1}H NMR (DMSO-d_6, ppm): 9.14 (s, 1H, PyrH), 5.84 (br, 1H, $NHNO_2$), 4.70 (s, 4H, CH_2). ^{13}C NMR (DMSO-d_6, ppm): d 151.2, 147.2, 134.9, 126.1, 51.4, 45.7. ^{15}N NMR (DMSO-d_6, ppm): – 17.68, – 24.91, – 25.92, – 26.05, – 27.97 [t (J = 1.77 Hz)], – 80.97 [t (J = 1.77 Hz)], – 153.65, – 157.93, – 173.36, – 176.91. IR (ν, cm^{-1}): 3633, 3449, 3136, 1580, 1538, 1521, 1487, 1452, 1416, 1362, 1339, 1309, 1258, 1228, 1156, 1120, 1062, 1036, 980, 922, 860, 808, 781, 748. Elemental analysis for $C_6H_6N_{10}O_6$ (314.18): calcd C 22.94, H 1.92, N 44.58%. Found: C 23.59, H 2.11, N 42.76%. IS: 34 J. FS: 240 N.

1-(2-(5-Bromo-3,4-dinitro-1H-pyrazol-1-yl) ethyl)-1H-tetrazol- 5-amine (**2f-45**): A solution was prepared by combining ammonium 3,4,5-trinitropyrazolate (0.90 g, 4.09 mmol) and tetraethylammonium bromide (TEAB, 0.09 g, 0.41 mmol) in 4 mL of DMF. To this solution, amino compound (0.79 g, 4.09 mmol) was added, and the reaction mixture was maintained at 80 °C for 16 h. After cooling, the mixture was poured into 10 mL of cold water. The resulting precipitate was collected by filtration, washed with 5 mL of water, and then dried in the air. As a result, compound **2f-45** was obtained as a yellow solid. The yield of the product was reported as 0.78 g, corresponding to a 55% yield. T_m: 172 °C (onset). T_{dec} (5 °C min^{-1}): 193 °C (onset). Density: 1.95 g cm^{-3}. ^{1}H NMR (DMSO-d_6, ppm): 6.76 (s, 2H, NH_2), 4.75–4.71 (m, 2H, CH_2), 4.65–4.61 (m, 2H, CH_2). ^{13}C NMR (DMSO-d_6, ppm): d 155.6, 147.9, 125.9, 120.4, 50.6, 43.3. IR (ν, cm^{-1}): 3341, 3157, 1645, 1590, 1558, 1542, 1503, 1433, 1411, 1359, 1333, 1298, 1126, 1111, 1048, 988, 935, 870, 847, 810, 767, 743, 723, 670. Elemental analysis for $C_6H_6BrN_9O_4$ (284.19): calcd C 20.70, H 1.74, N 36.22%. Found: C 21.11, H 1.67, N 36.41%. IS: > 40 J. FS: > 360 N.

1-(2-(3,5-Diazido-4-nitro-1H-pyrazol-1-yl) ethyl)-1H-tetrazol-5- amine (**2f-46**). Sodium azide (213 mg, 3.27 mmol) was added to a solution of **2f-45** (0.38 g, 1.09 mmol) in 3 mL of DMSO. The reaction mixture was stirred at room temperature for 8 h. Afterward, the mixture was poured into 15 mL of water. The resulting precipitate formed was collected by filtration, washed with water, and then dried under vacuum. As a result, compound **2f-46** was obtained as a pale-yellow solid. The yield of the product was reported as 0.28 g, corresponding to an 84% yield. T_{dec} (5 oC min^{-1}): 127 °C (onset). Density: 1.63 g cm^{-3}. ^{1}H NMR (DMSO-d_6, ppm): 6.72 (s, 2H, NH_2), 4.51 [t (J = 5.14 Hz), 2H, CH_2], 4.36 ppm [t (J = 5.07 Hz), 2H, CH_2]. ^{13}C NMR (DMSO-d_6, ppm): d 155.6, 142.5, 137.9, 116.2, 47.1, 43.2. 15N NMR (DMSO-d_6, ppm): 5.92, – 22.57, – 24.48 [t (J = 1.52 Hz)], – 92.09,

– 115.92 [t (J = 2.53 Hz)], – 143.33, – 143.64, – 147.01, – 150.67, – 180.13 [t (J = 1.27 Hz)], – 196.79, – 294.47, – 296.97, – 337.07 [t (J = 87.44 Hz)]. IR (ν, cm^{-1}): 3318, 3147, 2165, 2137, 1659, 1594, 1545, 1505, 1492, 1463, 1433, 1383, 1372, 1356, 1308, 1267, 1217, 1206, 1131, 1110, 963, 833, 790, 764, 647, 527. Elemental analysis for $C_6H_6N_{14}O_2$ (306.20): calcd C 23.53, H 1.98, N 64.04%. Found: C 23.74, H 1.85, N 63.05%. IS: 3 J. FS: 80 N.

1-(2-(5-Amino-3,4-dinitro-1H-pyrazol-1-yl) ethyl)-1H-tetrazol- 5-amine (**2f-47**). A solution was prepared by mixing ammonium 5-amino-3,4-dinitropyrazolate (0.30 g, 1.44 mmol) and tetraethylammonium bromide (TEAB, 0.03 g, 0.15 mmol) in 2 mL of DMF. To this solution, amino compound (0.28 g, 1.44 mmol) was added, and the reaction mixture was maintained at 80 °C for 16 h. After cooling, the mixture was poured into 5 mL of cold water. The resulting precipitate was collected by filtration, washed with 3 mL of water, and then dried in the air. As a result, compound **2f-47** was obtained as a yellow solid. The yield of the product was reported as 0.32 g, corresponding to a 78% yield. T_m: 192 °C (onset). T_{dec} (5 °C min^{-1}): 239 °C (onset). Density: 1.68 g cm^{-3}. ^1H NMR (DMSO-d_6, ppm): 7.99 (s, 2H, NH_2), 6.71 (s, 2H, NH_2), 4.47 (m, 4H, CH_2). ^{13}C NMR (DMSO-d_6, ppm): d 155.6, 147.7, 147.1, 108.3, 46.7, 42.6 ppm. ^{15}N NMR (DMSO-d_6, ppm): 5.82, – 22.78, – 23.74, – 24.28 [t (J = 1.52 Hz)], – 91.96, – 104.03 [t (J = 2.53 Hz)], – 179.75, – 212.52 (m), – 311.35 [t (J = 92.51 Hz)], – 336.67 [t (J = 87.19 Hz)]. IR (ν, cm^{-1}): 3449, 3349, 3151, 1638, 1585, 1547, 1522, 1481, 1456, 1384, 1361, 1332, 1298, 1255, 1121, 1090, 1058, 1031, 942, 868, 828, 786, 761. Elemental analysis for $C_6H_8N_{10}O_4$ (284.19): calcd C 25.36, H 2.84, N 49.29%. Found: C 25.43, H 2.71, N 48.72%. IS: > 40 J. FS: > 360 N.

3-(1H-1,2,3,4-Tetrazol-5-ylamino)-6-amino-1,2,4,5-tetrazine-1-N-oxide (**2f-49**): Trifluoroacetic anhydride (4 mL, 28 mmol) was added to a slurry of 50% hydrogen peroxide (1.3 mL, 25 mmol) in methylene chloride (20 mL) while stirring at a temperature below 10 °C. Next, 3-(1H-1,2,3,4-tetrazol-5-ylamino)-6-amino-1,2,4,5-tetrazine (7 mmol) was added to the mixture at 0 °C and stirred for 30 min. The stirring continued for several hours at room temperature. The solvent was then removed, and the resulting residue was washed with ether and air-dried. As a result, a red solid compound was obtained with a yield of 54%. IR (KBr): 3396, 3304, 3199, 2970, 2810, 1616, 1491, 1348, 1311, 1244, 1099, 1047, 833 cm^{-1}; ^1H NMR: d = 7.14 (s, 3 H), 7.65 ppm (s, 1H); ^{13}C NMR: d = 148.6, 154.3, 161.4 ppm; elemental analysis calcd (%) for $C_3H_4N_{10}O$ (196.13): C 18.37, H 2.06, N 71.42; found: C 18.56, H 2.41, N 69.38.

3-Amino-6-(3,5-dimethylpyrazol-l-yl)-1,2,4,5-tetrazine-2,4-di-N-oxide (**2f-50**): Trifluoroacetic anhydride (4 mL, 28 mmol) was added to a slurry of 50% hydrogen peroxide (1.3 mL, 25 mmol) in methylene chloride (20 mL) while stirring at a temperature below 10 °C. Next, 3-amino-6-(3,5-dimethylpyrazol-l-yl)-1,2,4,5-tetrazine (7 mmol) was added to the mixture at 0 °C and stirred for 30 min. The stirring continued for several hours at room temperature. The solvent was then removed, and the resulting residue was washed with ether and air-dried. As a result, a yellow solid compound was obtained with a yield of 74%. IR (KBr): 3283, 3234, 3148, 1674, 1499, 1416, 1326, 1304, 1268, 1224, 1083, 1003, 949 cm^{-1}; ^1H NMR: d = 2.21

(s, 3H), 2.44 (s, 3 H), 6.17 (s, 1H), 8.65 ppm (s, 2H); ^{13}C NMR: d = 11.6, 13.3, 109.2, 141.9, 145.7, 146.5, 150.5 ppm; elemental analysis calcd (%) for $C_7H_9N_7O_2$ (223.19): C 37.14, H 4.06, N 43.93; found: C 37.14, H 3.98, N 43.37.

3-Amino-6-(1,2-pyrazol-l-yl)-1,2,4,5-tetrazine-2,4-di-N-oxide (**2f-51**): Trifluoroacetic anhydride (4 mL, 28 mmol) was added to a slurry of 50% hydrogen peroxide (1.3 mL, 25 mmol) in methylene chloride (20 mL) with stirring at a temperature below 10 °C. Next, 3-amino-6-(1,2-pyrazol-l-yl)-1,2,4,5-tetrazine (7 mmol) was added to the mixture at 0 °C and stirred for 30 min. The stirring continued for several hours at room temperature. The solvent was then removed, and the resulting residue was washed with ether and air-dried. As a result, a yellow solid compound was obtained with a yield of 82%. IR (KBr): 3130, 1629, 1539, 1483, 1390, 1416, 1336, 1306, 1107, 1079, 1035, 1003, 974 cm^{-1}; ^{1}H NMR: d = 6.56 (dd, 1H, J = 2.6, 1.5 Hz), 7.82 (d, 1H, J = 1.5 Hz), 8.33 (d, 1H, J = 2.6 Hz), 8.52 ppm (s, 2H); ^{13}C NMR: d = 109.4, 129.4, 143.7, 145.9, 146.0 ppm; ^{15}N NMR: d = – 77.7, – 90.8, – 100.9, – 170.8, -311.9 ppm; elemental analysis calcd (%) for $C_5H_5N_7O_2$ (195.14): C 30.77, H 2.58, N 50.24; found: C 30.48, H 2.56, N 49.71.

3-Amino-6-(1,2,4-triazol-l-yl)-1,2,4,5-tetrazine-2,4-di-N-oxide (**2f-52**): Trifluoroacetic anhydride (4 mL, 28 mmol) was added to a slurry of 50% hydrogen peroxide (1.3 mL, 25 mmol) in methylene chloride (20 mL) while stirring at a temperature below 10 °C. Next, 3-amino-6-(1,2,4-triazol-l-yl)-1,2,4,5-tetrazine (7 mmol) was added to the mixture at 0 °C and stirred for 30 min. The stirring continued for several hours at room temperature. The solvent was then removed, and the resulting residue was washed with ether and air-dried. As a result, a yellow solid compound was obtained with a yield of 78%. IR (KBr): 3365, 3133, 3013, 1647, 1505, 1418, 1346, 1333, 1278, 1205, 1107, 1006, 946, 888 cm^{-1}; ^{1}H NMR: d = 8.29 (s, 1 H), 8.79 (s, 2 H), 9.24 ppm (s, 2H); ^{13}C NMR: d = 146.4, 148.9, 149.8, 159.3 ppm; elemental analysis calcd (%) for $C_4H_4N_8O_2{\cdot}H_2O$ (214.14): C 22.43, H 2.82, N 52.35; found: C 22.20, H 2.72, N 51.24.

3-Amino-6-(3-amino-5-nitro-1H-1,2,4-triazol-1-yl)-1,2,4,5-tetrazine-2,4- di-N-oxide monohydrate (**2f-53**): Trifluoroacetic anhydride (4 mL, 28 mmol) was added to a slurry of 50% hydrogen peroxide (1.3 mL, 25 mmol) in methylene chloride (20 mL) while stirring at a temperature below 10 °C. Next, 3-amino-6-(3,5-diamino-1H-1,2,4-triazol-1-yl)-1,2,4,5-tetrazine (7 mmol) was added to the mixture at 0 °C and stirred for 30 min. The stirring continued for several hours at room temperature. The solvent was then removed, and the resulting residue was washed with ether and air-dried. It was further purified on silica gel using a 1:1 hexane–EtOAc solvent system. As a result, a yellow solid compound was obtained with a yield of 25%. IR (KBr): 3442, 3314, 1638, 1572, 1527, 1474, 1341, 1311, 1105, 844 cm^{-1}; ^{1}H NMR: d = 8.62 (s, 2 H), 7.82 ppm (s, 2H); ^{13}C NMR: d = 144.0, 148.6, 151.0, 162.7 ppm; elemental analysis calcd (%) for $C_4H_4N_{10}O_4{\cdot}H_2O$ (274.05): C 17.52, H 2.21, N 51.09; found: C 17.87, H 2.53, N 51.87.

3-Amino-6-(4-nitro-3,5-dimethylpyrazol-l-yl)-1,2,4,5-tetrazine-2,4-di-Noxide (**2f-54**): Trifluoroacetic anhydride (4 mL, 28 mmol) was added to a slurry of 50% hydrogen peroxide (1.3 mL, 25 mmol) in methylene chloride (20 mL) with stirring at a temperature below 10 °C. Next, 3-amino-6-(4-nitro-3,5-dimethylpyrazol-l-yl)-1,2,4,5-tetrazine (7 mmol) was added to the mixture at 0 °C and stirred for 30 min. The stirring continued for several hours at room temperature. The solvent was then removed, and the resulting residue was washed with ether and air-dried. As a result, a yellow solid compound was obtained with a yield of 55%. IR (KBr): 3356, 1627, 1583, 1509, 1462, 1327, 1151, 1109, 1012, 806 cm^{-1}; ^{1}H NMR: d = 8.72 (s, 2H), 2.52 (s, 3H), 2.43 ppm (s, 3H); ^{13}C NMR: d = 12.3, 13.9, 132.2, 143.6, 144.5, 147.0, 147.1 ppm; elemental analysis calcd (%) for $C_7H_8N_9O_4$ (281.07): C 31.35, H 3.01, N 41.78; found: C 31.01, H 2.99, N 39.69.

3-Amino-6-(5-azido-1,2,4-triazol-l-yl)-1,2,4,5-tetrazine-2,4-di-N-oxide (**2f-55**): Trifluoroacetic anhydride (4 mL, 28 mmol) was added to a slurry of 50% hydrogen peroxide (1.3 mL, 25 mmol) in methylene chloride (20 mL) while stirring at a temperature below 10 °C. Next, 3-amino-6-(5-azido-1,2,4-triazol-l-yl)-1,2,4,5-tetrazine (7 mmol) was added to the mixture at 0 °C and stirred for 30 min. The stirring continued for several hours at room temperature. The solvent was then removed, and the resulting residue was washed with ether and air-dried. As a result, a yellow solid compound was obtained with a yield of 62%. IR (KBr): 3343, 2144, 1631, 1531, 1327, 1186, 1109, 1014, 734 cm^{-1}; ^{1}H NMR: d = 9.21 (s, 1 H), 8.78 ppm (s, 2H); ^{13}C NMR: d = 153.4, 145.6, 145.1, 143.6 ppm; elemental analysis calcd (%) for $C_4H_3N_{11}O_2$ (237.05): C 20.26, H 1.28, N 64.97; found: C 19.85, H 1.55, N 64.76.

3-Amino-6-(1H-tetrazol-5-yl)-1,2,4,5-tetrazine-2,4-di-N-oxide (**2f-56**): Trifluoroacetic anhydride (4 mL, 28 mmol) was added to a slurry of 50% hydrogen peroxide (1.3 mL, 25 mmol) in methylene chloride (20 mL) while stirring at a temperature below 10 °C. Next, 3-amino-6-(1H-tetrazol-5-yl)-1,2,4,5-tetrazine (7 mmol) was added to the mixture at 0 °C and stirred for 30 min. The stirring continued for several hours at room temperature. The solvent was then removed, and the resulting residue was washed with ether and air-dried. As a result, a yellow solid compound was obtained with a yield of 70%. IR (KBr): 3203, 3127, 2957, 1637, 1516, 1426, 1326, 1317, 1176, 1053, 950, 732 cm^{-1}; ^{1}H NMR: d = 9.12 ppm (s, 3H); ^{13}C NMR: d = 152.3, 147.6, 140.6 ppm; ^{15}N NMR: d = − 9.6, − 81.6, − 89.6, − 102.8, − 309.2 ppm; elemental analysis calcd (%) for $C_4H_3N_{11}O_2{\cdot}H_2O$ (255.06): C 16.75, H 2.34, N 58.60; found: C 17.05, H 2.29, N 58.84.

References

1. H. Wei, C. He, J. Zhang, J.M. Shreeve, Angew. Chem. Int. Ed. **54**, 9367–9371 (2015)
2. C. Zhang, J. Mol. Struct. THEOCHEM **765**, 77–83 (2006)
3. D. Fischer, T.M. Klapötke, M. Reymann, J. Stierstorfer, Chem. Eur. J. **20**, 6401–6411 (2014)
4. D. Fischer, T.M. Klapötke, J. Stierstorfer, Eur. J. Inorg. Chem. **34**, 5808–5811 (2014)
5. C. He, Y. Tang, L.A. Mitchell, D.A. Parrishc, J.M. Shreeve, J. Mater. Chem. A **4**, 8969–8973 (2016)
6. M.A. Epishina, A.S. Kulikov, N.N. Makhova, Russ. Chem. Bull. Int. Ed. **57**, 644–651 (2008)
7. Y. Tang, C. He, L.A. Mitchell, D.A. Parrish, J.M. Shreeve, Chem. Eur. J. **22**, 11846–11853 (2016)
8. J. Zhang, J.M. Shreeve, J. Am. Chem. Soc. **136**, 4437–4445 (2014)

9. M. Coburn, J. Hetercycl. Chem. **5**, 83–87 (1968)
10. G.D. Solodyuk, M.D. Boldyrev, B.V. Gidaspov, V.D. Nikolaev, Zh. Org. Khim. **17**, 861 (1980)
11. E.G. Francois, D.E. Chavez, M.M. Sandstrom, Propellants Explos. Pyrotech. **35**, 529–534 (2010)
12. Q. Yu, Z. Wang, B. Wu, H. Yang, X. Ju, C. Lu, G. Cheng, J. Mater. Chem. A **3**, 8156–8164 (2015)
13. Q. Yu, Z. Wang, H. Yang, B. Wu, Q. Lin, X. Ju, C. Lu, G. Cheng, RSC Adv. **5**, 27305–27312 (2015)
14. A.K. Zelenin, M.L. Trudell, R.D. Gilardi, J. Heterocycl. Chem. **35**, 151 (1998)
15. X. Zhao, S. Li, Y. Wang, Y. Li, F. Zhao, S. Pang, J. Mater. Chem. A **4**, 5495–5504 (2016)
16. Y. Tang, C. He, H. Gao, J.M. Shreeve, J. Mater. Chem. A **3**, 15576–15582 (2015)
17. Y. Tang, H. Gao, G.H. Imler, D.A. Parrishc, J.M. Shreeve, RSC Adv. **6**, 91477–91482 (2016)
18. Y. Tang, C. He, L.A. Mitchell, D.A. Parrish, J.M. Shreeve, Angew. Chem. Int. Ed. **55**, 5565–5567 (2016)
19. P.W. Leonard, D.E. Chavez, P.F. Pagoria, D.A. Parrish, Propellants Explos. Pyrotech. **36**, 233–239 (2011)
20. A.K. Sikderr, N. Sikderr, J. Hazard. Mater. **112**, 1–15 (2004)
21. J.Y. He, Y.H. Lu, Q. Lei, Y. Cao, Chin. J. Explos. Propellants **34**, 9–12 (2011)
22. J. Zhang, S. Dharavath, L.A. Mitchell, D.A. Parrish, J.M.J. Shreeve, Mater. Chem. A **4**, 16961–16967 (2016)
23. Q. Yu, G.H. Imler, D.A. Parrish, J.M. Shreeve, Dalton Trans. **47**, 12661–12666 (2018)
24. R.L. Willer, R.S. Day, R. Gilardi, C. George, J. Heterocycl. Chem. **29**, 1835–1839 (1992)
25. A.B. Sheremetev, N.S. Aleksandrova, N.V. Palysaeva, M.I. Struchkova, V.A. Tartakovsky, K.Y. Suponitsky, Chem. Eur. J. **19**, 12446–12457 (2013)
26. H. Gao, J.M. Shreeve, Chem. Rev. **111**, 7377–7436 (2011)
27. L. Zhai, B. Wang, K. Xu, H. Huo, N. Liu, Y. Li, H. Li, P. Lian, X. Fan, J. Energ. Mater. **34**, 92–102 (2016)
28. A.B. Sheremetev, S.E. Semenov, V.S. Kuzmin, Y.A. Strelenko, S.L. Ioffe, Chem. Eur. J. **4**, 1203–1206 (1998)
29. Y. Tang, H. Gao, L.A. Mitchell, D.A. Parrish, J.M. Shreeve, Angew. Chem. Int. Ed. **128**, 1159–1162 (2016)
30. Q. Yu, G. Cheng, X. Ju, C. Lu, Q. Lin, H. Yang, New J. Chem. **41**, 1202–1211 (2017)
31. R. Beaudegnies, S. Wendeborn, Heterocycles **60**, 2417–2424 (2003)
32. C. He, G.H. Imler, D.A. Parrishc, J.M. Shreeve, J. Mater. Chem. A **6**, 16833–16837 (2018)
33. Q. Wang, Y. Shao, M. Lu, Cryst. Growth Des. **18**, 6150–6154 (2018)
34. H. Xiong, G. Cheng, Z. Zhang, H. Yang, New J. Chem. **43**, 7784–7789 (2019)
35. E.C. Johnson, J.J. Sabatini, D.E. Chavez, R.C. Sausa, E.F.C. Byrd, L.A. Wingard, P.E. Guzmàn, Org. Process Res. Dev. **22**, 736–740 (2018)
36. T.M. Klapötke, N. Mayr, J. Stierstorfer, M. Weyrauther, Chem. Eur. J. **20**, 1410–1417 (2014)
37. M.A. Kettner, K. Karaghiosoff, T.M. Klapçtke, M. Suceska, S. Wunder, Chem. Eur. J. **20**, 7622–7631 (2014)
38. NASA, *Press release: STS-122 The Voyage of Columbus* (2008), pp. 82–84
39. D. Fischer, J.L. Gottfried, T.M. Klapötke, K. Karaghiosoff, J. Stierstorfer, D.T.G. Witkowski, Angew. Chem. Int. Ed. **55**, 16132–16135 (2016)
40. T. Lu, C. Wang, G. Wang, S. Wang, J. Song, H. Yin, G. Fan, F. Chen, New J. Chem. **43**, 13330–13333 (2019)
41. Y. Tang, H. Gao, L.A. Mitchell, D.A. Parrish, J.M. Shreeve, Angew. Chem. Int. Ed. **128**, 3252–3255 (2016)
42. V. Thottempudi, J. Zhang, C. He, J.M. Shreeve, RSC Adv. **4**, 50361–50364 (2014)
43. Y. Tang, C. He, L.A. Mitchell, D.A. Parrishb, J.M. Shreeve, J. Mater. Chem. A **3**, 23143–23148 (2015)
44. Q. Wang, Y. Shao, M. Lu, Cryst. Growth Des. **19**, 839–844 (2019)
45. J. Tian, H. Xiong, Q. Lin, G. Cheng, H. Yang, New J. Chem. **41**, 1918–1924 (2017)

46. M.A. Kettner, T.M. Klapçtke, T.G. Witkowski, F.V. Hundling, Chem. Eur. J. **21**, 4238–4241 (2015)
47. N. Fischer, D. Izsak, T.M. Klapötke, J. Stierstorfer, Chem. Eur. J. **19**, 8948–8957 (2013)
48. N. Fischer, D. Fischer, T.M. Klapötke, D.G. Piercey, J. Stierstorfer, J. Mater. Chem. **22**, 20418–20422 (2012)
49. T.M. Klapötke, D.G. Piercey, J. Stierstorfer, Dalton Trans. **41**, 9451–9459 (2012)
50. D. Fischer, T.M. Klapötke, J. Stierstorfer, Angew. Chem. Int. Ed. **53**, 8172–8175 (2014)
51. Y. Joo, J.M. Shreeve, Angew. Chem. Int. Ed. **49**, 7320–7323 (2010)
52. Q. Zhang, J. Zhang, D.A. Parrish, J.M. Shreeve, Chem. Eur. J. **19**, 11000–11006 (2013)
53. Y. Joo, B. Twamley, S. Garg, J.M. Shreeve, Angew. Chem. Int. Ed. **47**, 6236–6239 (2008)
54. Y. Joo, J.M. Shreeve, Angew. Chem. Int. Ed. **121**, 572–575 (2009)
55. T.M. Klapötke, P. Mayer, J. Stierstorfera, J.J. Weigand, J. Mater. Chem. **18**, 5248–5258 (2008)
56. Y. Guo, G. Tao, Z. Zeng, H. Gao, D.A. Parrish, J.M. Shreeve, Chem. Eur. J. **16**, 3753–3762 (2010)
57. T.M. Klapötke, D.G. Piercey, Inorg. Chem. **50**, 2732–2734 (2011)
58. Y. Tang, H. Yang, J. Shen, B. Wu, X. Ju, C. Lu, G. Cheng, New J. Chem. **36**, 2447–2450 (2012)
59. T.M. Klapötke, C.M. Sabaté, Chem. Mater. **20**, 3629–3637 (2008)
60. N. Fischer, K. Hüll, T.M. Klapötke, J. Stierstorfer, G. Laus, M. Hummel, C. Froschauer, K. Wurstb, H. Schottenbergerb, Dalton Trans. **41**, 11201–11211 (2012)
61. D. Fischer, T.M. Klapötke, D.G. Piercey, J. Stierstorfer, Chem. Eur. J. **19**, 4602–4613 (2013)
62. F. Li, X. Cong, Z. Du, C. He, L. Zhao, L. Meng, New J. Chem. **36**, 1953–1956 (2012)
63. D. Fischer, T.M. Klapötke, J. Stierstorfer, N. Szimhardt, Chem. Eur. J. **22**, 4966–4970 (2016)
64. P. He, H. Mei, J. Yang, J. Zhang, New J. Chem. **43**, 4235–4241 (2019)
65. T.M. Klapötke, N.K. Minar, J. Stierstorfer, Polyhedron **28**, 13–26 (2009)
66. Q. Wang, H. Lu, F. Pang, J. Huang, F. Nie, F. Chen, RSC Adv. **6**, 56827–56830 (2016)
67. T.M. Klapötke, P. Mayer, A. Schulz, J.J. Weigand, Propellants Explos. Pyrotech. **29**, 325–332 (2004)
68. Y. Tang, H. Yang, B. Wu, X. Ju, C. Lu, G. Cheng, Angew. Chem. Int. Ed. **52**, 4875–4877 (2013)
69. D. Chand, D.A. Parrishb, J.M. Shreeve, J. Mater. Chem. A **1**, 15383–15389 (2013)
70. M. Freis, T.M. Klapötke, J. Stierstorfer, N. Szimhardt, Inorg. Chem. **56**, 7936–7947 (2017)
71. A.A. Dippold, T.M. Klapötke, J. Am. Chem. Soc. **135**, 9931–9938 (2013)
72. J. Zhang, P. Yin, L.A. Mitchell, D.A. Parrishc, J.M. Shreeve, J. Mater. Chem. A **4**, 7430–7436 (2016)
73. Y. Luo, L. Zhai, F. Bi, C. Zhou, B. Wang, Chin. J Energ. Mater. **24**, 925–926 (2016)
74. A.A. Dippold, T.M. Klapötke, Chem. Eur. J. **18**, 16742–16753 (2012)
75. P. Yin, J.M. Shreeve, Angew. Chem. Int. Ed. **54**, 14513–14517 (2015)
76. T.M. Klapötke, M. Leroux, P.C. Schmid, J. Stierstorfer, Chem. Asian J. **11**, 844–851 (2016)
77. Q. Ma, H. Gu, J. Huang, F. Nie, G. Fan, L. Liao, W. Yang, New J. Chem. **42**, 2376–2380 (2018)
78. C. He, J.M. Shreeve, Angew. Chem. Int. Ed. **54**, 6260–6264 (2015)
79. Z. Xu, G. Cheng, S. Zhu, Q. Lin, H. Yang, J. Mater. Chem. A **6**, 2239–2248 (2018)
80. Y. Tang, C. He, L.A. Mitchell, D.A. Parrish, J.M. Shreeve, J. Mater. Chem. A **4**, 3879–3885 (2016)
81. Y. He, G. Fan, G. Zhang, M. Huang, Y. Liu, Chin. J Energ. Mater. **20**, 715–720 (2012)
82. H.H. Licht, H. Ritter, Propellants Explos. Pyrotech. **22**, 333–336 (1997)
83. S. Li, S. Pang, X. Li, Y. Yu, X. Zhao, Chin. Chem. Lett. **18**, 1176–1178 (2007)
84. C. Qi, S. Li, Y. Li, Y. Wang, X. Zhao, S. Pang, Chem. Eur. J. **18**, 16562–16570 (2012)
85. W. Liu, S. Li, Y. Li, Y. Yang, Y. Yu, S. Pang, J. Mater. Chem. A **2**, 15978–15986 (2014)
86. Y. Li, C. Qi, S. Li, H. Zhang, C. Sun, Y. Yu, S. Pang, J. Am. Chem. Soc. **132**, 12172–12173 (2010)
87. Y. Tang, J.M. Shreeve, Chem. Eur. J. **21**, 7285–7291 (2015)
88. D.L. Naud, M.A. Hiskey, H.H. Harry, J. Energ. Mater. **21**, 57–62 (2003)

89. A. Dippold, T.M. Klapötke, F.A. Martin, Z. Anorg, Allg. Chem. **637**, 1181–1193 (2011)
90. V. Thottempudi, H. Gao, J.M. Shreeve, J. Am. Chem. Soc. **133**, 6464–6471 (2011)
91. P. Yin, D.A. Parrish, J.M. Shreeve, Angew. Chem. Int. Ed. **53**, 12889–12892 (2014)
92. H. Yang, L. Huang, M. Xu, Y. Tang, B. Wang, G. Cheng, J. Org. Chem. **84**, 10629–10634 (2019)
93. M.V. Huynh, M.A. Hiskey, E.L. Hartline, D.P. Montoya, R. Gilardi, Angew. Chem. Int. Ed. **43**, 4924–4928 (2004)
94. Q. Zhang, C. He, P. Yin, J.M. Shreeve, Chem. Asian J. **9**, 212–217 (2014)
95. S.V. Chapyshev, D.V. Korchagin, Chem. Heterocycl. Compd. **53**, 791–796 (2017)
96. D.E. Chavez, M.A. Hiskey, R.D. Gilardi, Angew. Chem. Int. Ed. **39**, 1971–1973 (2000)
97. D.E. Chavez, M.A. Hiskey, D.L. Naud, Propellants Explos. Pyrotech. **29**, 209–215 (2004)
98. P. Yin, C. He, J.M. Shreeve, Chem. Eur. J. **22**, 2108–2113 (2016)
99. Y. Tang, G.H. Imler, D.A. Parrish, J.M. Shreeve, Org. Lett. **21**, 8201–8204 (2019)
100. P. Yin, Q. Zhang, J. Zhang, D.A. Parrish, J.M. Shreeve, J. Mater. Chem. A **1**, 7500–7510 (2013)
101. T.M. Klapötke, A. Preimesser, J. Stierstorfer, Z. Anorg, Allg. Chem. **638**, 1278–1286 (2012)
102. M.M. Breiner, D.E. Chavez, D.A. Parrish, Synlett **24**, 0519–0521 (2013)
103. I.L. Dalinger, I.A. Vatsadze, T.K. Shkineva, G.P. Popova, S.A. Sheveleve, Mendeleev Commun. **20**, 355–356 (2010)
104. C. Li, L. Liang, K. Wang, C. Bian, J. Zhang, Z. Zhou, J. Mater. Chem. A **2**, 18097–18105 (2014)
105. T.K. Shkineva, A.V. Kormanov, V.N. Boldinova, I.A. Vatsadze, I.L. Dalinger, Chem. Heterocycl. Compd. **54**, 703–709 (2018)
106. Y. Tang, C. He, G.H. Imler, D.A. Parrishb, J.M. Shreeve, Chem. Commun. **54**, 10566–10569 (2018)
107. D. Kumar, Y. Tang, C. He, G.H. Imler, D.A. Parrish, J.M. Shreeve, Chem. Eur. J. **24**, 17220–17224 (2018)
108. P. Yin, D.A. Parrish, J.M. Shreeve, Chem. Eur. J. **20**, 6707–6712 (2014)
109. P. Yin, J. Zhang, D.A. Parrish, J.M. Shreeve, Chem. Eur. J. **20**, 16529–16536 (2014)
110. R. Wang, Y. Guo, Z. Zeng, B. Twamley, J.M. Shreeve, Chem. Eur. J. **15**, 2625–2634 (2009)
111. H. Huang, Y. Shi, Y. Li, Y. Liu, J. Yang, RSC Adv. **6**, 64568–64574 (2016)
112. H. Wei, J. Zhang, C. He, J.M. Shreeve, Chem. Eur. J. **21**, 8607–8612 (2015)
113. L. Liang, K. Wang, C. Bian, L. Ling, Z. Zhou, Chem. Eur. J. **19**, 14902–14910 (2013)
114. T.P. Efimova, O.Y. Ozerova, T.A. Novikova, I.V. Belik, V.M. Berestovitskaya, Russ. J. Gen. Chem. **83**, 1779–1786 (2013)
115. Z. Xu, G. Cheng, H. Yang, X. Ju, P. Yin, J. Zhang, J.M. Shreeve, Angew. Chem. Int. Ed. **56**, 5877–5881 (2017)
116. J. Ma, J. Tang, H. Yang, Z. Yi, G. Wu, S. Zhu, W. Zhang, Y. Li, G. Cheng, ACS Appl. Mater. Interfaces **11**, 26053–26059 (2019)
117. A.A. Dippold, T.M. Klapötke, Chem. Asian J. **8**, 1463–1471 (2013)
118. Y. Joo, J.M. Shreeve, Eur. J. Org. Chem. **21**, 3573–3578 (2009)
119. G. Tao, B. Twamley, J.M. Shreeve, J. Mater. Chem. **19**, 5850–5854 (2009)
120. C. Bian, M. Zhang, C. Lia, Z. Zhou, J. Mater. Chem. A **3**, 163–169 (2015)
121. W. Fu, B. Zhao, M. Zhang, C. Li, H. Gao, J. Zhang, Z. Zhou, J. Mater. Chem. A **5**, 5044–5054 (2017)
122. Y. Tang, C. He, G.H. Imler, D.A. Parrish, J.M. Shreeve, J. Mater. Chem. A **4**, 13923–13929 (2016)
123. D. Kumar, L.A. Mitchell, D.A. Parrishc, J.M. Shreeve, J. Mater. Chem. A **4**, 9931–9940 (2016)
124. C. Li, M. Zhang, Q. Chen, Y. Li, H. Gao, W. Fu, Z. Zhou, Dalton Trans. **45**, 17956–17965 (2016)
125. M. Zhang, H. Gao, C. Li, W. Fu, L. Tang, Z. Zhou, J. Mater. Chem. A **5**, 1769–1777 (2017)
126. D. Kumar, C. He, L.A. Mitchell, D.A. Parrishc, J.M. Shreeve, J. Mater. Chem. A **4**, 9220–9228 (2016)
127. H. Wei, H. Gao, J.M. Shreeve, Chem. Eur. J. **20**, 16943–16952 (2014)

Chapter 3
Tricyclic-Based Energetic Materials

Structures-property analysis of these well-known explosives clearly suggests that the introduction of heterocyclic rings is a general approach to improve the thermal stability, heats of formation and energetic properties. On the basis of bicyclic-based energetic materials, a series of tricyclic-based nitrogen-rich heterocyclic compounds including bicyclic oxadiazole, tetrazole, triazole, pyrazole, imidazole, tetrazine and triazine were systematic studied including synthesis, physicochemical and energetic properties in this section.

3.1 Tricyclic Oxadiazole-Based Energetic Compound

Generally, oxadiazole have been considered as potential component of explosives and propellants due to high nitrogen contents, high energy density and good oxygen balance. Many backbone containing one and two oxadiazole rings have been applied for the design of energetic materials. And a series of compounds containing three oxadiazole rings show relatively increased crystal density compared to other simple furazan derivatives such as bis(nitrofurazano)furoxan (**BNFF**). As shown in Scheme 3.1, bis(aminofurazano)furoxan (**BAFF**) and bisaminofurazanodioxadiazine (**BAFOZ**) were selectively synthesized via 1,3-dipolar cycloaddition of 4-amino-N-hydroxy-1,2,5-oxadiazole-3-carbimidoyl chloride depending on the reaction conditions. Both diamino compounds were oxidized to give dinitro compounds **BNFF** and **BNFOZ** in moderate yield. **BNFF** exhibits a high density (1.90 g cm^{-3}) and detonation velocity ($D = 8930$ m s^{-1}), which is comparable to tranditional explosive **HMX**. Furthermore, the low melting point (110 °C) and relatively high thermal stability of **BNFF** make it easy to cast and process. Subsequently, **BNFF** was treated with H_2/Pd–C to give its deoxygenated product bis(nitrofurazano)furazan (**BNFF**^{R}) in moderate yield [1, 2]. Moreover, the reduction reaction of **BAFF** using $SnCl_2$/HCl/AcOH as reductant and then oxidation using Caro's acid is another synthetic

G. Cheng et al., *Nitrogen-Rich Polycyclic-Based Energetic Materials*,
https://doi.org/10.1007/978-981-97-4569-2_3

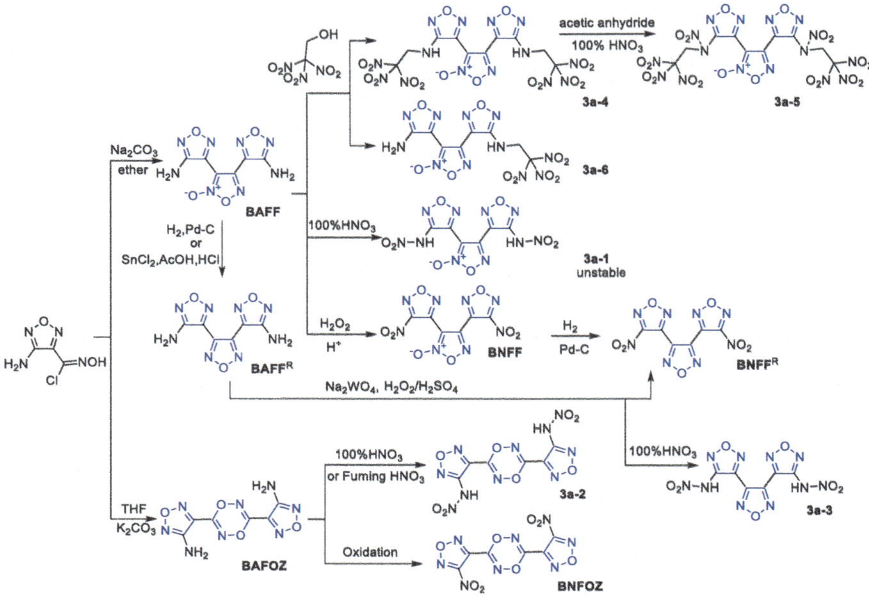

Scheme 3.1 Synthesis of C–C bonded tricyclic oxadiazole derivatives

method to **BNFF**^{R} [3]. In a typical procedure, the nitramino compound **3a-3** was obtained by the nitration of **BAFF**^{R} using 100% nitric acid as the nitrating reagent [4]. And the further nitration of **BAFF** and **BAFOZ** with fuming HNO_3 or 100% HNO_3 produced **3a-1** and **3a-2**, respectively. And compound **3a-1** was stable in diethyl ether solution, but decomposed after isolation at room temperature.

Considering the high energy properties of nitro and nitramino groups, the synthesis and characterization of a series of N-trinitroethylamino functionalizational compounds were reported. As shown in Scheme 3.1, the reaction of trinitroethanol and **BAFF** afforded target products containing one and two trinitroethyl fragments in yields of 30.1% and 68.0%, respectively. Compound **3a-4** was then treated with 100% nitric acid and acetic anhydride to afford the N-nitrated trinitroethylamino derivative **3a-4** [5]. The trinitroethylnitramino derivatives **3a-(4–6)** possess high detonation velocities (8673–9407 m s^{-1}) and relatively high densities (1.84–1.90 g cm^{-3}), as well as an acceptable impact sensitivity (3.6–21.1 J), which indicates that these trinitroethyl derivatives might be of interest for future applications as environmentally friendly and nitrogen-rich energetic materials.

There are two different synthetic routes to compound 3-[4-(1,2,4-oxadiazol-3-yl)-2-oxido-1,2,5-oxadiazol-3-yl]-1,2,4-oxadiazole (**3a-7**) shown in Scheme 3.2. Compound **3a-7** was synthesized through direct cyclization of 1,2,4-oxadiazole-3-chloroxime obtained through the oximation, cyclization and chlorination reaction starting from glyoxal in the presence of K_2CO_3 [6]. Another route to **3a-7** was the double cyclization reaction of bis-amidoxime by utilizing trimethyl orthoformate

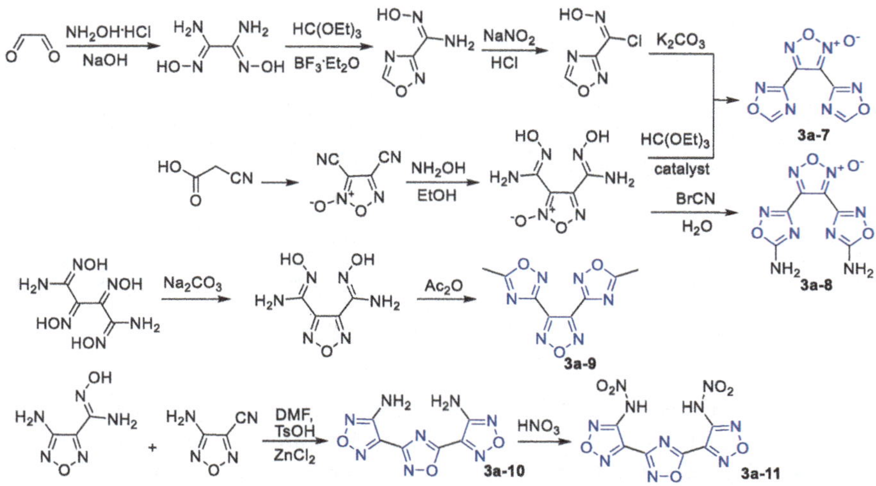

Scheme 3.2 Synthesis of C–C bonded tricyclic furazan-based 1,2,4-oxadiazole derivatives

and different calatysts [7, 8]. When the bis-amidoxime was reacted with cyanogen bromide, the amino groups were introduced into **3a-7** to form 3,4-bis-(5-amino-1,2,4-oxadiazolyl) furoxan (**3a-8**) [9]. The decomposition temperature of **3a-7** (T_{dec} = 214 °C) was increased to 273 °C after the introduction of amino groups. The methly group was also introduced to tricyclic oxadiazole backbone shown in Scheme 3.2. The diaminotetraoxime in aqueous sodium hydroxide was heated to be dehydrated into 3,4-dicarboxamidoxime-furazan. Then, this bis-amidoxime was reacted with acetic anhydride to give 3,4-bis(5-methyl-1,2,4-oxadiazol-3-yl)furazan (**3a-9**). In addition, the formation of 1,2,4-oxadiazole **3a-10** from amidoxime was proceeded using $ZnCl_2$ as the Lewis acid. Moreover, nitramino compound **3a-11** was obtained by the treatment of **3a-10** with excess fuming nitric acid [10].

When bifurazan or trifurazan rings are substituted with nitro groups such as 4,4′-dinitro-3,3′-bifurazan or **BNFF** above mentioned, their far greater stability makes them more favorable for practical applications. As shown in Scheme 3.3, the chlorination reaction of 3-hydroximoyl-4-aminofuroxan using $NaNO_2$ in concentrated HCl was firstly attempted to prepare 3-chlorohydroximoyl-4-aminofuroxan, which could undergo dimerization to give a trifuroxan structure (**3a-12**) by treating with dilute aqueous K_2CO_3 solution. However, the intermediate 3-chlorohydroximoyl-4-aminofuroxan was not obtained in the acid sodium nitrite solution probably due to the diazotization of amino group in the 4-position of 3-amidoxime-4-aminofuroxan as a starting material. Later, J. M. Shreeve's group reported the synthesis of **3a-12** by protecting the 4-amino group of 3-hydroximoyl-4-aminofuroxan. Then, **3a-12** was synthesized through the chlorination, cyclization and ammoniation starting from the protected 3-hydroximoyl-4-aminofuroxan. Further oxidation reaction of **3a-12** into dinitro compound **BNTFO-IV** using numerous oxidants such as H_2O_2/H_2SO_4 and $H_2O_2/H_2SO_4/Na_2WO_4$, was attempted. And the oxidation reaction of diamine **3a-12**

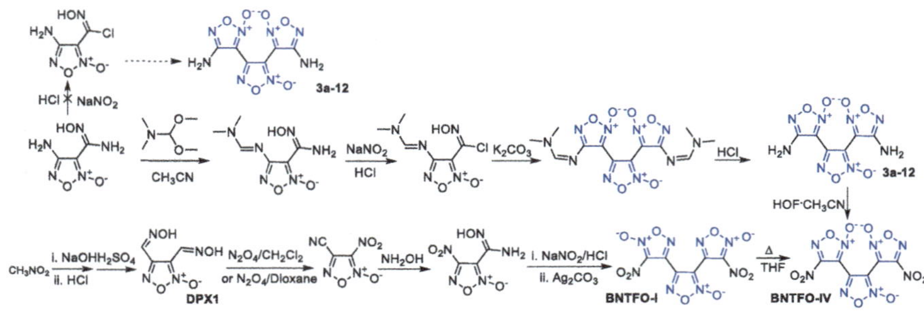

Scheme 3.3 Synthesis of C–C bonded tricyclic furoxan derivatives

into **BNTFO-IV** was successful only using HOF as an oxidant [11]. Another route to **BNTFO** starting from (E,E)-3,4-bis(oximomethyl)furoxan (**DPX1**) was reported by B. Wang's group. The **DPX1** was prepared in high purity by two-steps reaction from nitromethane (Scheme 3.3a-3). And the oxidation reaction of **DPX1** using N_2O_4 in CH_2Cl_2 or dioxane led to 3-nitro-4-cyanofuroxan, which was then treated with 50% aqueous hydroxylamine in ethanol to give 3-nitro-4-aminoximidofuoxan. And **BNTFO-I** was produced through diazotization of 3-nitro-4-aminoximidofuoxan with sodium nitrite in concentrated hydrochloric acid and subsequently cyclization using Ag_2CO_3 in THF. In an effort to obtain crystals suitable for X-ray diffraction of **BNTFO-I**, the isomer **BNTFO-IV** was formed in some solvents such CH_3CN and THF [12, 13]. Surprisely, **BNTFO-I** has a measured density of 1.98 g cm^{-3} slightly higher than **BNTFO-IV** (1.91 cm^{-3}) due to "inverse" molecule configuration, which results in higher detonation performance of **BNTFO-I** ($D = 9867$ m s^{-1}, $P = 45.0$ GPa) and **BNTFO-IV** ($D = 9503$ m s^{-1}, $P = 40.8$ GPa).

Apart from C–C bonded tricyclic oxadiazole derivatives above mentioned, a series of zero-hydrogen nitrofurazans containing an oxy, azo and azoxy bridge have also been synthesized. As shown in Scheme 3.4, dinitrofurazan was hydrolyzed in aqueous acetronitrile alkali followed by acidification with hydrochloric acid to give the 3,4-dihydroxyfurazan. The 3,4-dihydroxyfurazan was then converted to the bisalt by treatment with sodium ethoxide in ether. When dinitrofuran and 3-cyano-4-nitrofurazan was subjected to the bisalt in glyme, the desired trifurazan oxy-bridged compounds **3a-13** and **3a-14** were obtained, respectively. The addition reaction of compound **3a-14** with hydroxylamine led to pure aminoximidofurazan compound in a high yield. Then chloroximidofurazan compound is obtained by diazotization of aminoximidofurazan using sodium nitrite in concentrated hydrochloric acid. Finally, 3,4-bis(3-chlorodinitromethylfurazan-4-oxy)furazan **3a-15** was formed through the oxidation of chloroximidofurazan using N_2O_5 in chloroform [14, 15]. Encouraged by this, another trifurazan ether containing azo bridge (**3a-16**) was also synthesized from 3,3′-nitro-4,4′-azoxyfurazan following the same synthetic route to **3a-14**. Successful oxidation of **3a-16** into **3a-17** accompanied by the formation of a very slight amount of **3a-18** was finally achieved utilizing ammonium persulfate in oleum. 3-Nitro-4-amino-furazan and 3-nitro-3′-amino-4,4′-azoxyfurazan were readily prepared through the oxidation reaction of **DAAF** using H_2O_2 and $KMnO_4$, respectively. Then the oxidative coupling reaction of nitro-4-amino-furazan and 3-nitro-3′-amino-4,4′-azoxyfurazan was completed to give the product **3a-19**. Finally, the oxidation reaction of azo-bridged compound **3a-19** to azoxy-bridged compound was performed to give a mixture of positional isomers of **3a-20** via nonselective attack at four nitrogen atoms of both azo bridges.

In addition, the guanidine bridge was also introduced to trioxadiazole rings by the reaction of hydrazine groups with 3-methyl-4-furoxancarbaldehyde (Scheme 3.4). Starting from 3-methyl-4-furoxancarbaldehyde, the treatment of 1,3-diaminoguanidine monohydrochloride and triaminoguanidine nitrate salts led to the desired 1,3-bis-(3-methylfuroxan-4-methyleneamino) guanidine (**3a-22**) and 1,2,3-tris-(3-methylfuroxan-4-methyleneamino) guanidine (**3a-21**) in a high yield, respectively [16]. However, the detonation performance of these compounds was not ideal due to many methyl groups (Table 3.1).

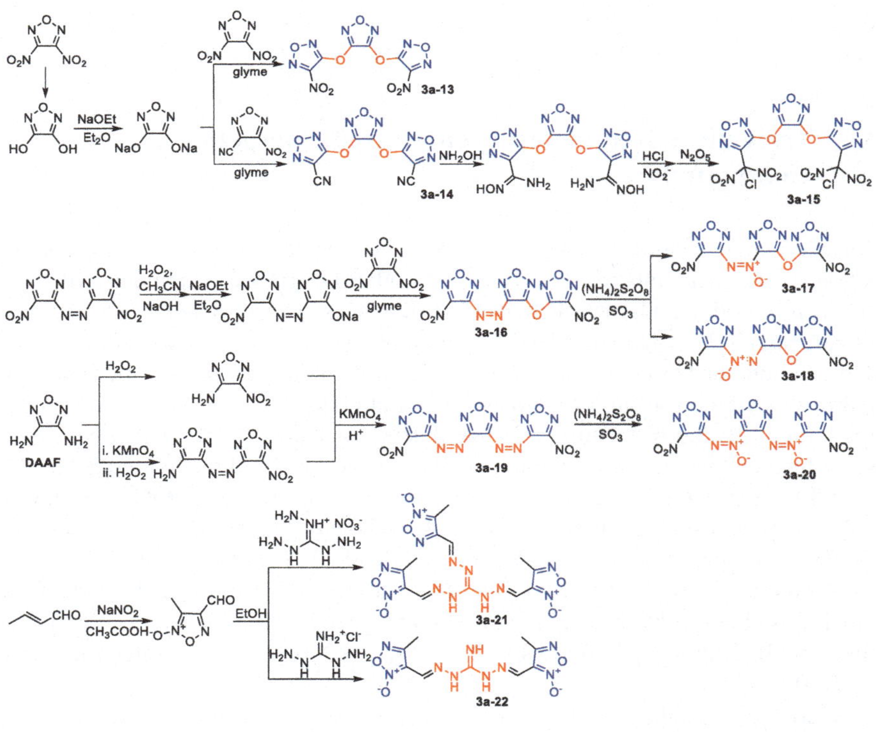

Scheme 3.4 Synthesis of oxy, azo, azoxy and guanidine bridged tricyclic oxadiazole derivatives

Table 3.1 Energetic properties of tricyclic oxadiazole-based energetic compounds

Compound	T_d (°C)	D (m s^{-1})	P (GPa)	$\Delta_f H$ (kJ mol^{-1})	IS (J)	FS (N)	d (g cm^{-3})
C–C bonded tricyclic oxadiazole derivatives							
3a-2	95	8490	32.2	592	< 1	32	1.82
BNFF	245	8930	/	645	3.1	/	1.90
3a-4	167	8935	35.9	923	14.2	240	1.84
3a-5	108	9407	40.5	1174	3.6	100	1.90
3a-6	169	8673	33.7	840	21.1	300	1.83
3a-7	214	8043	26.8	326	18	120	1.82
3a-12	184	8480	31.0	493	13	240	1.79
BNTFO-IV	146	9503	40.8	579	3	40	1.91
BNTFO-I	131	9867	45.0	668	3	35	1.98
Guanidine bridged tricyclic oxadiazole derivatives							
3a-21	232	7532	20.2	766	22	/	1.56
3a-22	215	7528	20.8	1214	18	/	1.57

3.2 Tricyclic Mixed Heterocyclic-Based Energetic Compounds

Considering the features of tetrazine above mentioned, it is expected that the combination of energetic tetrazine backbone with other heterocycle fragments will give rise to new compounds with better properties. As shown in Scheme 3.5, the 3,6-dihydrazinyl-1,2,4,5-tetrazine (**DHT**) can be prepared easily by the modified four-step reactions including ammoniation, cyclization, oxidation and substitution with hydrazine hydrate starting from guanidine nitrate [17]. Using NaH as a soft nucleophile in DMF, 3,6-bis(3-nitro-1,2,4-oxadiazole-5-guanyl)-1,2,4,5-tetrazine (**3b-1**) was prepared by the substitution reaction of 3,6-bis(3,5-dimethylpyrazol-1-yl)-1,2,4,5-tetrazine (**BT**) and 3-nitro-5-guanidino-1,2,4-oxadiazole (**NOG**), which was isolated by treatment of diaminoglycoluril (**AG**) with Oxone in acetone [18, 19]. Similar to synthetic route to **3b-1**, 3,6-bis(3′-azido-1′,2′,4′-oxadiazole-5′-guanyl)-1,2,4,5-tetrazine (**3b-2**) was also synthesized through the reaction of **BT** and **AOG**, which was obtained by the reaction of **NOG** and NaN_3 [20]. The target compounds **3b-1** and **3b-2** show high thermal stability (T_{dec} > 210 °C) and low the sensitivity (IS > 30 J).

Apart from oxadiazole, tetrazole fragment was also introduced to tetrazine backbone. There are two routes to prepare 3,6-bis(1H-1,2,3,4-tetrazol-5-ylamino)-1,2,4,5-tetrazine (**BTATz**) shown in Scheme 3.5. On one hand, nucleophilic substitution of the first 3,5-dimethylpyrazole leaving group with 5-aminotetrazole is easily proceeded to provide intermediate 3-(1H-1,2,3,4-tetrazol-5-ylamino)-6-(3,5-dimethyl-pyrazol-1-yl)-s-tetrazine monohydrate (**TADPTz**), and the second substitution was taken place to form **BATAz** in the presence of sulfolane at 135 °C. The

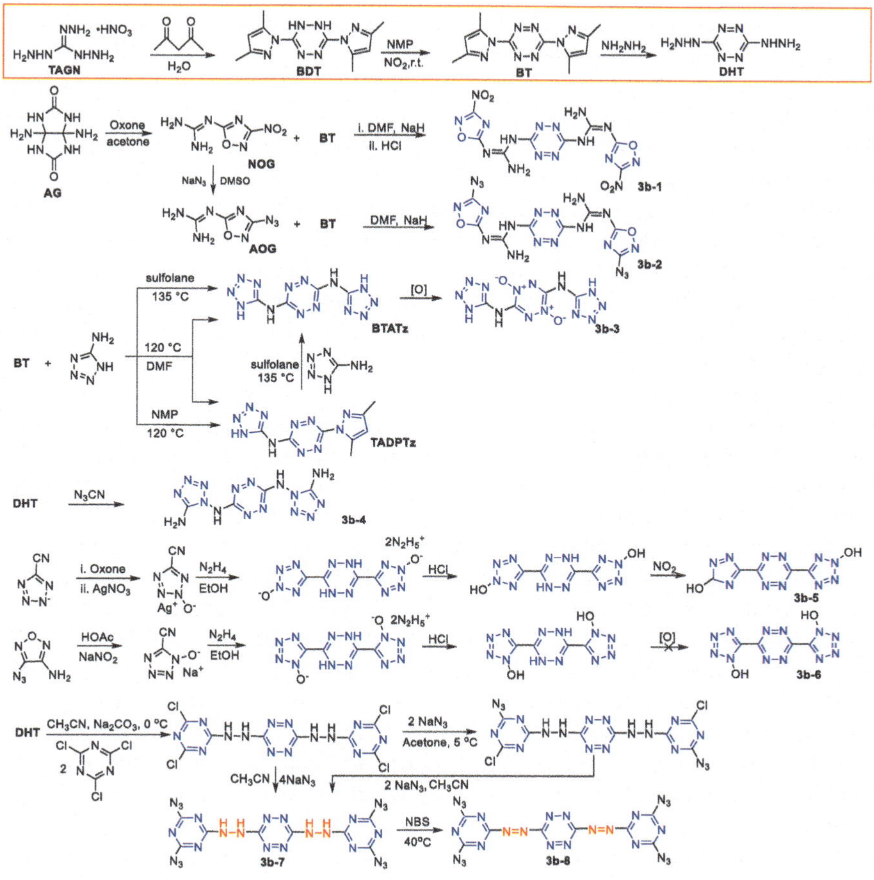

Scheme 3.5 Synthesis of tricyclic tetrazine derivatives

N-oxide derivative **3b-3** was obtained by the further oxidation of **BATAz** using Caro' acid [21]. On the other hand, the synthesis and characterization of **BATAz** with improved yield and simple details were demonstrated through the one-pot reaction starting from **BT** directly by A. K. Sikder. Further, one of the intermediate **TADPTz** was also obtained and characterized completely during the synthesis of **BTATz** [22]. DSC results revealed that **BTATz** showed an exothermic decomposition at 265–350 °C, higher than the exothermic decomposition peak of **TADPTz** in the temperature range of 245–275 °C. Similar to the synthesis route to **2b-(6–13)**, compound **3b-4** containing tetrazole and tetrazine backbones was synthesized through the cyclization of **DHT** using N_3CN [23]. As shown in Scheme 3.5, the cyanotetrazolate-2-oxide was synthesized by the oxidation of sodium 5-cyanotetrazolate using potassium peroxymonopersulfate (Oxone) in acetone. Another cyanotetrazolate-1-oxide anion was obtained from the reaction of azidoaminofurazan and sodium nitrite. As known, treatment of 5-cyanotetrazole with hydrazine in ethanol results in the formation of dihydrotetrazine. Cyanotetrazole 1- and 2-oxides were protonated with nitric acid yielding 5-cyano-2-hydroxytetrazole and 5-cyano-1-hydroxytetrazole, respectively. The subsequent cyclization reaction with hydrazine gave the corresponding dihydrotetrazines as their hydrazinium salts. These salts were then protonated with hydrochloric acid in water yielding the 1- and 2-hydroxy hydrates. When a dilute solution of NO_2 in acetonitrile was added to a slurry of these hydrates in acetonitrile, **3b-5** was obtained in a high purity. However, **3b-6** was not obtained probably because of the influence of steric hindrance on this oxidation reaction [24].

Based on the **DHT**, the metal-free polyazido compounds 3,6-bis(2-(4,6-diazido-1,3,5-triazin-2-yl)-hydrazinyl)-1,2,4,5-tetrazine (**3b-7**) and 3,6-bis-(2-(4,6-diazido-1,3,5-triazin-2-ly)-diazenyl)-1,2,4,5-tetrazine (**3b-8**) were synthesized by Cheng and Yang's group [25]. As shown in Scheme 3.5, 3,6-bis(2-(4,6-dichloro-1,3,5-triazin-2-yl)hydrazinyl)-1,2,4,5-tetrazine was obtained from the nucleophilic substitution of **DHT** and cyanuric chloride. Then the reaction of two equivalents of NaN_3 with one equivalent of bis(dichloro-1,3,5-triazin)-tetrazine was proceeded to give the bis(4-azido-6-chloro-1,3,5-triazin)-tetrazine in 57.6% yield, which was transformed to **3b-7** through further azidation by refluxing in acetonitrile. To enhance the yield, the addition of four equivalent of NaN_3 ensures bis(dichloro-1,3,5-triazin)-tetrazine was completely converted into azido in a 73.3% yield by refluxing in acetonitrile. Moreover, compound **3b-7** was further oxidized into azo compound **3b-8** using N-bromosuccinicmide (NBS) as an oxidation reagent in acetonitrile. And compound **3b-8** could be used as a kind of secondary explosive with excellent heats of formation and good explosive properties ($D = 8602\ m\ s^{-1}$, $P = 29.4$ GPa, $\Delta_f H = 2820\ kJ\ mol^{-1}$). It is noteworthy that compound **3b-7** can be used as a potential environmentally friendly candidate for lead-based-primary explosives, which was confirmed by perforated lead plate test.

The 1,3-dipolar cyclization of a nitrile oxide was one of the most valuable methods to obtain a furoxan. A new method to synthesize imidazolyl nitrile oxides starting from imidazole 2-nitrolic acid was reported by J. S. Kim's group shown in Scheme 3.6. The 1,4-dioxa-2,5-diazine derivative **3b-9** was obtained through the reaction of imidazole 2-nitrolic acid using $NaHCO_3$ as the main product followed by the formation of isomer **3b-10**. Also, the bisimidazolylfuroxan **3b-10** can formed through the dimerization of chloroxime, which was obtained by the treatment of oxime with NCS in aqueous K_2CO_3. Further nitration of **3b-10** with anhydrous HNO_3 gave the mixture of dinitrofuroxan (**3b-11**) and tetranitrobisimidazolylfuroxan (**3b-12**) [26]. The decomposition temperature of **3b-10** is about 234 °C, which is higher than that of diazine isomer **3b-9** ($T_d = 134$ °C). And compound **3b-12** also possesses a higher decomposition temperature ($T_d = 258$ °C) among polynitro compounds. By using the same method to **3b-10** from chloroxime, the target energetic compound 3,4-bis(1′,2′,4′-triazole-3′-yl)furoxan (**BTAF**) were designed via a three-step reaction including oxime formation, diazotization and cyclization starting from 3-cyano-1,2,4-triazole. And further nitration of **BTAF** into 3,4-bis-(1′-nitro-1′,2′,4′-triazole-3′-yl)furoxan (**BNTAF**) was proceeded using the mixture of acetic anhydride and concentrated nitric acid in acetic acid [27].

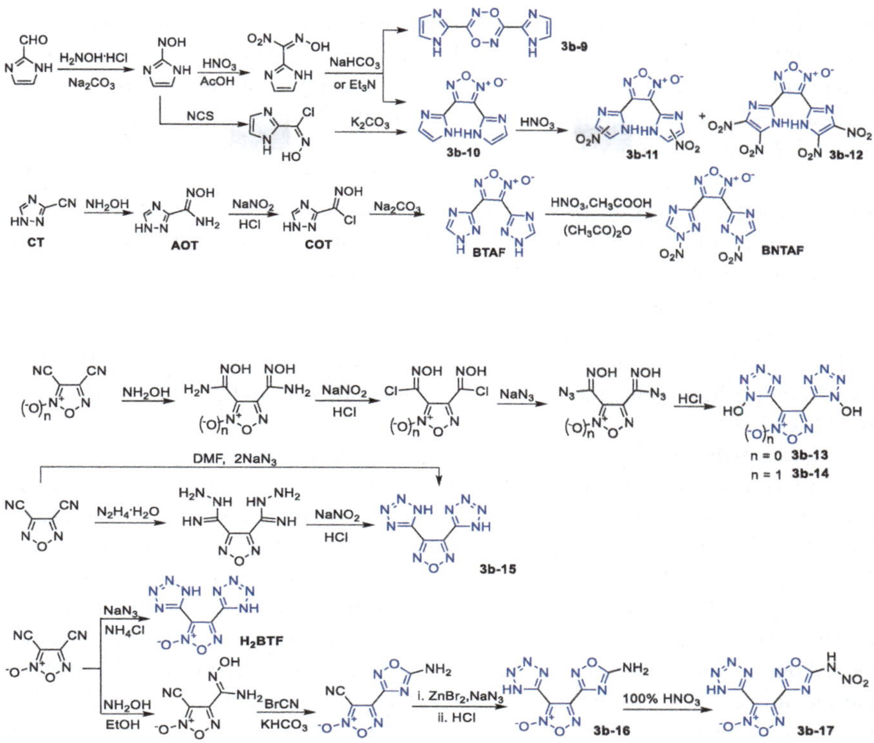

Scheme 3.6 Synthesis of tricyclic oxadiazole-imidazole, -triazole and -tetrazole derivatives

The cyclization reaction of hydroximoylazide derivative can proceed from chloroxime derivative to give a tetrazole ring. As shown in Scheme 3.6, hydroximoylamine-substituted oxadiazoles were obtained by exothermic addition of aqueous hydroxylamine in ethanol starting from nitriles. Then the diazotization of these hydroximoylamine derivatives in HCl results in the formation of chloroxime derivatives. Compounds **3b-13** and **3b-14** were obtained as slightly yellow sticky oils after the azidation of these chloroxime derivatives using sodium azide and then treatment with gaseous HCl starting from hydroximoyl chlorides [28]. As for these nitriles, the most common modification methodology is to constructe tetrazole ring by using of NaN_3. When the treatment 3,4-dicyanofuroxan or 3,4-dicyanofurazan with sodium azide in N,N-dimethylformamide (DMF), the correspondingly tetrazole products $\mathbf{H_2BTF}$ and **3b-15** can be obtained [29]. In addition, tetrazolylfurazan **3b-15** can also be prepared from 3,4-dicyanofurazan in two steps through intermediate amidrazone [30]. Also, the combination of tetrazolyl and (1,2,4-oxadiazol-5-yl)nitroamino group with furoxan was achieved through cyclization reaction by using of sodium azide and cyanogen bromide, respectively. As shown in Scheme 3.6, bishydroximoylamine-substituted furoxan was readily synthesized through the selective reaction of a cyano group of 3,4-dicyanofuroxan with an equivalent amount of aqueous hydroxylamine. Then the hydroximoylamine-substituted furoxan was reacted with cyanogen bromide in potassium bicarbonate solution to form 4-(5-amino-1,2,4-oxadiazol-3-yl)-3-cyano-1,2,5-oxadiazole 2-oxide. Subsequently, 4-(5-amino-1,2,4-oxadiazol-3-yl)-3-(1H-tetrazol-5-yl)-1,2,5-oxadiazole 2-oxide (**3b-16**) was formed by the cyclization reaction of 3-cyano-1,2,5-oxadiazole 2-oxide and sodium azide. And the further nitration of **3b-16** with 100% nitric acid as a nitration reagent was proceeded successfully to obtain the neutral nitroamine compound (**3b-17**) [31]. In Table 3.2, compounds **BTAF, BNTAF, H_2BTF** and **3b-17** containing furoxan backbone possess good detonation performance ($D > 7277$ m s^{-1}, $P > 20.1$ GPa) and acceptable sensitivity to impact ($IS > 7$ J) except **H_2BTF** ($IS < 2$ J). The decomposition temperatures of these compounds are in the range from 121 °C to 220 °C.

Apart from the nitrile groups on oxadiazole backbone, the nitrile groups on triazole and imidazole backbones can also be transformed to tetrazole ring by using sodium azide. As shown in Scheme 3.7, the 1,3-dipolar cycloaddition reaction of 4,5-dicyano-2H-1,2,3-triazole and sodium azide using zinc chloride as a catalyst yielded 4,5-bis-(1H-tetrazol-5-yl)-2H-1,2,3-triazole hydrate (**3b-18**). Similar to the synthetic route to compounds **3b-13**, the hydroxytetrazole **3b-19** can also be prepared from 4,5-dicyano-2H-1,2,3-triazole. It is noteworthy that the final cyclization reaction was carried out in concentrated hydrochloric acid, the hydroxytetrazole did not obtained and the amide groups formed instead [32]. Moreover, when 4,5-dicyano-2H-1,2,3-triazole was treated with sodium hydroxide, 2-nitro-4,5-dicyano-2H-1,2,3-triazole was obtained followed by adding bromonitromethane in a high yield. Then 5,5′-(2-(nitromethyl)-2H-1,2,3-triazole-4,5-diyl)bis(1H-tetrazole) (**3b-20**) was prepared using sodium azide in the presence of zinc chloride [33]. Also, the nitrile groups on imidazole backbone were converted easily to tetrazole rings in high yields using sodium azide and iodine in DMF to obtain 4,5-di(1H-tetrazol-5-yl)-1H-imidazol-2-amine (**3b-21**) [34]. As shown in Table 3.2, compounds **3b-(18–21)** with tetrazole rings possess high decomposition temperature (above 227 °C) and high heats of formation (above 750 kJ mol^{-1}) but a high sensitivity to impact ($IS < 10$ J).

The advantages of polynitropyrazole have been discussed in the section of biscyclic-based energetic material. Herein, a series of compounds containing 1,2,4-triazole and polynitropyrazoles were synthesized by J. M. Shreeve. As shown in Scheme 3.7, the substitution reaction of 4-amino-3,5-dichloro-triazole and a series of ammonium salts in DMF was proceeded to give the tricyclic mixed heterocyclic-based compounds containing polynitropyrazole and triazole (**3b-22**, **3b-27** and **3b-29**). In order to enhance the performance, the amino group of **3b-22** was converted to nitroamine group by the treatment of **3b-22** with 100% nitric acid to form 3,5-bis(3,5-dinitro-4-nitroamino-1H-pyrazol-1-yl)methyl)-1H-1,2,4-triazole (**3b-26**). And 4-amino-3,5-bis(3,4,5-trinitro-1H-pyrazol-1yl)methyl)-1,2,4-triazole (**3b-23**) was obtained through the oxidation of **3b-22** using 30% hydrogen peroxide in concentrated sulfuric acid with a reaction time below 12 h. However,

Table 3.2 Energetic properties of tricyclic mixed heterocyclic-based energetic compounds

Compound	T_d (°C)	D (m s^{-1})	P (GPa)	Δ_fH (kJ mol^{-1})	IS (J)	FS (N)	d (g cm^{-3})
Tricyclic tetrazine derivatives							
3b-1	329 (peak)	8002	28.5	957	> 70	/	1.81
3b-2	210	7592	25.0	1564	30	/	1.73
3b-4	209	8331	25.0	1289	1.5	/	1.62
3b-7	194	8365	26.8	2114	7.4	35	1.75
3b-8	189	8602	29.4	2820	5.3	29	1.76
Tricyclic oxadiazole derivatives							
BTAF	189	7277	20.1	420	38	8	1.75
BNTAF	139	8490	32.4	842	7	100	1.84
H_2BTF	220	7778	23.9	728	< 2	/	1.62
3b-17	121	8639	31.8	441	11	160	1.86
Tricyclic triazole-tetrazole and imidazole-tetrazole derivatives							
3b-18	277	8360	24.8	801	2	240	1.69
3b-19	246	8277	26.0	823	1	240	1.67
3b-20·H_2O	281	8636	29.5	946	10	240	1.79
3b-21	227	6330	16.9	750	/	/	1.65
Tricyclic triazole-pyrazole derivatives							
3b-22	287	8052	25.8	521	> 40	> 360	1.73
3b-23	213	8561	31.9	620	12	120	1.81
3b-24	218	8522	31.6	509	10	120	1.82
3b-25	134	8797	34.1	583	4	80	1.84
3b-26	122	8681	33.2	532	9	120	1.86
3b-27	309	7848	26.2	528	15	240	1.86
3b-28·H_2O	308	8329	27.9	1254	18	240	1.75
3b-29	181	7990	25.8	584	20	240	1.72
3b-30	354	8024	23.1	555	> 40	> 360	1.77
3b-31	238	8747	33.0	738	30	360	1.87
3b-32	134	9008	35.9	792	20	270	1.92
Tricyclic tetrazole derivatives							
3b-34	141	8122	27.5	682	> 40	/	1.69
3b-35	182	8185	30.0	1650	10	/	1.62

when the reaction time was increased to 24 h, another product 3,5-bis(3,4,5-trinitro-1Hpyrazol-1yl)methyl)-1H-1,2,4-triazole (**3b-24**) was obtained due to the cleavage of the N-NH_2 bond in acid conditions. Compound **3b-24** can also be synthesized by further nitration of **3b-23** in 100% nitric acid. When **3b-23** was nitrated with a mixture of 100% nitric acid and acetic anhydride, nitration of the 1,2,4-triazole

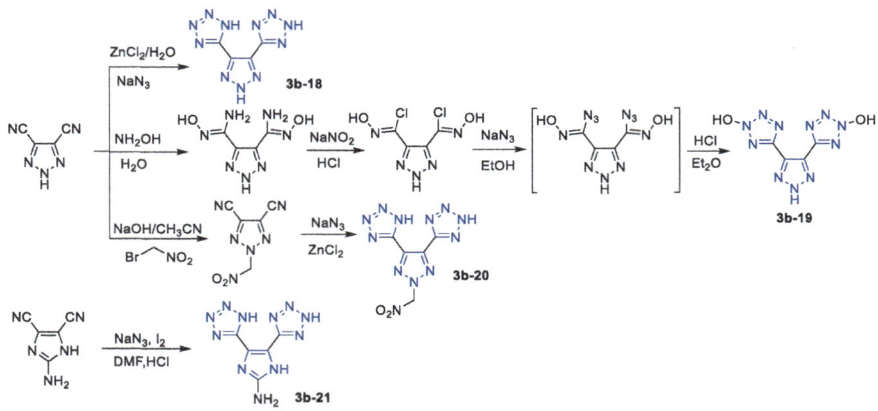

Scheme 3.7 Synthesis of tricyclic triazole-tetrazole and imidazole-tetrazole derivatives

ring with concomitant loss of the NH_2 moiety occurred to give 3,5-bis(3,4,5-trinitro-1H-pyrazol-1yl)methyl)-1-nitro-1,2,4-triazole (**3b-25**). This reaction process may include two steps, the cleavage of N-NH_2 bond in acid conditions and then the nitration of NH in the triazole ring. This is also confirmed by the nitration of compound **3b-24** in the mixture of 100% nitric acid and acetic anhydride into product **3b-25** similarly. The neutral compound **3b-28** was isolated as the monohydrate from the cyclization of **3b-27** using hydrazine in methanol and then acidification with diluted HCl [35].

According to above mentioned synthetic route to **2c-8** by condensation of carboxyl compounds and aminoguanidine hydrochloride, this cyclization reaction was also attempted using other substrates to construct 3-amino-1,2,4-triazole by Yang's group [36]. As shown in Scheme 3.8, 4-nitro-1*H*-pyrazole-3,5-dicarboxylate was prepared by a three-step reaction (nitration, oxidation and esterification) from 3,5-dimethylpyrazole. Then compound 3,5-bis(1*H*-1,2,4-triazol-3-amine)-1*H*-pyrazole (**3b-30**) was successfully synthesized through a one-pot cyclization of the carboxylate and aminoguanidinium sulphate using NaOMe as a nucleophilic reagent. The further oxidation product 4-nitro-3,5-bis(1H-1,2,4-triazol-3-nitro)-1H-pyrazole (**3b-31**) and nitration product 4-nitro-3,5-bis(1H-1,2,4-triazol-3-nitramino)-1H-pyrazole (**3b-32**) were acquired by Sandmeyer reaction using sodium nitrite in sulfuric acid and nitration in the mixture of 100% HNO_3 and 98% H_2SO_4 from **3b-30**, respectively.

These compounds (**3b-(22–32)**) provide detonation pressures and velocities in the range 23.1–35.9 GPa and 7848–9008 m s^{-1}, respectively. They also possess moderate sensitivities with IS values above 9 J and FS values above 120 N except compound **3b-25** (IS = 4 J, FS = 80 N). This result demonstrates that the incorporation of triazole with pyrazole enhances the balance of energy and safety of these polynitropyrazoles.

Similar to the synthetic route to **2b-6**, the tricyclic tetrazole-based compound **3b-33** was prepared from triamine using cyanogen azide (Scheme 3.9). Further introduction of trinitroethyl was carried out through the condensation of aminotetrazoles

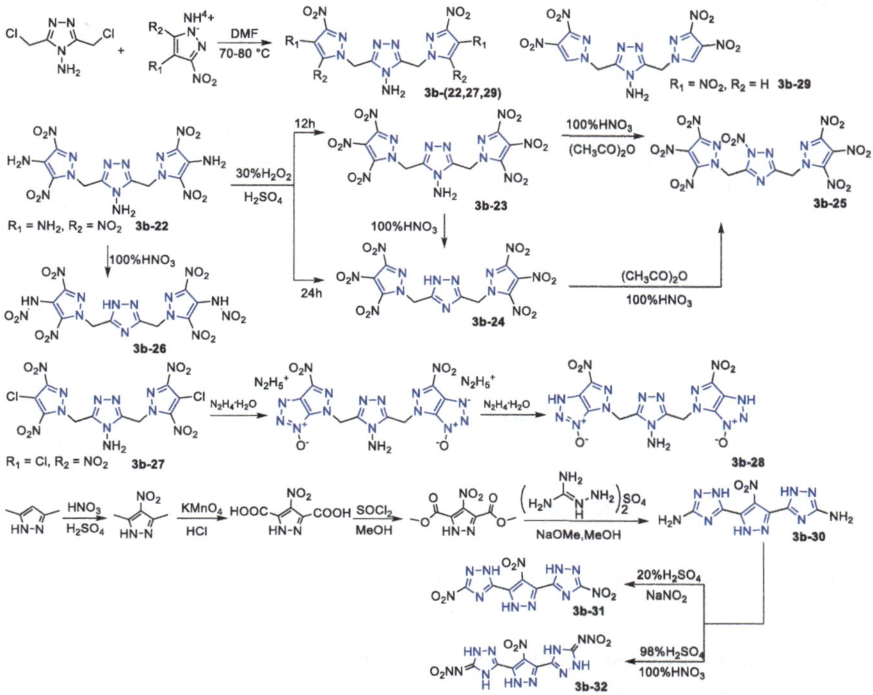

Scheme 3.8 Synthesis of tricyclic triazole-pyrazole derivatives

with formaldehyde followed by the reactions of 5-(1-hydroxymethyl)-aminotetrazole intermediate with nitroform to obtain trinitroethyl compound **3b-34** [37]. And nitramino compound **3b-35** was also synthesized by treatment of triamine with 100% nitric acid [38]. Another tricyclic tetrazole-based compound **3b-36** was prepared through the reaction of tricarbonitrile and sodium azide in the presence of zinc chlomide directly [39].

3.3 Experimental Section

3,4-Bis(4-amino-1,2,5-oxadiazol-3-yl)-1,2,5-oxadiazole N-oxide (**BAFF**): A suspension of 0.01 mol of the compound in 20 mL of diethyl ether was prepared. The suspension was cooled to a temperature range of 0–3 °C. Under vigorous stirring, a 3% aqueous solution of Na_2CO_3 was slowly added to the suspension while maintaining the temperature within the range of 0–3 °C. The mixture was kept at a temperature below 10 °C for 2 h. Subsequently, the solvent was removed, and the solid residue was washed with a small amount of cold water. Finally, the residue was dried in air. The overall yield of the compound obtained was 70%.

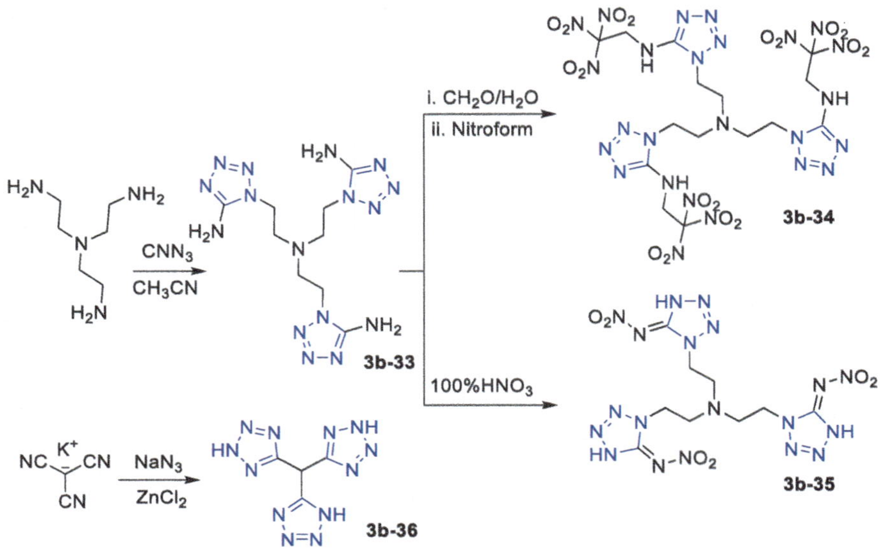

Scheme 3.9 Synthesis of tricyclic tetrazole derivative

T_m, 167–168 °C. IR (KBr) 3450, 3330, 1650, 1620, 1470, 1160, 1110, 1050, 970, 890, 850. ^1H NMR: 6.6 s ppm. ^{13}C NMR: 104.33, 133.41, 136.29, 146.66, 155.20, 156.12 ppm. Found, %:C 28.41; H 1.75; N 44.56. M^+. 252. $C_6H_4N_8O_4$. Calculated, %: C 28.57; H 1.58; N 44.44. M 252.

Bis(aminofurazano)furazan (**BAFF^R**): A solution of **BAFF** (0.25 g, 1.0 mmol) in 10 mL of methanol was prepared. To the solution, 10% Pd/C (0.06 g) was added. The mixture was then stirred for 30 min under a hydrogen atmosphere. Afterwards, the suspension was filtered through Celites and washed with diethyl ether. The resulting filtrate was concentrated under vacuum, resulting in a crude product that appeared as a brown liquid. To purify the crude product, flash silica gel chromatography was employed using a solvent mixture of THF and DCM in a ratio of 5:18. This process yielded a white solid (0.142 g, 60.7% yield). T_m, 178–180 °C; IR (KBr) 3458, 3351, 1643, 1609, 1533, 1264, 974 cm^{-1}; ^1H NMR (acetone-d_6, 400 MHz) δ 8.44; ^{13}C NMR (acetone-d_6, 100 MHz) δ 155.6, 142.4, 140.1; ESI–MS (pos.): 237.93 ($[M+2H]^{2+}$, C6H4N8O3$^+$; calc. 236.04).

3,4-Bis(nitrofurazano)furoxan (**BNFF**): A solution of 60% H_2O_2 (3.3 mL, 0.07 mol) in 26 mL of methylene chloride was prepared. Trifluoroacetic anhydride (7.4 mL, 0.05 mol) was added slowly to the solution while maintaining a temperature below 5 °C. Then, BAFF (1.0 g, 4.0 mmol) was added gradually. The mixture was stirred at room temperature for 1 h and then heated to reflux for 3 h. After the reaction, the mixture was washed twice with water and then twice with a 10% $NaHCO_3$ solution. The organic layer was dried over $MgSO_4$, filtered, and concentrated under reduced pressure. The resulting residue was subjected to purification using flash silica gel chromatography with a solvent mixture of methylene chloride and hexane in a 1:1 ratio. This process yielded a white solid (0.63 g, 51% yield). T_m, 108–110 °C (T_d); ^{13}C NMR (DMSO-d_6, 100 MHz) δ 160.2, 160.0, 143.0, 139.5, 137.0, 103.5.

Bis(nitrofurazano)furazan (**BNFF^R**): A solution of **BNFF** (0.3 g, 1.0 mmol) in 10 mL of methanol was prepared. To this solution, 10% Pd/C (0.06 g) was added. The mixture was stirred for 30 min under a hydrogen atmosphere. After the stirring period, the suspension was filtered through Celites and washed with diethyl ether. The resulting filtrate was concentrated under vacuum, resulting in a crude product that appeared as a brown liquid. To purify the crude product, flash silica gel chromatography was conducted using a solvent mixture of THF and DCM in a ratio of 5:18. This process yielded a white solid (0.138 g, 48.6% yield). T_m, 160–162 °C (melt); IR (KBr) 2360, 2340, 1579, 1547, 1384, 1357, 1196 cm^{-1}; ^{13}C NMR (Acetone-d_6, 100 MHz) δ 160.8, 141.8, 139.5; ESI–MS (pos.): 320.16 ($[M+Na]^+$, $C_6N_8O_7^+$; calc. 296.02.

3,6-Bis(4-nitroamino-1,2,5-oxadiazol-3-yl)-1,4,2,5-dioxadiazine (**3a-2**): In a reaction flask, 2 g (8 mmol) of diamine was added in portions under vigorous stirring to 20 mL of concentrated nitric acid. The nitric acid solution was cooled to a temperature range of – 5 to – 15 °C, and the addition of the diamine was done while maintaining the temperature below – 5 °C. The mixture was stirred for 10 min at – 5 °C. The reaction mixture was then allowed to slowly warm up to room temperature. Subsequently, it was diluted with 70–100 mL of trifluoroacetic acid. The resulting precipitate was filtered off, thoroughly washed with trifluoroacetic acid, and dried in air. This process yielded colorless crystals with a yield of 2.6 g (95%). T_m = 95 °C (with vigorous decomposition). IR spectrum, 3280, 1696, 1644, 1632, 1612, 1532, 1440, 1312, 1140, 1120, 1052, 1032, 1024, 1016, 944. ^1H NMR spectrum ($CDCl_3$-DMSO-d_6), δ, ppm: 10.25 (NH). Found, %: C 20.84; H 1.00; N 41.05. $C_6H_2N_{10}O_8$. Calculated, %: C 21.06; H 0.59; N 40.94.

4″-Amino-4-((2,2,2-trinitroethyl)amino)-[3,3′:4′,3″-ter(1,2,5-oxadiazole)] 2′-oxide (**3a-6**) and 4,4″-bis((2,2,2-trinitroethyl)amino)-[3,3′:4′,3″-ter(1,2,5-oxadiazole)] 2′-oxide (**3a-4**): First, 3,4-bis(aminofurazano)furoxan (0.504 g, 2.0 mmol) was dissolved in 50 mL of methanol at ambient temperature. Then, a solution of 2,2,2-trinitroethanol (0.905 g, 5.0 mmol) was added to the reaction solution. The resulting mixture was refluxed for 1 h. After refluxing, the solvent was removed under vacuum, leaving behind a crude residue. This residue was then purified using flash column chromatography on silica gel, employing a solvent mixture of 1:2 CH_2Cl_2 to petroleum ether. Through this purification process, 0.250 g of pure product **3a-6** was obtained as a white solid with a yield of 30.1%. Additionally, 0.786 g of product **3a-4** was separated with a yield of 68.0%.

3a-6, ^1H NMR (300 MHz, Acetone-d_6): δ = 7.01 (t, J = 6.0 Hz, 1H, –NH–), 6.18 (s, $–NH_2$), δ = 5.47 (d, J = 6.0 Hz, 2H, $–CH_2–$) ppm; ^{13}C NMR (75 MHz, Acetone-d_6): δ = 155.76, 154.70, 146.44, 135.48, 132.88, 125.50, 103.65, 47.58 ppm; IR: 3499, 3390, 3322, 2921, 1574, 1462, 1302, 1184, 1150, 1113, 993, 963, 872, 822, 803, 779, 695 cm^{-1}; elemental analysis calcd. (%) for $C_8H_5N_{11}O_{10}$ (415.19): calcd. C 23.14, H 1.21, N 37.11; found: C 23.24, H 1.16, N 38.3.

3a-4, ^1H NMR (300 MHz, Acetone-d_6): δ = 6.96 (t, J = 6.0 Hz, 1H, –NH–), 6.76 (t, J = 6.0 Hz, 1H, –NH–), δ = 5.57 (d, J = 6.0 Hz, 2H, $–CH_2–$), δ = 5.47

(d, J = 6.0 Hz, 2H, –CH_2–) ppm; ^{13}C NMR (75 MHz, Acetone-d_6): δ = 155.66, 154.81, 145.86, 135.50, 132.59, 125.42, 103.91, 47.77, 47.53 ppm; IR: ῡ = 3409, 3349, 1588, 1566, 1478, 1428, 1298, 1274, 1113, 1028, 998, 837, 806, 781, 696, 660, 635 cm^{-1}; elemental analysis calcd. (%) for $C_{10}H_6N_{14}O_{16}$ (578.24): calcd. C 20.77, H 1.05, N 33.91; found: C 20.60, H 1.14, N 33.53.

4,4″-Bis(nitro(2,2,2-trinitroethyl)amino)-[3,3′:4′,3″-ter(1,2,5-oxadiazole)] 2′-oxide (**3a-5**): A nitration reagent consisting of 100% nitric acid and acetic anhydride (20 mL: 16 mL) was vigorously stirred. To this mixture, **3a-4** (0.578 g, 1.0 mmol) was added at 0 °C. The temperature was strictly controlled to remain below 5 °C throughout the addition process. After the addition of acetic anhydride was complete, the reaction was allowed to proceed at room temperature. It was important to monitor the reaction temperature and provide timely cooling if the temperature rose sharply during the nitrification process. The reaction should continue for two hours until the temperature stabilizes. At the end of the reaction, the system was quenched by lowering the temperature with crushed ice. This caused a white solid to precipitate out. The precipitate was then washed three times with water (30 mL each) and dried in the air. The final product, **3a-5**, was obtained as a yellow solid with a yield of 0.603 g (90.2%). ^{1}H NMR (300 MHz, Acetone-d_6): δ = 6.67 (s, 2H, –CH_2–) ppm; ^{13}C NMR (75 MHz, Acetone-d_6): δ = 151.74, 151.42, 143.53, 142.50, 139.16, 123.16, 104.70, 53.40, 52.88 ppm; IR: 3009, 1589, 1546, 1464, 1398, 1350, 1276, 1192, 1113, 1042, 1022, 929, 854, 806, 775, 728, 625, 607 cm^{-1}; elemental analysis calcd. (%) for $C_{10}H_4N_{16}O_{20}$ (668.23): calcd. C 17.97, H 0.60, N 33.54; found: C 17.87, H 0.69, N 33.43.

3-[4-(1,2,4-Oxadiazol-3-yl)-2-oxido-1,2,5-oxadiazol-3-yl]-1,2,4-oxadiazole (**3a-7**). A suspension of 1,2,4-oxadiazol-3-carbohydroximoyl chloride (0.295 g, 2 mmol) in diethyl ether (10 mL) was prepared. Dropwise addition of K_2CO_3 (0.157 g, 1.13 mmol) in water (4 mL) to the suspension took place over a period of 30 min. After stirring for 2 h at 10 °C, the resulting solution was extracted with ethyl ether twice, using 20 mL each time. The combined organic layer was then dried over $MgSO_4$ and concentrated under vacuum. As a result, a white solid (0.098 g, 44% yield) was obtained.^{1}H NMR (600 MHz, $CDCl_3$): 8.99 (1H, s), 9.03 (1H, s) ppm; ^{13}C NMR (151 MHz, $CDCl_3$): 105.59, 144.68, 156.13, 158.23, 166.06, 166.10 ppm; IR (KBr): ν = 3118, 1610, 1573, 1552, 1532, 1440, 1378, 1367, 1275, 1102, 1093, 1051, 1022, 956, 944, 907, 895, 823, 613 cm^{-1}; elemental analysis calcd (%) for $C_7H_4N_6O_3$ (220.15): C 38.19, H 1.83, N 38.18, found: C 38.06, H 1.85, N 38.08.

3,4-Bis-(5-amino-1,2,4-oxadiazolyl) furoxan (**3a-8**): An aqueous solution of sodium hydroxide (0.72 g, 17.9 mmol) was prepared. Into this solution, 3,4-bis-aminooxime furoxan (1.44 g, 7.1 mmol) was added. The resulting mixture was heated under reflux for 10 min, leading to the formation of a clear solution. After cooling to 40 °C, cyanogen bromide (1.89 g, 17.85 mmol) was added in portions. The reaction mixture was stirred overnight at room temperature. Subsequently, the mixture was acidified by adding 2 M hydrochloric acid until reaching a pH of 2. The resulting precipitate was filtered off and thoroughly washed with water. This process yielded 0.93 g (54% yield) of product **3a-8**, which appeared as a brownish powder. DSC (5 °C min^{-1}): 273 °C (T_d). IR (ATR): 3472 (w), 3340 (w), 3249 (w), 3177 (w), 1662 (s), 1608 (s), 1566 (m), 1529 (m), 1449 (m), 1416 (m), 1359 (w), 1259 (w), 1220 (w), 1112 (m), 1073 (w), 1050 (m), 1008 (w), 976 (w), 946 (m), 910 (w), 824 (s), 761 (m), 724 (m), 683 (m). Raman (1064 nm, 500 mV, 25 °C) 2859 (20), 1661 (19), 1606 (7), 1569 (77), 1530 (24), 1509 (42), 1455 (8), 1425 (25), 1378 (22), 1220 (58), 1108 (16), 1099 (3), 1010 (8), 977 (59), 911 (11), 827 (13), 757 (42), 546 (6), 498 (12), 464 (11), 421 (18), 362 (28), 314 (3), 304 (4), 283 (9), 272 (3), 244 (2), 214 (21), 126 (100), 95 (8). ^{1}H NMR (DMSO-d$_6$) $\delta = 8.4$ (br, NH_2), 8.3 (br, NH_2). ^{13}CNMR (DMSO-d$_6$) $\delta = 173.0$ (CNH_2), 159.6 (CNH_2), 156.9 (C_{ring}), 155.9 (C_{ring}), 144.7 (CNO), 106.7(CN(O)O). EA (C6H4N8O4, 252.15 g mol^{-1}) Calc.: C 28.58, H 1.60, N 44.44%, found: C 27.73, H 2.03, N 42.28%.

3,4-Bis(5-Methyl-1,2,4-oxadiazol-3-yl)furazan (**3a-9**): A mixture of furazan-3,4-dicarboxamidoxime (1.86 g, 0.01 mol) and acetic anhydride (25.5 g, 0.25 mol) was refluxed for 6 h. After this reflux period, the mixture was poured into 150 mL of water. The excess acetic anhydride in the solution was hydrolyzed. The resulting solution was then cooled using an ice bath, while sodium hydroxide (16.0 g, 0.40 mol) was added with stirring. The mixture was stored at 0 °C overnight. The product was collected by filtration and subjected to recrystallization twice using a mixture of methanol and water. This process yielded 0.52 g (22% yield) of the compound **3a-9**. T_m, 79 °C. Anal. Calcd. for $C_8H_6N_6O_3$: C, 41.03; H, 2.58; N, 35.89. Found: C, 40.63; H, 2.61; N, 35.59.

3,4-Bis(4′-aminofuroxano)furoxan (**3a-12**): A suspension of 1.0 g (6 mmol) of 4-aminofuroxan-3-carboxamide oxime in 8 mL of acetonitrile was prepared. To this suspension, 1.1 g (9 mmol) of dimethylformamide dimethyl acetal was added. The initial product dissolved, and after a short period of time, a precipitate started to form. After 30 min, the precipitate was filtered off. This resulted in the isolation of 1.04 g (77% yield) of the desired product, 4-dimethylaminomethyleneaminofuroxan-3-carboxamide oxime. T_m, 214–215 °C (from a 1:1 mixture of ethanol and acetonitrile). NMR spectrum (δ, ppm): 2.98 and 3.13 [3H each, s, $(CH_3)_2N$]; 6.04 (2H, c, NH2); 8.24 (1H, s, CH); 10.2 (1H, s, OH). IR spectrum, cm^{-1}: 3455 and 3318 (NH_2), 3200 (OH), 1640 (C = N), 1605 (furoxan).

A suspension of 4-dimethylaminomethyleneaminofuroxan-3-carboxamide oxime (1.07 g, 5 mmol) in water (6 mL) was prepared. Concentrated HCl (6 mL) was then added to the suspension. While maintaining the reaction temperature between − 5 and 0 °C, a solution of sodium nitrite (0.52 g, 7.5 mmol) in water (5 mL) was added dropwise to the reaction mixture. The solution was stirred for one hour and allowed to reach ambient temperature over another hour. The resulting white precipitate was collected by filtration and air dried. This yielded 1.01 g (86.7% yield) of the product, which is 3-chlorohydroximoyl-4-dimethylaminomethyleneamino-furoxan. T_d 144.6 °C; IR (cm^{-1}) $\nu = 3262$, 2954, 2920, 2820, 1629, 1605, 1581, 1526, 1478, 1432, 1406, 1362, 1292, 1254, 1113, 1077, 1018, 1981, 981, 918, 837, 751, 719, 603, 588; ^{1}H NMR (DMSO-d$_6$): 13.40 (1H, s, N–OH), 8.27 (1H, s, N = CH), 3.44 (3H, s, –CH_3), 3.11 (3H, s, –CH_3); ^{13}C NMR (DMSO-d$_6$): δ 160.23, 156.52, 120.75, 108.61, 40.21, 34.18. EA ($C_6H_8Cl_1N_5O_3$, 223.61): calcd, C, 30.85; H, 3.45; N, 29.98; found, C: 30.82; H: 3.43; N: 29.98.

3-Chlorohydroximoyl-4-dimethylaminomethyleneamino-furoxan (0.94 g, 4 mmol) was suspended in diethyl ether (20 mL). A solution of 3% potassium carbonate (2 mmol) was added to the suspension at a temperature of 0–5 °C. The reaction mixture was stirred for 2 h at 0 °C, followed by stirring overnight at ambient temperature. The diethyl ether solvent was then removed by blowing air, resulting in the formation of a white solid. The white solid was filtered and washed with 10% aqueous ammonia (5 mL). Afterward, it was air dried, yielding 0.53 g (67.3% yield) of the 3,4-bis(4′-dimethylaminomethyleneaminofuroxano-3′) furoxan. T_{dec}: 170.2 °C; IR (cm^{-1}) $\nu = 3153$, 2929, 1630, 1518, 1480, 1433, 1390, 1362, 1336, 1254, 1209, 1110, 1064, 987, 956, 931, 894, 835, 769, 754, 728, 680, 582. ^1H NMR (CD_3CN): 8.07 (2H, s, CH); 3.05 (3H, s, $-CH_3$); 3.04 (3H, s, $-CH_3$); 2.80 (3H, d, $-CH_3$); 2.77 (3H, d, $-CH_3$); ^{13}C NMR (CD_3CN): δ 161.66, 161.57, 157.01, 156.97, 144.15, 105.64, 104.11, 103.58, 41.41, 41.37, 34.74, 34.67. EA ($C_{12}H_{14}N_{10}O_6$, 394.30): calcd, C, 36.55; H, 3.58; N, 35.52; found, C: 36.70; H: 3.60; N: 35.16.

First, 3,4-Bis(4′-dimethylaminomethyleneaminofuroxano-3′) furoxan (0.20 g, 0.5 mmol) was added to a mixture of concentrated HCl (1 mL) and water (3 mL). The resulting suspension was stirred at 40 °C for 8 h. After the reaction, the solid product was collected by filtration and washed with ice water (2 mL). The collected solid was then subjected to recrystallization from acetonitrile, resulting in the formation of a white solid, which is referred to as **3a-12**. The yield of the compound was determined to be 0.08 g (55.6%). T_d: 183.6 °C; IR (cm^{-1}) $\nu = 3444$, 3411, 3333, 1624, 1587, 1560, 1523, 1464, 1432, 1401, 1264, 1213, 1185, 1111, 1056, 963, 935, 827, 765, 731, 697, 649. ^1H NMR (CD_3CN): 5.45 (2H, s, NH2); 5.15 (2H, s, NH2); ^{13}C NMR (CD_3CN): δ 156.63, 156.03, 144.66, 104.28, 103.58, 100.38 ppm. ^{15}N NMR (ppm): − 7.58; − 23. 83; − 33.94; − 34.10; − 50.68; − 51.47; − 292.45; − 329.85 ppm. EA ($C_6H_4N_8O_6$, 284.15): calcd, C, 25.36; H, 1.42; N, 39.44; found, C: 25.14; H: 1.23; N: 37.37.

Bis-3,4-(3-nitrofurazan-4-oxy)furazan (**3a-13**): In a dry atmosphere at 30 °C, a solution of salt (1,2,5-oxadiazolidine-3,4-dione, disodium salt) weighing 2.92 g (20 mmol) was dissolved in glyme (25 ml). Then, a solution of 3,4-dinitrofurazan weighing 1.6 g (10 mmol) in glyme (25 ml) was added to the above mixture. The resulting mixture was stirred at 30 °C for 0.5–1 h. TLC (thin-layer chromatography) was used to monitor the progress of the reaction. After cooling, CH_2Cl_2 (250 ml) was added to the mixture. The resulting mixture was washed with water, dried using $MgSO_4$, filtered, and then evaporated. The residue obtained was a mixture. To separate the product, silica gel flash chromatography was performed using CH_2Cl_2/pentane as the eluent. The first fraction collected contained difurazano[3,4-b;3′,4′-e]dioxine. The second fraction contained 3,3-dinitrodifurazanyl ether. The third fraction yielded 1.41 g (43%) of colorless crystals identified as **3a-13**. T_m 38–40 °C; IR 1600, 1560, 1490, 1365, 1265, 1205, 1190, 1040 (cm^{-1}). Anal. Calcd for $C_6N_8O_9$ (328.11): C, 21.96; N, 34.15. Found: C, 21.89; N, 34.09 (%).

Bis-3,4-(3-cyanofurazan-4-oxy)furazan (**3a-14**): In a dry atmosphere at 30 °C, a solution of salt (1,2,5-oxadiazolidine-3,4-dione, disodium salt) weighing 2.92 g

(20 mmol) was dissolved in glyme (25 ml). Then, a solution of 3-cyano-4-nitrofurazan weighing 1.4 g (10 mmol) in glyme (25 ml) was added to the above mixture. The resulting mixture was stirred at 30 °C for 0.5–1 h. TLC (thin-layer chromatography) was used to monitor the progress of the reaction. After cooling, CH_2Cl_2 (250 ml) was added to the mixture. The resulting mixture was washed with water, dried using $MgSO_4$, filtered, and then evaporated. The sole isolable product, diether **3a-14**, was obtained as colorless crystals weighing 2.56 g. (89% yield). m.p. 70–72 °C; IR 2210, 1615, 1590, 1575, 1510, 1470, 1410, 1290, 1270, 1240, 1035 (cm^{-1}); ^{13}C NMR ($CDCl_3$): δ = 162.1 (NC–C–C), 153.8 (O–C–C–O), 128.7 (NC–C), 106.3 (CN). –MS m/z 288 (M^+), 258 (M^+-NO). Anal. Calcd for $C_8N_8O_5$ (288.14): C, 33.35; N, 38.89. Found: C, 33.21; N 38.76 (%).

3,4-Bis(3-chlorodinitromethylfurazan-4-oxy)furazan (**3a-15**): To a mixture of water (55 mL) and isopropanol (25 mL), **3a-14** (5.76 g, 20 mmol) was added. Then, sodium carbonate anhydrous (2.23 g, 21 mmol) and hydroxylamine hydrochloride (2.92 g, 42 mmol) were added at ambient temperature. The reaction mixture was stirred for an additional 2 h at 50 °C. The resulting precipitate was filtered, washed with ice-cold water, and dried under vacuum over P_2O_5 to obtain 3,4-bis(3-aminoximidofurazan-4-oxy)furazan as a white solid. The yield of the product was 6.74 g (91.2%). T_m 219–220 °C; ^{1}H NMR (DMSO-d_6, 500 MHz) δ 10.61 (s, 2H, OH), 6.35 (s, 4H, NH_2); ^{13}C NMR (DMSO-d_6, 125 MHz) δ 159.25, 154.20, 142.29, 141.19; IR (cm^{-1}, KBr) 3503, 3487, 3382, 3276, 2893, 1675, 1595, 1576, 1553, 1523, 1439, 1254, 1175, 1034, 967; anal. calcd for $C_8H_6N_{10}O_7$ (%): C 27.13, N 39.55, H 1.71, found C 26.64, N 39.28, H 1.70.

In a mixture of concentrated hydrochloric acid (55 mL) and water (30 mL), 3,4-bis(3-aminoximidofurazan-4-oxy)furazan (5.31 g, 15 mmol) was dissolved. Then, saturated sodium nitrite (2.16 g, 31 mmol) in water (12 mL) was added dropwise to the stirred solution of compound **3a-14** at 0 °C. The reaction mixture was stirred for 2 h and then heated to 20 °C for an additional 1.5 h until N_2 evolution ceased. The resulting white precipitate was filtered, washed twice with cold water, and dried under vacuum. This yielded 3,4-bis(3-chloroximidofurazan-4-oxy)furazan as a white solid, weighing 5.40 g (86.3%). T_m 82–83 °C; ^{1}H NMR (DMSO-d_6, 500 MHz) δ 13.74 (s, 2H, OH); ^{13}C NMR (DMSO-d_6, 125 MHz) δ 158.26, 153.41, 143.23, 123.30; IR (cm^{-1}, KBr) 3583, 3505, 3346, 2832, 2759, 1701, 1605, 1589, 1560, 1528, 1512, 1423, 1301, 1252, 1028, 943; anal. calcd for $C_8H_2N_8O_7Cl_2$ (%): C 24.45, N 28.51, H 0.51, found C 24.23, N 28.19, H 0.85.

A suspension of 3,4-bis(3-chloroximidofurazan-4-oxy)furazan (0.75 g, 1.9 mmol) in $CHCl_3$ (50 mL) at 15 °C was prepared. N_2O_5 (1.63 g, 23 mmol) was then added to the suspension. The mixture was heated to 45 °C and maintained at this temperature for 40 min. Afterward, the solvent was evaporated, leaving behind a residue. The residue was subjected to column chromatography on silica gel using ethyl acetate-pentane (v/v = 1:6) as the eluent. This process allowed for the isolation of pure colorless crystals identified as compound **3a-15**. The yield of the product was 0.27 g (28.0%). mp 79–80 °C; ^{13}C NMR (DMSO-d_6, 125 MHz) δ 159.45, 152.73, 139.27, 113.06; IR (cm^{-1}, KBr) 1602, 1584, 1544, 1501, 1275, 1234, 1030, 981, 816, 786, 621; anal. calcd for $C_8N_{10}O_{13}Cl_2$ (%): C 18.66, N 27.19, found C 18.40, N 26.95.

3-(3-Nitrofurazan-4-azo)-4-(3-nitrofurazan-4-oxy)furazan (**3a-16**): In a dry atmosphere at 50 °C, a solution of Na-salt 4-hydroxy-4′-nitroazofurazan (2.49 g, 10 mmol) in diglyme (30 ml) was prepared. Then, a solution of 3,4-dinitrofurazan (1.6 g, 10 mmol) in glyme (15 ml) was added to the above mixture. The resulting mixture was stirred for 3–4 h at 70 °C. TLC (thin-layer chromatography) was used to monitor the progress of the reaction. After cooling, CH_2Cl_2 (350 ml) was added to the mixture. The resulting mixture was washed with water, dried using $MgSO_4$, filtered, and then evaporated. The residue obtained was a mixture of compound **3a-16** and 3,3-dinitrodifurazanyl ether. To separate the products, silica gel flash chromatography was performed using CH_2Cl_2/CCl_4 (1:1) as the eluent. The first fraction collected contained 3,3-dinitrodifurazanyl ether. The second fraction yielded an orange oil identified as compound **3a-16** with a 51% yield. MS m/z 340 (M^+), 294 ($M^+–NO_2$); IR 1650, 1585, 1340, 1315, 1210, 1010, 980 (cm^{-1}); Anal. Calcd for $C_6N_{10}O_8$ (340.13): C, 21.19; N, 41.18. Found: C, 21.26; N, 41.05 (%).

3-(3-Nitrofurazan-4-NNO-azoxy-4-(3-nitrofurazan-4-oxy)furazan (**3a-17**) and 3-(3-Nitrofurazan-4-ONN-azoxy-4-(3-nitrofurazan-4-oxy)furazan (**3a-18**): A mixture of $(NH_4)_2S_2O_8$ (11.4 g, 50 mmol) and 20% oleum (30 ml) was prepared. To this mixture, **3a-16** (0.34 g, 1 mmol) was added all at once with stirring. The resulting orange suspension was stirred at 70 °C for 2 h, then cooled to 0 °C and poured over crushed ice. The mixture was extracted with CH_2Cl_2 (4 × 25 ml), and the combined organic layers were subsequently washed with water, dried using $MgSO_4$, and concentrated. The products were isolated through silica gel chromatography, using a mixture of CH_2Cl_2 and pentane as eluants. It was observed that under these conditions, a greater amount of **3a-17** was formed, with only a small amount of **3a-18** present. The quantity of **3a-18** was not sufficient for complete characterization. The remaining oil was recrystallized from hexane, resulting in the isolation of **3a-17** (0.29 g, 82% yield). mp 41–42 °C; MS m/z 356 (M^+), IR: 1650, 1585, 1340, 1315, 1210, 1010, 980 (cm^{-1}); Anal. Calcd for $C_6N_{10}O_9$ (356.13): C, 20.24; N, 39.33. Found: C, 20.30; N, 39.19 (%).

3,4-Bis(3-nitrofurazan-4-azo-)furazan (**3a-19**): A suspension of 4-amino-3-nitrofurazan (1.3 g, 10 mmol) and 4-amino-4′-nitroazofurazan (2.26 g, 10 mmol) in concentrated hydrochloric acid (50 mL) was prepared. With stirring, a 0.1N aqueous solution of $KMnO_4$ (250 ml) was added dropwise to the suspension at 45–50 °C. The resulting mixture was stirred for 2 h. The resulting black suspension was extracted with $CHCl_3$ (4 × 25 ml), and the combined orange organic layers were washed with water, dried using $MgSO_4$, and then concentrated. The residue obtained was a mixture of compound **3a-19** along with 4,4′-dinitroazofurazan and 4,4′-bis(4-nitrofurazanyl-3-azo)azo-furazan. To separate the products, silica gel flash chromatography was performed using CH_2Cl_2/pentane as the eluent. The first fraction collected contained 4,4′-dinitroazofurazan, the second fraction contained **3a-19**, and the third fraction contained 4,4′-bis(4-nitrofurazanyl-3-azo)azo-furazan. **3a-19,** MS m/z 352 (M^+). IR 1580, 1350, 1245, 1180, 1040, 830 (cm^{-1}); Anal. Calcd for $C_6N_{12}O_7$ (352.14): C, 20.47; N, 47.73. Found: C, 20.56; N, 47,69 (%).

3-(4-Nitrofurazan-3-NNO-azoxy)-4-(4-nitrofurazan-3-ONN-azoxy)-furazan (**3a-20**): A mixture of $(NH_4)_2S_2O_8$ (22.8 g, 0.1 mol) and **3a-19** (0.352 g, 1 mmol) in 20% oleum (50 ml) was stirred at 80–90 °C for 0.5 h. After cooling to 0 °C and pouring over crushed ice, the mixture was extracted with $CHCl_3$ (4 × 25 ml). The combined organic layers were then washed with water, dried using $MgSO_4$, and concentrated. The residue obtained from the concentration step was a mixture. To isolate the desired product, silica gel flash chromatography was performed using CH_2Cl_2/pentane as the eluent. MS m/z 384 (M^+), 338 (M^+–NO_2), 292 (M^+–$2NO_2$), 278, 254 (M^+-O-nitrofurazanyl), 238 (M^+-2O-nitrofurazanyl), 226, 210, 196, 180, 162, 158 (nitrofurazan), 144, 142 (nitrofurazan), 122, 68; IR 1580, 1510, 1350, 1300, 1215, 1160, 1130, 1030, 980, 940, 820 (cm^{-1}); Anal. Calcd for $C_6N_{12}O_9$ (384.14): C, 18.76; N, 43.75. Found: C, 18.77; N, 43.66 (%).

1,2,3-Tris-(3-methylfuroxan-4-methyleneamino)guanidine (**3a-21**). In a solution of triaminoguanidium nitrate (0.835 g, 5.0 mmol) and 3-methyl-4-furoxancarbaldehyde (3.84 g, 30.0 mmol) in 30 mL of ethanol, refluxing was carried out for 24 h. Afterward, the reaction mixture was cooled to room temperature and the resulting precipitate, which was the target product, was collected by filtration. A yield of 1.84 g of yellow solid was obtained, corresponding to an 85% yield. ^1H NMR (300 MHz, DMSO): δ (ppm): 11.20 (s, 2H, –NH–), 8.38 (s, 3H, –CH=N–), 2.41 (s, 9H, –CH_3); ^{13}C NMR (75 MHz, DMSO): δ (ppm): 155.6, 153.7, 134.6, 113.4, 10.7 ppm. IR (KBr): 3422 (w), 3330 (m), 2361 (m), 1638 (m), 1604 (vs), 1576 (s), 1544 (s), 1489 (m), 1457 (s), 1409 (m), 1377 (m), 1303 (m), 1284 (w), 1152 (s), 1102 (m), 1033 (m), 963 (w), 929 (w), 869 (w), 820 (m), 728 (vw), 666 (w), 611 (w), 496 (w). $C_{13}H_{14}N_{12}O_6$ (434.33): calcd C 35.95, H 3.25, N 38.70; found C 36.39, H 3.29, N 38.65.

1,3-Bis-(3-methylfuroxan-4-methyleneamino)guanidine (**3a-22**). In a solution of 1,3-diaminoguanidine monohydrochloride (0.628 g, 5.0 mmol) and 3-methyl-4-furoxancarbaldehyde (2.56 g, 20.0 mmol) in 30 mL of ethanol, refluxing was conducted for 24 h. After the reflux period, the reaction mixture was cooled to room temperature, and the resulting precipitate, which represented the target product, was collected through filtration. The remaining filtrate was subjected to column chromatography on silica gel to separate and isolate the additional target product. This process allowed for the purification of the compound. Ultimately, a yield of 0.85 g of white solid was obtained, corresponding to a 55% yield. ^1H NMR (300 MHz, DMSO): δ (ppm): 12.63 (br, –NH–), 8.66 (s, 2H, –CH=N–), 8.62 (s, 2H, –NH_2), 2.43 (s, 6H, –CH_3); ^{13}C NMR (75 MHz, DMSO): δ (ppm): 154.9, 140.0, 113.3, 10.7. IR (KBr): 3453 (s), 3210 (s), 2812 (s), 2361 (m), 1669 (vs), 1611 (vs), 1465 (s), 1433 (s), 1378 (s), 1346 (m), 1314 (w), 1280 (w), 1189 (w), 1118 (s), 1033 (m), 954 (w), 927 (w), 826 (m), 798 (w), 739 (w), 678 (m), 625 (m), 600 (w), 486 (s). Elemental analysis: $C_9H_{11}N_9O_4$ (309.24): calcd C 34.96, H 3.59, N 40.76; found C 34.90, H 3.56, N 39.98.

3,6-Bis(4-amino-1,2,5-oxadiazol-3-yl)-1,4,2,5,-dioxadiazine(**BAFOZ**): A solution of 3-amino-4-chloroximidofurazan (0.33 g, 2 mmol) in THF (10 mL) was prepared. To this solution, a solution of K_2CO_3 (0.15 g) in water (10 mL) was added dropwise over a period of 30 min. The reaction mixture was allowed to stand for 2 h below 10 °C and then filtered. The filtration process yielded a solid product weighing

0.175 g, corresponding to a 70% yield. T_m 230–233 °C (T_d); ^{1}H NMR (DMSO-d_6, 400 MHz) δ 6.54 (bs); ^{13}C NMR (DMSO-d_6, 100 MHz) δ 155.0, 153.5, 136.0.

3,6-Bis(4-nitro-1,2,5-oxadiazol-3-yl)-1,4,2,5-dioxadiazine (**BNFOZ**): A solution of BAFOZ (1.0 g, 4.0 mmol) and ammonium persulfate (3.75 g, 16.4 mmol) in concentrated sulfuric acid (16 mL) was prepared. Dropwise addition of 60% hydrogen peroxide (10.0 mL, 0.256 mol) was carried out. The resulting mixture was stirred at temperatures below 10 °C for 1 h, followed by stirring at 50 °C for 4 h. The formed precipitate was subsequently filtered to obtain a white solid product weighing 0.74 g, corresponding to a 60% yield. T_m 102–104 °C; IR (KBr) 1587, 1556, 1359, 1273, 1088, 1030 cm^{-1}; ^{13}C NMR (DMSO-d_6, 100 MHz) δ 160.4, 153.7, 140.4. Anal Calcd: C, 23.09; N, 35.90. Found: C, 23.37; N, 34.69.

(E,E)-3,4-Bis(oximomethyl)furoxan (**DPX1**): A total of 13 g (0.12 mol) of freshly-prepared nitroacetaldehyde oxime was divided into three equal portions. The first portion was added slowly to 15 mL of concentrated sulfuric acid while maintaining the temperature below 10 °C. To this solution, a mixture of 6 mL of 65% oleum and 7 mL of concentrated sulfuric acid was added under the same temperature control. The second portion of nitroacetaldehyde oxime was added in a similar manner, followed by the remaining 6.5 mL of the oleum solution. Finally, the third portion of nitroacetaldehyde oxime was added while keeping the temperature below 10 °C. The resulting solution was stirred for 10 min at this temperature, after which the cooling bath was removed. The solution was allowed to self-warm while stirring, reaching a temperature of 30 °C, which was maintained for one hour. The solution was then carefully poured over 215 g of crushed ice and refrigerated for 12 h. After filtration and rinsing with cold water, a yield of 8.04 g (75%) of **DPX1** was obtained. T_m = 168 °C (T_d) (DSC, 5 °C min^{-1}). IR (cm^{-1}): 3365 (m), 3294 (m), 2360 (w), 2340 (w), 1626 (m), 1601(s), 1498 (m), 1422 (m), 1342 (w), 1320 (w), 1273 (m), 1262 (w), 1131 (w), 1054 (w), 1007 (s), 996 (s), 966 (m), 919 (m), 903 (s), 868 (w), 854 (w), 796 (w), 730 (s), 702 (m). Raman (200 mW, cm^{-1}): 3057 (2), 3002 (2), 1626 (64), 1615 (7), 1592 (100), 1513 (39), 1488 (24), 1438 (11), 1337 (33), 1271 (20), 1129 (32), 1052 (9), 988 (9), 923.^{1}H NMR (acetone-d_6, 400.18 MHz, 25 °C, TMS): 11.65 (s, 1H), 11.47 (s, 1H), 8.23 (s, 1H), 8.17 (s, 1H).^{13}C NMR (acetone-d_6, 100.63 MHz, 25 °C, TMS): 150.3 (1C), 138.9 (1C), 135.7 (1C), 108.7 (1C).^{15}N NMR (acetone-d_6, 40.55 MHz, 25 °C, CH3NO2) 9.01 (s, broad 2N), – 7.83 (s, 1N), – 26.36 (s, 1N) EA: Calc. for $C_4H_4N_4O_4$ (MW = 172.10 g mol^{-1}): Calc. C 27.92, H 2.34, N 32.55; Found C27.45, H 2.37, N 32.21. MS DEI^+ m/z (%) 172.1 (54) [M], 142.1 (18) [M-30], 126.1 (12), 112.1 (100) [M-60], 64.1(44), 30.1 (41.6). Impact sensitivity: 10 J. Friction sensitivity: 192 N. Electrostatic discharge (ESD) sensitivity: 0.35 J.

3,4-Bis(3-nitrofuroxan-4-yl)furoxan (**BNTFO-I**): To a mixture of 3-nitro-4-aminoximidofuroxan (1.9 g, 10 mmol), concentrated hydrochloric acid (9 mL), and water (16 mL) at room temperature, sodium nitrite (0.8 g, 11 mmol) in water (4 mL) was added dropwise under stirring at 0 °C. The resulting solution was stirred for 3 h. Next, diethyl ether (60 mL) was added to the mixture. The diethyl ether solution was separated and the solvent was removed under reduced pressure. The residue obtained was dissolved in 20 mL of THF. Ag_2CO_3 (1 g, 3.6 mmol) was added to the

THF reaction mixture at 0–5 °C and stirred for 2 h. Insoluble solid was removed by filtration. The filtrate was then concentrated, and the resulting residue was subjected to column chromatography on silica gel using ethyl acetate-pentane (v/v = 1:10) as the eluent. This process allowed for the isolation of colorless crystals, referred to as BNTFO-I, with a yield of 1.13 g (67%). T_m = 127.6 °C, T_d (onset) = 130.7 °C. IR (KBr): 1657, 1616, 1557, 1527, 1477, 1447, 1381, 1356, 1335, 1275, 1133, 1085, 1030, 984, 950, 860, 831, 792, 761, 744, 712, 692, 646, 592, 511, 477 cm^{-1}. ^{13}C NMR (DMSO-d_6): δ = 143.72, 139.65, 137.27, 126.03, 125.52, 104.44. Elemental analysis calcd (%) for $C_6N_8O_{10}$: C 20.94, N 32.56. Found: C 21.09, N 32.33. HRMS (EI, m/z) calcd. for $C_6N_8O_{10}$ [M]+ 343.9737, found 343.9738.

3,4-Bis(4-nitrofuroxan-3-yl)furoxan (**BNTFO-IV**): **BNTFO-I** (0.5 g) was dissolved in 20 mL of THF. The resulting solution was heated at reflux for 3 h. Afterward, the solvent was evaporated, and the residue was subjected to recrystallization from $CHCl_3$. This process yielded **BNTFO-IV** as a colorless solid, with a yield of 0.21 g (42%). T_m = 145.4 °C, T_d (onset) = 146.5 °C. IR (KBr): 1661, 1636, 1617, 1566, 1544, 1488, 1453, 1425, 1414, 1335, 1302, 1244, 1202, 1102, 1081, 1052, 983, 950, 834, 797, 767, 753, 740, 701, 673, 637, 537, 476, 455 cm^{-1}. ^{13}C NMR (DMSO-d_6): δ = 157.22, 156.95, 141.41, 102.12, 100.05, 99.11. Elemental analysis calcd (%) for $C_6N_8O_{10}$: C 20.94, N 32.56. Found: C 21.05, N 32.41. HRMS (EI, m/z) calcd. for $C_6N_8O_{10}$ [M]+ 343.9737, found 343.9741.

Synthesis of 3,6-bis(3,4-dimethyl-1H-pyrazol-1-yl)-1,4-dihydro-1,2,4,5-tetrazine **(BDT)**: In a 250 mL round bottom flask, **TAGN** (10 g, 71.1 mmol) and a magnetic bar were added along with 80 mL of water. The mixture was dissolved at room temperature and then cooled to 0 °C using an ice bath. Next, 2,4-pentanedione (14.6 mL, 142.3 mmol) was added dropwise. The solution gradually turned yellow, and a white precipitate formed. Once all the 2,4-pentanedione was added, the mixture was heated to 70 °C until no further precipitation was observed. During this heating period, the white solid dissolved, and a yellow precipitate appeared. It is important not to exceed a reaction time longer than 2 h as prolonged reaction times can result in an unsuccessful reaction, leading to the formation of a sticky yellow solid. It should also be noted that if improperly dried starting material I is used, it can result in the formation of a sticky solid. The resulting yellow solid was then subjected to hot filtration, followed by washing with water (3 × 50 mL). Finally, the solid was dried under high vacuum. This process yielded 8.1 g (83%) of a yellow solid. ^1H NMR (500 MHz, $CDCl_3$, δ ppm): 2.24 (s, 6H,), 2.50 (s, 6H,), 5.98 (s, 2H), 8.11 (br. s., 2H). ^{13}C NMR (126 MHz, $CDCl_3$, δ ppm): 13.6, 14.0, 110.1, 142.7, 145.7, 150.1.

Synthesis of 3-(3,4-dimethyl-1H-pyrazol-1-yl)-6-(3,4-dimethyl-4,5-dihydro-1H-pyrazol1-yl)-1,2,4,5-tetrazine **(BT)**: In a 1 L round bottom flask, $NaNO_2$ (8.0 g, 115.7 mmol) was placed. The round bottom flask was equipped with a large magnetic stirrer, and the gas outlet of the dropping funnel was connected to a trap containing a 1 M NaOH solution. 3,6-Bis(3,4-dimethyl-1H-pyrazol-1-yl)-1,4-dihydro-1,2,4,5-tetrazine (10.5 g, 38.6 mmol) was added to the flask and dissolved in 430 mL of a 1:2 mixture of DCM/H_2O. The mixture was then cooled to 0 °C. Acetic acid (11 mL, 192.8 mmol) was added dropwise to the mixture. During the addition, the color changed from yellow to bright red, accompanied by the generation of an orange gas. After the addition was completed, the mixture was vigorously stirred for 2 h until gas evolution ceased, and the organic phase was separated. The aqueous layer was extracted with DCM (2 × 100 mL), and the combined organic layers were washed with saturated $NaHCO_3$ (2 × 100 mL) and dried over Na_2SO_4. The solution was then filtered and concentrated under vacuum. The resulting dark red solid was suspended in water, filtered under vacuum, and washed with water (1 × 50 mL) followed by cold MeOH (2 × 25 mL). This process yielded 8.3 g (80%) of a bright red powder. ^{1}H NMR (500 MHz, $CDCl_3$, δ ppm): 2.39 (s, 6H), 2.71 (s, 6H), 6.19 (s, 2H). ^{13}C NMR (125 MHz, $CDCl_3$, δ ppm):14.0, 14.8, 112.1, 144.0, 154.6, 159.5.

Synthesis of 3,6-dihydrazineyl-1,2,4,5-tetrazine (**DHT**): A solution of 3-(3,4-dimethyl-1H-pyrazol-1-yl)-6-(3,4-dimethyl-4,5-dihydro-1H-pyrazol-1-yl)-1,2,4,5-tetrazine (8.3 g, 30.7 mmol) in acetonitrile (150 mL) was prepared. Hydrazine monohydrate (4.5 mL, 1 mmol) was added to the solution, resulting in an immediate color change from red to maroon. The mixture was refluxed for 2 h, then cooled to 25 °C and filtered. The obtained solid was washed with acetonitrile (2 × 50 mL) and dried under high vacuum. This process yielded 4.3 g (99%) of a maroon powder. ^{13}C NMR (125 MHz, $CDCl_3$, δ ppm): 168.3.

3,6-Bi(3-nitro-1,2,4-oxadiazole-5-guanyl)-1,2,4,5-tetrazine (**3b-1**): To a solution of diaminoglycoluril (**AG**, 1.384 g, 8 mmol) in a mixture of water (150 mL), acetone (50 mL), and acetonitrile (30 mL), a solution of Oxone (80 g, 0.13 mol) in H_2O (500 mL) was added dropwise. The Oxone solution had been neutralized with $NaHCO_3$ (25.2 g, 0.3 mol) at 0–5 °C over a period of 30 min. The addition was performed while maintaining the temperature at 0–5 °C. After stirring for 10 h at 0 °C, the reaction mixture was filtered, and the filtrate was subjected to extraction with ethyl acetate (150 mL). The combined organic layers were washed with brine, dried over $MgSO_4$, and then filtered. The filtrate was concentrated under reduced pressure. The resulting residue was purified by column chromatography using an ethyl acetate-petroleum ether mixture (1:3 ratio) as the eluent. This purification process yielded the product, 3-nitro-5-guanidino-1,2,4-oxadiazole (**NOG**), as a pale yellow solid, with a yield of 550 mg (40%).

In an oven-dried, 50 mL round-bottomed flask equipped with a stir bar, **NOG** (860 mg, 5 mmol) was placed. The compound was then dissolved in anhydrous dimethylformamide (20 mL). The resulting solution was cooled to 0 °C, and NaH (60% dispersion in oil, 280 mg, 7 mmol) was added in portions over a period of 5 min. The reaction mixture was stirred for 30 min at 0 °C. Next, **BT** (540 mg, 2 mmol) was added to the reaction mixture in one portion. The reaction was continued to be stirred at 0 °C for 1 h and then allowed to warm to room temperature, where it was stirred for an additional 4 h. The reaction mixture was then poured into ice water (20 mL) and acidified by the addition of 3N HCl until the pH reached 1. The resulting orange precipitate was filtered, thoroughly washed with water, and dried. This process yielded **3b-1** (800 mg, 94.7%) as the final product. T_{dec}: 329 °C; ^{1}H NMR (500 MHz, DMSO-d_6): δ 12.36 (bs, NH), 9.27 (bs, 2H), 8.74 (bs, 2H) ppm. ^{13}C NMR (125 MHz, DMSO-d_6): δ 155.8, 159.2, 169.0, 175.3 ppm. IR (cm^{-1}, KBr):

3365, 3264, 3216, 1633, 1568, 1538, 1477, 1405, 1386, 1302, 1052, 1026, 980, 955, 828, 787, 755, 695, 599. MS (421.0584): m/z negative mode, 421.05796 $[M-H]^-$.

3,6-Bis(3′-azido-1′,2′,4′-oxadiazole-5′-guanyl)- 1,2,4,5-tetrazine (**3b-2**): In a solution containing DMSO (4 mL), H_2O (1 mL), and NaN_3 (195 mg, 3 mmol), **NOG** (344 mg, 2 mmol) is added. The resulting mixture is stirred using an overhead stirrer and heated to 50 °C for 5 h. After the reaction time has elapsed, the mixture is poured into ice water (10 mL). The formed pale yellow precipitate is then filtered, washed thoroughly with water, and dried. This process yields the product solid, 3-azido-5-guanidino-1,2,4-oxadiazole (**AOG**), with a yield of 120 mg (36%). $T_{dec.}$: 196 °C. ^1H NMR (500 MHz, DMSO-d_6): δ 7.14(NH_2). ^{13}C NMR (125 MHz, DMSO-d_6): δ 174.6, 163.5, 159.8. IR (cm^{-1}, KBr): 3433, 3391, 3230, 3103, 2160, 1673, 1580, 1537, 1420, 1214, 974, 766. Anal. calcd for $C_3H_4N_8O$ (%): C 21.43; H2.37; N 66.65; Found: C 21.60; H 2.37; N 66.52.

In an oven-dried, 50 mL round-bottomed flask equipped with a stir bar, **AOG** (252 mg, 1.5 mmol) is placed. The compound is then dissolved in anhydrous dimethylformamide (20 mL). The resulting solution is cooled to 0 °C, and NaH (60% dispersion in oil, 92 mg, 3.3 mmol) is added portion-wise over a period of 5 min. The reaction mixture is stirred for 30 min at 0 °C. Next, **BT** (162 mg, 0.6 mmol) is added to the reaction mixture in one portion. The reaction is continued to be stirred at 0 °C for 1 h and then allowed to warm to room temperature, where it is stirred for an additional 4 h. The reaction mixture is then poured into ice water (20 mL) and acidified by the addition of 3 N HCl until the pH reaches 1. The resulting brown–red precipitate is filtered, thoroughly washed with water, and dried. This process yields **3b-2** (240 mg, 96.7%) as the final product. T_{dec}: 210 °C. ^1H NMR (500 MHz, DMSO-d6): δ 12.14 (bs, 2H), 9.05 (bs, 2H), 8.80(bs, 2H). ^{13}C NMR (125 MHz, DMSO-d6): δ 154.9, 158.6, 164.0, 174.0. IR (cm^{-1}, KBr): 3372, 3262, 3212, 2153, 1635, 1538, 1367, 1294, 1194, 1050, 774, 610.

3,6-Bis-(1H-1,2,3,4-tetrazol-5-ylamino)-s-tetrazine (**BTATz**): In a liter flask, dry sulfolane (2700 mL), anhydrous 5-amino-1H-tetrazole (197 g, 2.32 mol), and compound **BT** (250.0 g, 0.926 mol) are thoroughly mixed using a high-torque mechanical stirrer. Initially, the reaction mixture forms a thick, orange slurry, likely due to the formation of a complex between **BT** and 5-amino-1H-tetrazole. However, as the mixture is slowly heated over a period of 6 h to 135 °C, it gradually converts into a dark solution. The heating process is carefully controlled and maintained at 135 °C using a programmable temperature controller. Within several hours, the product begins to precipitate. After 18 h at 135 °C, the resulting slurry is cooled to 35 °C, and DMF (200 mL) is added to prevent the sulfolane from freezing. The product is then filtered, thoroughly washed with ethanol, and oven dried at 100 °C for several days, yielding the crude material (170 g, 74% yield). To further purify the product, it is triturated with boiling ethanol (3 L) for 18 h. The mixture is then cooled, filtered, and washed again with ethanol. Subsequently, the material is triturated with DMF at 110 °C for 18 h, followed by cooling, filtration, and additional washing with ethanol. It is recommended to perform all triturations with vigorous stirring, especially when purifying large amounts of **BTATz**. Finally, the purified material is dried in a vacuum oven for 24 h at 110 °C, resulting in the production of pure bright orange **BTATz** (142 g, 62% yield). Longer drying times may be necessary to remove any traces of DMF solvent. ^{1}H NMR (500 MHz, DMSO-d_6): δ 12.5 (bs, 4H); ^{13}C NMR (125 MHz, DMSO-d_6): δ 158.15, 151.04.

3-(1H-1,2,3,4-tetrazol-5-ylamino)-6-(3,5-dimethyl-pyrazol-1-yl)-s-tetrazine monohydrate **(TADPTz)**: In a flask, anhydrous 5-amino-1H-tetrazole (3.14 g, 0.0365 mol) and 3,6-bis(3,5-dimethyl pyrazol-1-yl)-1,2,4,5-tetrazine (5.0 g, 0.0185 mol) were combined. Approximately 50 mL of dry dimethylformamide was added to the flask. The reaction mixture was then slowly heated to 120 °C and maintained at this temperature for 4 h. The progress of the reaction was monitored using thin-layer chromatography (TLC). Once the reactant BPT was completely consumed (as indicated by its absence on TLC), the hot solution was poured over crushed ice and stirred. The resulting mixture was then filtered to obtain a rosy red precipitate. The precipitate was thoroughly washed with water and dried. The filtered solid was refluxed with acetone and dissolved in hot methanol. The methanol solution was allowed to cool, leading to the formation of crystals of the desired product. The final product weighed 3.33 g, corresponding to a yield of 65%. $T_{m.}$: 275 °C. ^{1}H NMR (500 MHz, DMSO-d_6): δ 2.24, 2.50, 6.27, 6.45, 14.5. ^{13}C NMR (125 MHz, DMSO-d_6): δ 13.03, 13.46, 109.76, 142.06, 150.38, 151.26, 157.85, 158.69. IR (cm^{-1}, KBr): 3372, 3262, 3212, 2153, 1635, 1538, 1367, 1294, 1194, 1050, 774, 610.

N1,N1′-(1,2,4,5-tetrazine-3,6-diyl)bis(1H-tetrazole-1,5-diamine) **(3b-4)**: To begin the reaction, cyanogen bromide (1.30 g, 12.3 mmol) is dissolved in dry acetonitrile (20 mL) at 0 °C. Sodium azide (3.82 g, 58.8 mmol) is then added to the solution. The resulting reaction mixture is stirred at 0–15 °C for 4 h. The mixture is then filtered to remove any inorganic salt present. The filtered salt should be quickly dissolved in cold water. Next, the resulting solution is added to a suspension containing 3,6-dihydrazinyl-1,2,4,5-tetrazine (**DHT**) (100 mg, 0.704 mmol) in water (10 mL) at 0 °C. The mixture is stirred at ambient temperature for a period of 4 days. After the reaction has taken place, the suspension is filtered to obtain an orange filter cake. The filtrate is then dried in air to yield the final product. IR (cm^{-1}, KBr): 3385, 3298, 3256, 3192, 3026, 2936, 2843, 1655, 1558, 1483, 1431, 1332, 1112, 1061, 953, 560; ^{1}H NMR (500 MHz, DMSO-d_6): δ 7.06 (s, 4H; NH_2), 11.98 ppm (brs, 2H; NH); ^{13}C NMR (125 MHz, DMSO-d_6): δ 154.9, 161.1 ppm; ^{15}N NMR (DMSO-d_6): δ 332.24, 279.6, 171.7, 93.0, 30.0, 19.7, 6.2 ppm; Anal. calcd for $C_4H_6N_{16}$ (%): C 17.27, H 2.17, N 80.56; found: not determinable (explodes).

1,4-Bis(2-hydroxytetrazole)-1,2,4,5-tetrazine **(3b-5)**: To start the reaction, sodium cyanotetrazolate·1.5H_2O (5 g, 34.7 mmol) is dissolved in 70 mL of distilled water, and potassium acetate (29.8 g, 304 mmol) is added. The resulting solution is heated to 40 °C, and Oxone (66.8 g, 98 mmol) is slowly added. The suspension is stirred for 20 h at 40 °C. Afterward, the solution is poured into 1.6 L of acetone, causing precipitated salts to form. These salts are then filtered off. The solvent is evaporated under reduced pressure, leaving behind solid residue. This residue is dissolved in ethanol. Insoluble residues are filtered out, and the filtrate is evaporated under vacuum. Next, the colorless solid obtained from the previous step is dissolved in 100 mL of distilled water. A solution of silver nitrate (5.9 g, 34.7 mmol) in 100 mL of distilled water is added to the suspension in the dark while stirring. The resulting colorless suspension is further stirred in the dark for 15 min. The suspension is then filtered and washed with 20 mL of distilled water. After drying in the dark, a white solid known as silver cyanotetrazolate-2-oxide (5.5 g, 73%) is obtained. T_{dec}: 224 °C; IR (cm^{-1}, KBr): 3627, 3464, 2921, 2242, 1596, 1441, 1428, 1403, 1365, 1340, 1267, 1225, 1090, 1013, 812, 710; ^{13}C NMR (125 MHz, DMSO-d_6): δ 132.0, 112.3 ppm.

To prepare the compound, sodium-5-cyano-tetrazolate $1.5H_2O$ (5.00 g, 34.7 mmol, 1.0 equiv) is dissolved in 25 mL water. The solution is cooled to 0 °C, and nitric acid (18.5 mL, 2 M, 37.0 mmol, 1.1 equiv) is added. The mixture is stirred for 30 min before allowing it to warm up to room temperature. The solvent is then removed under reduced pressure at room temperature, leaving behind a residue that is dissolved in ethanol. Any remaining solids are filtered out, and the ethanol is evaporated at room temperature under reduced pressure, resulting in crude 5-cyanotetrazole. The crude product is dissolved in 66 mL of water. Next, potassium hydroxide (17.5 mL, 35.0 mmol, 1.0 equiv), potassium acetate (29.80 g, 303.6 mmol, 8.7 equiv), and Oxone (66.03 g, 107.4 mmol, 3.1 equiv) are added to the solution. The mixture is stirred at 40 °C overnight. Afterward, the solution is poured into acetone (1.60 L) and stirred, followed by filtration to remove precipitated salts. The acetone is then evaporated under reduced pressure, and the remaining solids are dissolved in hot ethanol. Insoluble remains are filtered off, and the ethanol is removed under reduced pressure. The residue is dissolved in 15 mL of water. The solution is cooled to 0 °C, and 2 M nitric acid (17.4 mL, 34.7 mmol, 1.0 equiv) is added. The mixture is stirred for 30 min, and then the solvent is evaporated under high vacuum at room temperature. The solid residue is dissolved in dry ethanol, and any insoluble remains are filtered off. Hydrazine hydrate (2.9 mL, 3.06 g, 60.9 mmol, 1.8 equiv) is added to the mixture, which is then refluxed at about 85 °C for 2 h. Additional hydrazine hydrate (2.96 mL, 3.06 g, 60.9 mmol, 1.8 equiv) is added, and the mixture is refluxed for another 22 h. Finally, the precipitate is filtered off and rinsed with dry ethanol, resulting in the formation of a white solid known as hydrazinium 1,4-bis(2-N-oxidotetrazolate)-dihydro-1,2,4,5-tetrazine (4.38 g, 13.9 mmol, 79.8% relative to the amount of sodium 5-cyanotetrazolate sesquihydrate). T_{dec}: 176 °C; IR (cm^{-1}, KBr): 3306, 3272, 3187, 3148, 3079, 2945, 2877, 2649, 2520, 2151, 2119, 1637, 1613, 1545, 1512, 1461, 1448, 1406, 1396, 1372, 1302, 1246, 1193, 1100, 1038, 1014, 993, 972; ^{13}C NMR (125 MHz, DMSO-d_6): δ 148.10, 140.54 ppm.

A suspension of hydrazinium 1,4-bis(2-N-oxidotetrazolate)-dihydro-1,2,4,5-tetrazine (100 mg, 0.35 mmol, 1.0 equiv) is prepared in 100 mL of acetonitrile. A solution of nitrogen dioxide in acetonitrile is then added dropwise to the suspension until it turns red and no more solid dissolves. The remaining solid is filtered off, and the resulting solution is slowly evaporated under a stream of nitrogen at room temperature. The solid obtained from the evaporation is subsequently dried under high vacuum at 40 °C. ^{13}C NMR (125 MHz, DMSO-d_6): δ 158.4, 154.9 ppm.

Hydrazinium 1,4-bis(1-N-oxidotetrazolate)-dihydro-1,2,4,5-tetrazine **(3b-6)**: Sodium cyanotetrazolate-1-oxide (2.00 g, 14.1 mmol, 1.0 equiv) is dissolved in 12 mL of water. The resulting solution is then cooled to 0 °C. Next, 2 M nitric acid (7.04 mL, 14.1 mmol, 1.0 equiv) is added to the solution. The mixture is stirred for

30 min before the solvent is evaporated under reduced pressure at room temperature. The solid residue obtained from the evaporation step is dissolved in dry ethanol while insoluble remains are filtered off. Following that, hydrazine hydrate (1.20 mL, 1.24 g, 24.7 mmol, 1.8 equiv) is added to the mixture. The mixture is refluxed at 85 °C for 2 h. Another portion of hydrazine hydrate (1.20 mL, 1.24 g, 24.7 mmol, 1.8 equiv) is then added, and the refluxing process continues for another 22 h. Afterward, the precipitate formed is filtered off and rinsed with dry ethanol. This yields **3b-6** as a white solid (1.84 g, 82.5%). T_{dec}: 180 °C; IR (cm^{-1}, KBr): 3589, 326, 3183, 3111, 2968, 2909, 2726, 2635, 2298, 2202, 2056, 1988, 1954, 1903, 1653, 1615, 1538, 1514, 1457, 1445, 1411, 1346, 1318, 1243, 1167, 1150, 1103, 1082, 1005, 988, 951, 853, 782, 737; ^1H NMR (500 MHz, DMSO-d_6): δ 10.04 (s), 7.07 ppm; ^{13}C NMR (125 MHz, DMSO-d_6): δ 140.44, 134.26 ppm.

3,6-Bis(2-(4,6-diazido-1,3,5-triazin-2-yl)hydrazineyl)-1,2,4,5-tetrazine (3b-7): 11.06 g (60.0 mmol) of cyanuric chloride and 100 mL of acetonitrile are added to a three-necked round-bottom flask. The mixture is stirred for 30 min at 0 °C. Subsequently, 4.26 g (30.0 mmol) of **DHT** (Dihydrotetrazine) and 7.63 g (72.0 mmol) of Na_2CO_3 (Sodium carbonate) are added. The stirring process continues for 18 h until completion of the reaction, which is confirmed by TLC (Thin-layer chromatography) using CH_2Cl_2:CH_3OH (12:1) as the eluent. The resulting precipitate is then centrifuged and dissolved in a large amount of acetone to remove Na_2CO_3. The solution is further condensed by rotary evaporation, resulting in the formation of an orange product, namely 3,6-bis(2-(4,6-dichloro-1,3,5-triazin-2-yl)hydrazinyl)-1,2,4,5-tetrazine. The obtained yield of the product is 7.75 g (59%). ^1H NMR (500 MHz, DMSO-d_6): δ 11.19 (2H, s, NH), 10.39 (2H, s, NH). ^{13}C NMR (125 MHz, DMSO-d_6): δ 162.31, 168.30, 169.77, 170.93 ppm.

0.876 g (2.0 mmol) of 3,6-bis(2-(4,6-dichloro-1,3,5-triazin-2-yl)hydrazinyl)-1,2,4,5-tetrazine is suspended in 10 mL of acetonitrile and stirred for 30 min at room temperature. Next, 0.52 g (8.0 mmol) of NaN_3 (Sodium azide) is added to the reaction mixture. The mixture is then refluxed for 16 h and subsequently cooled to room temperature. The resulting product is centrifuged, washed with cool water to remove excess NaN_3, and dried in air. This yields **3b-7** as a brown solid, with a mass of 0.68 g and a yield of 73.3%. T_{dec}: 193.9 oC (DSC, 5 °C min^{-1}); IR (cm^{-1}, KBr): 2200, 2147, 1788, 1700, 1533, 1473, 1397, 1238, 1135, 1008), 957, 611, 577); Elemental Analysis ($C_8N_{26}H_4$, 464.4): Calcd: C 20.69, H 0.87, N 78.44%; found: C 20.63, H 0.76, N 77.89%; ^1H NMR (500 MHz, DMSO-d_6): δ 10.39 (2H, s, NH), 11.18 (2H, s, NH). ^{13}C NMR (125 MHz, DMSO-d_6): δ 162.31, 168.30, 169.77, 170.93 ppm.

3,6-Bis((4,6-diazido-1,3,5-triazin-2-yl)diazenyl)-1,2,4,5-tetrazine (3b-8): In a 25 mL flask, 0.712 g (4.0 mmol) of NBS (N-bromosuccinimide) is added to 15 mL of acetonitrile and stirred until completely dissolved. Then, 0.464 g (1.0 mmol) of **3b-7** is added to the mixture at 40 °C after 30 min. After the reaction, the resulting solution is poured into 60 mL of ice water. An orange-yellow precipitate forms, which is then filtered, washed with water, and dried. This yields a yellow–brown powder with a mass of 0.14 g, resulting in a yield of 30.4%. T_{dec}: 188.6 oC (DSC, 5 °C min^{-1}); IR (cm^{-1}, KBr): 2349, 2165, 2143, 1759, 1713, 1546, 1537, 1352, 1205, 1128, 979, 896,

805, 548; Elemental Analysis (C_8N_{26}, 460.3): Calcd: C 20.87, N 79.13%; found: C 20.56, N 78.41%;^{1}H NMR (500 MHz, DMSO-d_6): no hydrogen signal. ^{13}C NMR (125 MHz, DMSO-d_6): δ 163.22, 169.20, 170.68, 171.83 ppm.

3,6-Bis(imidazol-2-yl)-1,4,2,5-dioxadiazine **(3b-9)**: Firstly, a solution of hydroxylamine hydrochloride (1.44 g, 20.7 mmol) in water (5 mL) was prepared and adjusted to a pH of 7. Sodium carbonate (1.10 g, 10.4 mmol) was then added to this solution. Next, imidazole-2-carboxaldehyde (1.00 g, 10.4 mmol) was added slowly and incrementally to the resulting reaction mixture. The mixture was stirred for 1 h at 70 °C. After completion of the reaction, the product was filtered and washed with cold water. The resulting solid was then dried under vacuum conditions, leading to the formation of a white solid known as imidazole-2-carboxaldehyde oxime. The yield of the product was 0.83 g (72%). T_{dec}: 178 °C; ^{1}H NMR (500 MHz, DMSO-d_6): δ 7.97 (1H, s), 7.14 (1H, s), 7.09 (1H, s); ^{13}C NMR (125 MHz, DMSO-d_6): δ 140.9, 140.3, 138.6, 136.9.

Imidazole-2-carboxaldehyde oxime (0.25 g, 2.25 mmol) was dissolved in acetic acid (1.5 mL). To this solution, a dropwise addition of 98% nitric acid (0.3 mL, 7.20 mmol) was performed at a temperature range of 0–4 °C. The resulting mixture was stirred for 3 h at room temperature. After completion of the reaction, the mixture was filtered and the solid product was obtained. This solid was washed with diethyl ether and dried under vacuum conditions, resulting in the formation of a yellow solid known as imidazole-2-nitrolic acid. The yield of the product was determined to be 0.32 g (91%). T_{dec}: 116 °C; ^{1}H NMR (500 MHz, DMSO-d_6): δ 7.60 (2H, s); ^{13}C NMR (125 MHz, DMSO-d_6): δ 149.8, 129.2, 124.1.

To carry out the reaction, imidazole-2-nitrolic acid (0.62 g, 8.33 mmol) was added gradually to a solution of 5% sodium bicarbonate (14 mL, 8.33 mmol) at 4 °C. The resulting mixture was stirred for 30 min in the temperature range of 0–4 °C. Afterwards, the reaction mixture was filtered and the solid product obtained was washed with cold water. The solid was then dried under vacuum conditions, resulting in the formation of a brown solid. The yield of the product was determined to be 0.32 g (91%). T_{dec}: 130 °C; ^{1}H NMR (500 MHz, DMSO-d_6): δ 7.35 (4H, s); ^{13}C NMR (125 MHz, DMSO-d_6): δ 153.1, 131.2, 126.2; ESI–MS (pos.): 240.94 ($[M+Na]^+$, $C_8H_6N_6O_2^+$; calc. 218.14).

3,4-Bis(imidazol-2-yl)furoxan (**3b-10**): In the reaction, a solution of N-chlorosuccinimide (3.40 g, 24 mmol) in 50% aqueous methanol (150 mL) was prepared. Imidazole-2-carboxaldehyde oxime (**3b-9**, 2.24 g, 20.2 mmol) dissolved in methanol (100 mL) was then added dropwise to the solution at a temperature range of 0–4 °C. The mixture was stirred for 2 h at the same temperature range. Following this, a slow addition of 3% aqueous potassium carbonate (100 mL, 25 mmol) was made to the reaction mixture while maintaining a temperature of 0–4 °C. The resulting mixture was stirred for an additional 2 h at ambient temperature. Subsequently, the reaction mixture was concentrated under vacuum and extracted four times with diethyl ether (50 mL each time). The organic layer obtained from the extraction was washed with distilled water, dried using $MgSO_4$, and then further concentrated under vacuum. This process yielded a crude product in the form of a brown solid. To purify the crude product, it was subjected to flash silica gel chromatography using a solvent mixture of THF/Chloroform (1:9). As a result, a white solid known as **3b-10** was obtained with a mass of 2.5 g and a yield of 58%. T_{dec}: 224 °C; ^{1}H NMR (500 MHz, DMSO-d_6): δ 7.87 (2H, s), 7.05 (2H, s); ^{13}C NMR (125 MHz, DMSO-d_6): δ 142.0, 139.7, 130.4, 126.2, 115.5; ESI–MS (pos.): 241.91 ($[M+Na]^+$, $C_8H_6N_6O_2^+$; calc. 218.14).

3,4-Bis(4,5-dinrtoimidazol-2-yl)furoxan (**3b-12**): 3,4-Bis(imidazol-2-yl)furoxan (5, 0.5 g, 23 mmol) was added gradually to 100% nitric acid (7 mL) at 0 °C. The resulting mixture was stirred for 2 h at ambient temperature. After this period, the reaction mixture was quenched by adding acetic acid (70 mL). The mixture was then concentrated under vacuum, resulting in the formation of a crude yellow liquid. To purify the crude product, it was subjected to recrystallization from a mixture of methanol and MC (an organic solvent). As a result, an ivory solid was obtained with a mass of 0.62 g and a yield of 68%. T_m: 147 °C, T_{dec}: 257; ^{13}C NMR (125 MHz, DMSO-d_6): δ 163.2, 159.5, 137.1, 129.6, 118.1; ESI–MS (pos.): 437.18 ($[M+K]^+$, $C_8H_2N_{10}O_{10}^+$; calc. 398.16).

Synthesis of 3,4-bis(1′-nitro-1′,2′,4′-triazole-3′-yl)-furoxan (**BNTAF**): 3-Cyano-1,2,4-triazole weighing 10.00 g (0.106 mol) was added to a three-necked round-bottom flask equipped with a mechanical stirrer and a dropping funnel. Then, 100 mL of water was added to the flask. Hydroxylamine hydrochloride was added dropwise to the reaction flask at room temperature until all of it was completely added. The pH value of the reaction mixture was then adjusted to 8 using a 30% sodium hydroxide solution. The reaction was carried out at a temperature range of 60–65 °C for an additional 2.5 h. During this time, the desired product, 3-amino oxime-1,2,4-triazole (**AOT**), precipitated as a white solid. The white precipitate was filtered, resulting in the isolation of 6.60 g of white solid **AOT**. The yield of the isolated compound was determined to be 48.8%. The purity of the product was found to be 99.8% according to High-Performance Liquid Chromatography analysis. The melting point of the compound was measured to be in the range of 176.8–177.8 °C using the capillary method. ^{1}H NMR (500 MHz, DMSO-d_6): δ 5.83 (s, 2H, NH_2), 8.32 (s, 1H, NH), 9.96 (s, 1H, CH), 14.28 (s, 1H, OH); ^{13}C NMR (125 MHz, DMSO-d_6): δ 143.79, 147.76, 151.84; IR (cm^{-1}, KBr): 3472, 3453, 3360, 3336, 3143, 1655, 932, 1472, 1390, 1292, 1151. Anal. calcd for $C_3H_5N_5O$: C 28.35, N 55.10, H 3.96; found C 28.97, N 55.49, H 4.107; MS m/z: 127 (M^+).

3-Amino oxime-1,2,4-triazole (**AOT**) weighing 4.00 g (0.031 mol) was transferred into a three-necked round-bottom flask equipped with a mechanical stirrer and a dropping funnel. A solution containing 9.5 mL of concentrated hydrochloric acid in 30 mL of water was then added to the flask. The resulting solution was cooled to 5 °C after the solid **AOT** was completely dissolved. A sodium nitrite solution was then added dropwise to the reaction flask while maintaining a temperature of 5 °C. Once the sodium nitrite solution was added completely, the reaction mixture was stirred for an additional 3 h at the same temperature. Afterward, the reaction temperature was raised to 25 °C, and the reaction mixture was stirred for another 1.5 h. During this time, a white precipitate formed. The white precipitate was filtered, resulting in the isolation of 2.80 g of a white solid known as 3-chloride oxime-1,2,4-triazole (**COT**). The yield of the isolated compound was determined to be 60.7%.

The melting point of the product was measured to be in the range of 194.7–195.8 °C using the capillary method. ^{1}H NMR (500 MHz, DMSO-d_6): δ 8.60 (s, 1H, CH), 12.61 (s, 1H, NH), 14.41 (s, 1H, OH); ^{13}C NMR (125 MHz, DMSO-d_6): δ 128.36, 145.95, 154.79; IR (cm^{-1}, KBr): 3237, 1619, 937, 748, 1477, 1370, 1279, 1040. Anal. calcd for $C_3H_3N_4OCl$: C 24.59, N 38.23, H 2.064; found C 24.79, N 37.70, H 2.050.

3-Chloride oxime-1,2,4-triazole (**COT**) weighing 2.50 g (0.017 mol) was transferred into a three-necked round-bottom flask equipped with a mechanical stirrer and a dropping funnel. Diethyl ether (30 mL) was added to the flask. The solution was then cooled to a temperature range of 0–3 °C. Next, a 3% sodium carbonate solution (33 mL) was added dropwise to the reaction flask while maintaining a temperature of 0–3 °C. After the addition of the sodium carbonate solution was complete, the mixture was stirred for an additional 4 h in the temperature range of 2–10 °C. The diethyl ether was subsequently removed under reduced pressure. This resulted in the formation of a white precipitate. The precipitate was filtered, yielding 1.45 g of white solid known as 3,4-bis(1′,2′,4′-triazole-3′-yl)furoxan (**BTAF**). The yield of the isolated compound was determined to be 71.2%. T_m: 188.8 °C. ^{1}H NMR (500 MHz, DMSO-d_6): δ 8.86 (s, 2H); ^{13}C NMR (125 MHz, DMSO-d_6): δ 146.37, 149.26, 157.27; IR (cm^{-1}, KBr): 3423, 3095, 1625, 1468, 991, 1494, 1360, 1274, 1024. Anal. calcd for $C_6H_4N_8O_2$: C 32.73, N 50.90, H 1.83; found C 32.23, N 50.80, H 1.86; MS m/z: 220 (M^+).

3,4-Bis(1′,2′,4′-triazole-3′-yl)furoxan (**BTAF**) weighing 2.00 g (0.009 mol) was transferred into a three-necked round-bottom flask equipped with a mechanical stirrer and a dropping funnel. Then, 12 mL of acetic acid was added to the flask. A mixing solution containing 2 mL of concentrated nitric acid and 6 mL of acetic anhydride was prepared. This mixing solution was then added to the flask containing **BTAF**. The addition of the mixing solution was done completely and the resulting mixture was stirred at 25 °C for an additional 2 h. Afterward, the reaction mixture was poured into ice-water. As a result, a white precipitate formed. The white precipitate was filtered, yielding 2.10 g of white solid with a yield of 74.5%. T_m: 99.2 °C; ^{1}H NMR (500 MHz, DMSO-d_6): δ 8.80 (d, 2H); ^{13}C NMR (125 MHz, DMSO-d_6): δ 108.82, 145.75, 145.85, 146.67, 149.79, 149.80; IR (cm^{-1}, KBr): 3151, 3115, 1621, 1479, 963, 1553, 1383, 1346, 1273, 1026. Anal. calcd for $C_6H_2N_{10}O_6$: C 23.24, N 45.16, H 0.65; found C 23.36, N 45.05, H 0.69; MS m/z: 310 (M^+).

Bis(1-hydroxytetrazolyl)furazane (**3b-13**): 10.8 g (90 mmol) of 3,4-Dicyanofuroxan was dissolved in 45 mL of ethanol. The solution was then added within 15 min to a mixture of 22.2 g (336 mmol) of 50% hydroxylamine solution and 90 mL of ethanol. The solvent was subsequently removed under reduced pressure until crystallization began. This process resulted in the formation of a yellowish powder. Upon filtering the reaction mixture, 9.3 g (50 mmol) of bisaminohydroximoylfurazane was obtained as the final product with a yield of 55%. T_{dec}: 198 °C. ^{1}H NMR (500 MHz, DMSO-d_6): δ 10.33, 6.20 ppm. ^{13}C NMR (125 MHz, DMSO-d_6): δ 148.7, 142.1 ppm.

6.9 g (37 mmol) of Bisaminohydroximoylfurazane was dissolved in 200 mL of semi-concentrated hydrochloric acid. A solution of 6.3 g (92 mmol) of sodium nitrite in 30 mL of water was then added dropwise to the mixture over the course of one hour. Throughout the addition, the temperature was maintained below 0 °C. After the complete addition of the sodium nitrite solution, the resulting mixture was stirred for one hour and allowed to come to ambient temperature. It was then diluted with 200 mL of water and extracted with 35 mL of diethyl ether. The organic phase (diethyl ether layer) was separated and dried over magnesium sulfate. The solvent was subsequently removed under reduced pressure, resulting in the isolation of 7.97 g (35 mmol) of bischlorohydroximoylfurazane. The compound obtained was in the form of an oily liquid that partially started to crystallize. The yield of the product was determined to be 96%. T_{dec}:115 °C. IR (cm^{-1}, KBr): 3502, 3388, 2991, 2877, 1731, 1607, 1561, 1507, 1499, 1397, 1390, 1376, 1359, 1343, 1265, 1193, 1094, 1057, 1032, 999, 962, 900, 887, 863, 818, 795, 749, 664. ^{1}H NMR (500 MHz, DMSO-d_6): δ 13.61 ppm. ^{13}C NMR (125 MHz, DMSO-d_6): δ 148.7, 122.9 ppm. Anal. calcd for $C_4H_2N_4O_3Cl_2$ (%): C 21.35, H 0.90, N 24.90, found C 23.19, H 1.53, N 22.76.

7.96 g (35 mmol) of Bischlorohydroximoylfurazane was dissolved in 50 mL of ethanol. Then, a solution of 6.53 g (100 mmol) of sodium azide in 50 mL of water was added to the mixture while maintaining a temperature of 0–5 °C. The resulting suspension was stirred for 1 h on ice. Afterward, it was diluted with 100 mL of water and the pH was adjusted to 2 using 2 M hydrochloric acid. The product was then extracted with 30 mL of diethyl ether. The organic phase (diethyl ether layer) was separated and dried over magnesium sulfate. It's worth noting that the product does not need to be isolated at this stage for the subsequent steps. If the solvent is removed, a yellowish oil known as bisazidohydroximoylfurazane is obtained with an approximate yield of 80%. This compound may partially start to crystallize upon standing. IR (cm^{-1}, KBr): 3258, 3035, 2981, 2855, 2361, 2325, 2132, 1733, 1614, 1558, 1516, 1445, 1389, 1340, 1272, 1223, 1108, 1093, 1026, 976, 936, 899, 855, 819, 781, 753, 668. ^{1}H NMR (500 MHz, DMSO-d_6): δ 12.85 ppm. ^{13}C NMR (125 MHz, DMSO-d_6): δ 147.1, 132.6 ppm.

The ethereal solution of bisazidohydroximoylfurazane was saturated with gaseous hydrogen chloride (HCl) below 5 °C. The reaction flask was sealed and then allowed to come to ambient temperature, where it was stirred overnight. After the reaction period, the solvent was removed under reduced pressure. This process resulted in the formation of bis(1-hydroxytetrazolyl)furazane as a yellowish oily liquid. T_{dec}: 91 °C. IR (cm^{-1}, KBr): 3404, 2255, 2128, 1713, 1660, 1463, 1344, 1246, 1197 1103, 1053, 1022, 1005, 982, 922, 895, 819, 758, 729, 709, 686, 673. ^{1}H NMR (500 MHz, DMSO-d_6): δ 9.02 ppm. ^{13}C NMR (125 MHz, DMSO-d_6): δ 142.1, 136.6 ppm. ^{15}N NMR (500 MHz, DMSO-d_6), δ 38.4, 2.1, 17.4, 51.1, 111.1 ppm.

Bis(1-hydroxytetrazolyl)furoxane (**3b-14**): 1.4 g of Dicyanofuroxane (10 mmol) was dissolved in 30 mL of ethanol. Then, 1.3 g (10 mmol) of 50% hydroxylamine solution in 10 mL of ethanol was added to the mixture. The resulting solution was stirred for 30 min. After the stirring period, the solvent was removed, leaving behind a residue. This residue was then suspended in 20 mL of diethyl ether. The solid formed was filtered, yielding 1.7 g (8.4 mmol) of bisaminohydroximoylfuroxane as a yellowish product. The yield of the isolated compound was determined to be 84%. T_{dec}: 180 °C. IR (cm^{-1}, KBr): 3463, 3371, 3309, 1668, 1647, 1579, 1539, 1504, 1418,

1360, 1311, 1229, 1082, 1021, 951, 929, 858, 810, 744, 688. ^{1}H NMR (500 MHz, DMSO-d_6): δ 10.64, 10.08, 6.98, 6.08 ppm. ^{13}C NMR (125 MHz, DMSO-d_6): δ 151.6, 142.5, 139.8, 109.9 ppm. Anal. calcd for $C_4H_6N_6O_4$ (%): C 23.77, H 2.99, N 41.58, found C 23.99, H 2.86, N 41.45; IS: 10 J, FS: 240 N, ESD: 0.25 J.

Bisaminohydroximoylfuroxane weighing 34.8 g (0.2 mol) was dissolved in 500 mL of 34% hydrochloric acid, which corresponded to 595 g (6.0 mol) of hydrochloric acid. The solution was cooled using a salt-ice bath, and an additional 500 g of ice was added to the solution. Separately, 31.1 g (0.5 mol) of sodium nitrite was dissolved in a small amount of water. This sodium nitrite solution was then added dropwise to the reaction mixture over the course of one hour while maintaining the temperature below 0 °C. After the complete addition of the sodium nitrite solution, the reaction mixture was allowed to warm to ambient temperature and further diluted by adding 1 L of ice water. The product was extracted three times with 200 mL of diethyl ether. The organic phase (diethyl ether layer) was separated, and it was dried over magnesium sulfate. As a result, bischlorohydroximoylfuroxane was obtained as a slightly yellow oily liquid. This compound partially started to crystallize, leading to a total yield of 41.1 g (170 mmol) with a yield of 81%.^{1}H NMR (500 MHz, DMSO-d_6): δ 13.78, 13.58 ppm. ^{13}C NMR (125 MHz, DMSO-d_6): δ 150.6, 124.8, 120.3, 110.2 ppm.

In this step, bischlorohydroximoylfuroxane weighing 3.9 g (16 mmol) was dissolved in 20 mL of ethanol. The solution was cooled using an ice bath. Meanwhile, an aqueous solution of sodium azide weighing 2.6 g (40 mmol) was added to the mixture in small portions. After the addition of sodium azide, the reaction mixture was stirred for 1 h. The resulting yellowish solution was then diluted with 70 mL of ice water and the pH was adjusted to 1 by adding concentrated hydrochloric acid. The mixture was extracted with 30 mL and 10 mL of diethyl ether. The organic phase (diethyl ether layer) was separated and dried over magnesium sulfate. Finally, the solvent was removed under reduced pressure, resulting in the formation of bisazidohydroximoylfuroxane as a yellowish oil. The approximate yield of the product was around 80%. ^{1}H NMR (500 MHz, DMSO-d_6): δ 13.00, 12.75 ppm. ^{13}C NMR (125 MHz, DMSO-d_6): δ 149.4, 133.7, 130.4, 107.7 ppm.

The oily compound bisazidohydroximoylfuroxane weighing 3.3 g (13 mmol) was dissolved in 100 mL of diethyl ether. Gaseous hydrogen chloride (HCl) was then passed through the reaction mixture until saturation was reached. This process was conducted at 0–5 °C and the reaction flask was sealed. The sealed reaction mixture was stirred overnight at room temperature. After the stirring period, the solvent was removed, resulting in the formation of bis(1-hydroxytetrazolyl)furoxane. This compound remained as a yellowish resinous substance. IR (cm^{-1}, KBr): 3423, 2460, 1607, 1461, 1402, 1369, 1301, 1259, 1223, 1194, 1135, 1091, 1000, 965, 816, 762, 744, 728, 696. ^{1}H NMR (500 MHz, DMSO-d_6): δ 6.67 ppm. ^{13}C NMR (125 MHz, DMSO-d_6): δ 143.4, 136.8, 134.4, 103.4 ppm.

3,4-Bis(1H-5-tetrazolyl)furoxan (**3b-15**): 3,4-Dicyanofuroxan weighing 1.000 g (7.3 mmol), NaN_3 weighing 1.147 g (17.6 mmol), and NH_4Cl weighing 0.786 g (14.7 mmol) were combined in a 25 mL round-bottomed flask containing 4 mL of DMF. The reaction mixture was then heated at 80 °C for 4 h with vigorous stirring.

After the heating period, the mixture was cooled to room temperature, and 4 mL of water was added. To adjust the pH to a range of 1–2, the solution was acidified using 2% HCl. The resulting solution was then extracted with ethyl acetate. The organic phase was collected, and it was washed with brine to remove any remaining DMF. The organic phase was further dried using magnesium sulfate ($MgSO_4$). Finally, the solvent was evaporated, leading to the formation of a light yellow solid. The yield of the product obtained was 0.972 g, corresponding to a yield of 60%. T_{dec}: 229 °C. ^{13}C NMR (125 MHz, DMSO-d_6): δ 149.0, 146.3, 144.5, 106.8 ppm. IR (cm^{-1}, KBr): 3135, 2908, 2446, 1620, 1579, 1457, 1418, 1386, 1282, 1232, 1203, 1181, 1128, 1093, 1069, 1025, 1003, 960, 932, 823, 780, 735, 704, 542, 486, 390. HRMS (ESI): m/z calcd. for [$C_4H_2N_{10}O_2$-H]: 221.0278 [M-H]; found: 221.0281.

3,4-Bis(1H-5-tetrazolyl)furoxan (**H_2BTF**): 3,4-Dicyanofuroxan weighing 1.000 g (7.3 mmol), NaN_3 weighing 1.147 g (17.6 mmol), and NH_4Cl weighing 0.786 g (14.7 mmol) were combined in a 25 mL round-bottomed flask containing 4 mL of DMF. The reaction mixture was then heated at 80 °C for 4 h with vigorous stirring. This allowed the reaction to proceed.

3-(5-Amino-1,2,4-oxadiazol-3-yl)-4-(1H-tetrazol-5-yl)-1,2,5-oxadiazole 2-oxide (**3b-16**): 3,4-Dicyanofuroxan weighing 5.2 g (43.3 mmol) was dissolved in ethanol (25 mL). The solution was then added over a period of 20 min to a 50% hydroxylamine solution weighing 3.14 g (47.6 mmol) that had been diluted with ethanol (30 mL). After the addition, the solvent was removed under reduced pressure over a period of 2 h. This process involved evaporating the solvent to leave behind the desired compound. Upon filtering the reaction mixture, the product 3-cyano-4-(N'-hydroxycarbamimidoyl)-1,2,5-oxadiazole 2-oxide was obtained as a yellowish solid. T_{dec}: 124 °C; ^{1}H NMR (500 MHz, DMSO-d_6): δ 11.00 (s, 1H), 6.34 (s, 2H) ppm. ^{13}C NMR (125 MHz, DMSO-d_6): δ 149.9, 141.2, 108.4, 96.6 ppm. IR (cm^{-1}, KBr): 3298, 3021, 2266, 1617, 1417, 1171, 1023, 913, 831, 753, 585, 516; elemental analysis for $C_4H_3N_5O_3$ (%):C 28.41, H 1.79, N 41.42. Found: C 28.27, H 1.65, N 40.97.

A solution of potassium hydrogen carbonate weighing 3.6 g (36 mmol) in water (40 mL) was added to a suspension of 3-cyano-4-(N′-hydroxycarbamimidoyl)-1,2,5-oxadiazole 2-oxide weighing 5.07 g (30 mmol) in alcohol (30 mL). Cyanogen bromide weighing 3.8 g (36 mmol) was then added to the mixture in portions. The resulting mixture was stirred for 24 h at 20 °C. During this period, a precipitate formed. The precipitate was collected by filtration and washed with water. As a result, 3-(5-amino-1,2,4-oxadiazol-3-yl)-4-cyano-1,2,5-oxadiazole 2-oxide was obtained as a white solid. The yield of the product obtained was 4.3 g, corresponding to a yield of 74%. T_m: 202 °C. T_{dec}: 239 °C; ^{1}H NMR (500 MHz, DMSO-d_6): δ 8.55 (s, 1H) ppm. ^{13}C NMR (125 MHz, DMSO-d_6): δ 174.2, 159.1, 147.6, 107.5, 98.9 ppm. IR (cm^{-1}, KBr): 3447, 3263, 3193, 2245, 1664, 1605, 1518, 1464, 1378, 1355, 1273, 1104, 1034, 962, 829, 768, 660, 532, 465; elemental analysis for $C_5H_2N_6O_3$ (%): calcd C 30.94, H 1.04, N 43.30%. Found: C 31.10, H 1.23, N 43.24%.

3-(5-Amino-1,2,4-oxadiazol-3-yl)-4-cyano-1,2,5-oxadiazole 2-oxide weighing 2.32 g (12.0 mmol) was dissolved in water (60 mL). Sodium azide weighing 1.56 g (24.0 mmol) and zinc bromide weighing 3.24 g (14.4 mmol) were then added to the solution. The resulting mixture was clear and heated under reflux for 7 h. Refluxing involves heating the mixture with a condenser attached so that any volatile components can condense and return to the reaction mixture. After the reflux period, the solution was cooled to ambient temperature, and 2N HCl (30 mL) was added. Upon addition of the hydrochloric acid, a colorless precipitate was formed. The colorless precipitate was then collected by filtration. T_{dec}: 251 °C; ^{1}H NMR (500 MHz, DMSO-d_6): δ 12.85 (s, 1H), 8.38 (s, 2H) ppm. ^{13}C NMR (125 MHz, DMSO-d_6): δ 174.0, 15 °C 9.7, 148.0, 146.3, 107.3 ppm. IR (cm^{-1}, KBr): 3412, 3288, 3211, 3057, 3012, 1785, 1677, 1629, 1522, 1371, 1287, 1220, 1109, 1056, 993, 830, 541; elemental analysis for $C_5H_3N_9O_3$ (%): calcd C 25.32, H 1.28, N 53.16%. Found: C 25.30, H 1.41, N 52.72%.

3-(5-(Nitroamino)-1,2,4-oxadiazol-3-yl)-4-(1H-tetrazol-5-yl)-1,2,5-oxadiazole 2-oxide (**3b-17**): Compound **3b-16** weighing 1.1 g (5.0 mmol) was added in portions to a mixture of concentrated nitric acid (HNO_3) with stirring and cooling at 0 °C. The resulting mixture was stirred for 2 h at a temperature range of − 10 to 0 °C. After the stirring period, the solvent was removed by blowing air, which likely involved evaporating the solvent. Then, trifluoroacetic acid (CF_3COOH) was added to the flask, likely to dissolve or react with the remaining components in the reaction mixture. The addition of CF_3COOH led to the formation of a precipitate. The precipitate was collected by filtration and washed with CF_3COOH. This washing step helps to purify the solid product. As a result, compound **3b-17** weighing 0.9 g was obtained, corresponding to a yield of 63%. Compound **3b-17** was obtained as a white solid. T_d (onset): 121 °C; 1 H NMR (d_6-DMSO) δ 13.64 (s, 2H) ppm. 13C NMR (d_6-DMSO) δ 176.7, 158.8, 148.0, 145.8, 107.0 ppm. IR (KBr): 3154, 3056, 2962, 1642, 1511, 1401, 1270, 1107, 1058, 986, 843, 745 cm^{-1} elemental analysis (%) for $C_5H_2N_{10}O_5$ (282.02): calcd C 21.29, H 0.71, N 49.65%. Found: C 21.28, H 0.97, N 49.26%.

4,5-Bis(1H-tetrazol-5-yl)-2H-1,2,3-triazole (**3b-18**): 4,5-Dicyano-1,2,3-triazole weighing 2.227 g (18.70 mmol) was dissolved in water (100 mL). Sodium azide weighing 2.930 g (45.10 mmol) and zinc chloride weighing 3.469 g (25.45 mmol) were added to the solution. The resulting mixture was clear and heated under reflux for 3 h. Refluxing involves heating the mixture with a condenser attached so that any volatile components can condense and return to the reaction mixture. After the reflux period, a colorless precipitate formed after several minutes. The solution was then cooled to ambient temperature, and concentrated hydrochloric acid weighing 50 mL was added while cooling to 0 °C. The precipitate was removed by filtration under suction, likely using a Buchner funnel setup. The crude product was then washed with water, ethanol, and diethyl ether, each with a volume of 10 mL. This washing step helps to purify the solid product. As a result, 4,5-bis(1H-tetrazol-5-yl)-2H-1,2,3-triazole hydrate was obtained as a colorless solid. The yield of the product obtained was 3.532 g, corresponding to a yield of 85%. T_{dec}: 277 °C; ^{13}C NMR (125 MHz, DMSO-d_6): δ 148.1, 131.8 ppm; MS (DCI$^+$): m/z (%): 206.2 (9) [M+H$^+$]; Anal. calcd for $C_4H_3N_{11}\cdot H_2O$ (%): C 21.53, H 2.26, N 69.04; found: C 21.74, H 2.36, N 67.73; Anal. calcd for $C_4H_3N_{11}$ (%): C 23.42, H 1.47, N 75.11; found: C 24.12, H 1.81, N 73.97.

4,5-Bis(1-hydroxytetrazol-5-yl)-2H-1,2,3-triazole dihydrate (**3b-19**): 4,5-Dicyano-2H-1,2,3-triazole weighing 2.111 g (17.73 mmol) was dissolved in water (100 mL). Hydroxylamine, with a volume of 2.2 mL (35.97 mmol), was then added to the solution. The resulting mixture was heated under reflux for 2 h. Refluxing involves heating the mixture with a condenser attached so that any volatile components can condense and return to the reaction mixture. After the reflux period, the solution was concentrated under reduced pressure to reduce its volume to approximately 50 mL. This process likely involved using a rotary evaporator to remove the solvent under reduced pressure. As a result of concentration, a precipitate formed. The precipitate was then filtered off, likely using a Buchner funnel setup. The isolated product obtained from the filtration was **3b-19**, which weighed 3.013 g (16.27 mmol). This corresponds to a yield of 92%. Compound **3b-19** is referred to as 2H-1,2,3-triazole-4,5-bis(carboxamidoxime) and it exists as a colorless solid. T_{dec}: 252 °C. ^{1}H NMR (500 MHz, DMSO-d_6): δ 9.91 (s, 2H), 6.67 ppm (s, 4H); ^{13}C NMR (125 MHz, DMSO-d_6): δ 146.7, 133.6 ppm; Anal. calcd for $C_4H_7N_7O_2$ (%): C 25.95, H 3.81, N 52.96; found: C 26.41, H 3.68, N 52.56; sensitivities (grain size: 40 J, FS > 360 N, ESD 0.75 J).

2H-1,2,3-Triazole-4,5-bis(carboxamidoxime) weighing 2.01 g (10.84 mmol) was dissolved in a solution of concentrated hydrochloric acid (25 mL) and water (25 mL). The resulting solution was then cooled down to 10 °C. Sodium nitrite weighing 3.73 g (54.09 mmol) dissolved in water (15 mL) was added dropwise to the solution while maintaining the temperature at 10 °C. After the complete addition of sodium nitrite, the solution was stirred at ambient temperature overnight, allowing the reaction to proceed. The reaction mixture was then extracted three times with diethyl ether (25 mL). The organic layer from each extraction was combined. The combined organic layer was washed with brine (20 mL) to remove any impurities or remaining aqueous components. Next, the organic layer was dried over magnesium sulfate, which helps remove any residual water present. Insolubles were filtered off, likely using a Buchner funnel setup. The solvent was evaporated from the filtrate to yield 2H-1,2,3-triazole-4,5-bis(carboxoxim dichloride), which weighed 2.155 g (9.62 mmol). This corresponds to a yield of 89%. Compound 2H-1,2,3-triazole-4,5-bis(carboxoxim dichloride) exists as a colorless solid. T_{dec}: 114 °C. ^{1}H NMR (500 MHz, DMSO-d_6): δ 15.74 (s, 1H), 12.73 ppm (s, 2H); ^{13}C NMR (125 MHz, DMSO-d_6): δ 137.9, 126.1 ppm; Anal. calcd for $C_4H_3Cl_2N_5O_2$ (%): C 21.45, H 1.35, N 31.26; found: C 22.40, H 1.40, N 30.78; sensitivities (grain size: 360 N, ESD 0.4 J.

A solution of 2H-1,2,3-Triazole-4,5-bis(carboxoxim dichloride) weighing 2.817 g (12.58 mmol) in ethanol (35 mL) was cooled to 0 °C. Sodium azide weighing 5.747 g (88.40 mmol) dissolved in water (25 mL) was then added dropwise to the cooled solution while maintaining the temperature below 5 °C. The resulting solution was clear and yellow in color. It was stirred overnight, allowing the reaction to proceed. After the stirring period, the solution was acidified with a 2 M hydrochloric acid solution (30 mL). This step helps to neutralize any remaining base and convert it into the corresponding chloride salt. The acidified solution was then extracted four times with diethyl ether (100 mL) to separate the organic phase from the aqueous phase. The combined organic layers were washed with brine to remove any impurities, and then dried over magnesium sulfate. The drying agent (magnesium sulfate) was then filtered off, leaving behind a solution of compound **3b-18** in diethyl ether. The solution of compound **3b-18** in diethyl ether was cooled to 0 °C, and hydrogen chloride gas was bubbled through intermittently while ensuring that the temperature did not exceed 10 °C. This step likely involved passing hydrogen chloride gas through the solution using a suitable apparatus. The saturated solution obtained after bubbling hydrogen chloride gas was stirred for 3 days, allowing the reaction to proceed further. After the stirring period, the solvent was evaporated in air, leaving behind a colorless residue. This residue was then recrystallized in water, which involved dissolving the residue in hot water followed by cooling to induce crystallization. The resulting crystals were collected by suction filtration, yielding compound **3b-19** weighing 1.126 g (4.75 mmol). This corresponds to a yield of 38%. Compound **3b-19** is obtained as a colorless solid. T_{dec}: 246 °C; ^{1}H NMR (500 MHz, DMSO-d_6): δ 9.11 ppm (s, 2H, OH); ^{13}C NMR (125 MHz, DMSO-d_6): δ 138.9, 129.1 ppm; ^{15}N NMR (DMSO-d_6): δ 2.6, 18.9, 53.5, 54.9, 100.3, 116.1 ppm; Anal. calcd for $C_4H_3N_{11}O_2 \cdot 2H_2O$ (%): C 17.59; H 2.58, N 56.40; found: C 18.06, H 2.56, N 55.94; sensitivities (grain size: 360 N, ESD 1.50 J.

5,5′-(2-(Nitromethyl)-2H-1,2,3-triazole-4,5-diyl)bis(1H-tetrazole) (**3b-20**): T_m: 132 °C; T_{dec}: 281 °C; ^{1}H NMR (500 MHz, DMSO-d_6): δ 9.8 (s, 4H), ^{13}C NMR (125 MHz, DMSO-d_6): δ 150.2, 95.06 ppm; IR (cm^{-1}, KBr): 3655, 3622, 3429, 3105, 3070, 3040, 3601, 2976, 1911, 1625, 1578, 1509, 1479, 1465, 1449, 1387, 1365, 1331, 1294, 1261, 1210, 1201, 1168, 1068, 1033, 1013, 972, 898, 835, 753, 744, 699, 657, 607, 565, 538, 513; elemental analysis calcd for $C_5H_4N_{12}O_2H_2O$ (%): C 21.28, H 2.14, N 59.57; found: C 21.61, H 2.511, N 59.55.

4,5-Di(1H-tetrazol-5-yl)-1H-imidazol-2-amine (**3b-21**): Sodium azide weighing 0.395 g (6 mmol) and iodine weighing 0.06 g (0.47 mmol) were added to a solution of 2-amino-1H-imidazole-4,5-dicarbonitrile weighing 0.266 g (2 mmol) in DMF (10 mL). The resulting mixture was refluxed for 6 h with stirring. Refluxing involves heating the mixture with a condenser attached so that any volatile components can condense and return to the reaction mixture. After the reflux period, the reaction mixture was cooled to room temperature and hydrochloric acid (10 mL, 1 M) was added. This acidification step helps neutralize any remaining base and convert it into the chloride salt. The reaction mixture was then extracted with ethyl acetate. This extraction step allows the desired product to be separated from other impurities or side products. The crude product obtained from the extraction was then purified by column chromatography over silica gel using a solvent system such as n-hexane–EtOAc. This technique helps separate different compounds present in the mixture based on their polarity. The purified product was dried over sodium sulfate to remove any residual moisture. Sodium sulfate is commonly used as a drying agent to absorb water. The solvent was removed under reduced pressure, likely using a rotary evaporator, leaving behind the isolated product as an orange solid. The product obtained had satisfactory purity after purification. (0.225 g, 51.3%). T_{dec}: 265 °C. IR (KBr): 3136, 2934, 2241, 2150, 1651, 1523, 1437, 1388, 1344, 1307, 1251, 1197, 667, 642, 505. ^{1}H NMR (500 MHz, DMSO-d_6): δ 7.21 (s, 2H). ^{13}C NMR (125 MHz, DMSO-d_6): δ 162.7, 153.8, 142.3.

General procedure for syntheses of compounds **3b-22** and **3b-27**: In this reaction, a solution of 3,5-dichloromethyl-1,2,4-triazole weighing 0.36 g (2.0 mmol) and either the ammonium salt of 4-chloro-3,5-dinitropyrazole or 4-amino-3,5-dinitropyrazole (at a molar equivalent of 1.1, weighing 2.2 mmol) were prepared in DMF (10 mL). The resulting solution was then heated to 80 °C for a duration of 8 h. After the reaction period, the solution was poured into 75 g of ice water. The addition of ice water causes the reaction mixture to cool rapidly and facilitates the precipitation of the product. The formed precipitate was then filtered off and washed with water to remove any soluble impurities and then with CH_2Cl_2 (dichloromethane) to further purify the solid. Finally, the collected solid was dried in air to remove any remaining moisture. This could involve leaving it in an open container to allow the solvent to evaporate completely.

4-Amino-3,5-bis(4-amino-3,5-dinitro-pyrazol-1yl)methyl)-1,2,4-triazole (**3b-22**): 0.68 g of **3b-22** (yield: 75%) was obtained as a yellowish solid. T_{dec}: 287 °C; ^{1}H NMR (500 MHz, DMSO-d_6): δ 7.34 (s, 4H), 6.11 (s, 2H), 6.01 ppm (s, 4H); ^{13}C NMR (125 MHz, DMSO-d_6): δ 150.2, 140.4, 131.8, 130.3, 48.7 ppm; IR (cm^{-1}, KBr): 3484, 3467, 3366, 3348, 3203, 3028, 1645, 1571, 1512, 1473, 1422, 1301, 1227, 1150, 1084, 961, 890, 827, 795, 664; Anal. calcd for $C_{10}H_{10}N_{14}O_8$ (%): C 26.44, H 2.22, N 43.17; found: C 26.25, H 2.37, N 41.92; IS: > 40 J; FS: > 360 N.

4-Amino-3,5-bis(4-chloro-3,5-dinitro-pyrazol-1yl)methyl)-1,2,4-triazole (**3b-27**): Compound **3b-27** (0.69 g; yield 70%) was obtained as a white solid. T_{dec}: 309 °C; ^{1}H NMR (500 MHz, DMSO-d_6): δ 6.12 (s, 4H), 5.34 ppm (s, 2H); ^{13}C NMR (125 MHz, DMSO-d_6): δ 160.7, 152.3, 149.7, 128.7, 50.6 ppm; IR (cm^{-1}, KBr): 3149, 3123, 1763, 1610, 1552, 1535, 1414, 1385, 1327, 1268, 1204, 1103, 1046, 955, 874, 838, 731, 611; Anal. calcd for $C_{10}H_6Cl_2N_{12}O_8$ (%): C 24.36, H 1.23, N 34.08; found: C 24.64, H 1.60, N 33.46.

4-Amino-3,5-bis(3,4,5-trinitro-1H-pyrazol-1yl)methyl)-1,2,4-triazole (**3b-23**): A cooled and vigorously stirred suspension of compound **3b-22** weighing 0.91 g (2.0 mmol) in sulfuric acid (98%; 4 mL) at 0 °C was prepared. Hydrogen peroxide (30%, 2 mL) was then added slowly to the suspension while maintaining the temperature at 0 °C. Once the addition of hydrogen peroxide was complete, the reaction mixture was allowed to warm up to room temperature and left stirring for 12 h. After the reaction time, the reaction mixture was poured onto crushed ice weighing 150 g. A white solid formed upon pouring the reaction mixture onto crushed ice. This solid was separated by filtration, followed by washing with water to remove any impurities or residual reactants. The collected solid residue was then dried under vacuum (in vacuo) to remove any remaining moisture and solvents. After drying, compound **3b-23** was obtained as a yield of 0.70 g (68%). Compound **3b-23** refers to the product obtained from the reaction and it exists as a white solid. T_{dec}: 213 °C; ^{1}H NMR (500 MHz, DMSO-d_6): δ 6.06 (s, 4H), 5.18 ppm (s, 2H); ^{13}C NMR (125 MHz, DMSO-d_6): δ 152.5, 150.9, 144.6, 122.5, 50.4 ppm; IR (cm^{-1}, KBr): 3381, 3337, 3273, 3208, 1626, 1479, 1339, 1250, 1080, 906, 845, 804, 770, 670 Anal. calcd for $C_{10}H_6N_{14}O_{12}$ (%): C 23.36, H 1.18, N 38.13; found: C 23.83, H 1.22, N 37.01; IS: 12 J; FS: 120 N.

3,5-Bis(3,4,5-trinitro-1H-pyrazol-1yl)methyl)-1H-1,2,4-triazole (**3b-24**): Method A: A similar procedure to that of **3b-23** was followed with the exception of extending the reaction time to 24 h. Subsequently, the reaction mixture was poured into ice water and the resulting suspension was extracted using ethyl acetate (30 mL). The combined organic phases were then separated using column chromatography, utilizing a hexane/ethyl acetate solvent system. This process yielded **3b-24** as the final product, with a yield of 21%.

Method B: Compound **3b-23** weighing 0.26 g (0.5 mmol) was slowly added to nitric acid (3 mL; 100%) at 0 °C. After stirring for 30 min, the resulting mixture was poured into ice water. The formed white precipitate was collected by filtration, washed with water, and then dried in air. This yielded compound **3b-24** weighing 0.20 g with a yield of 80%. T_m: 168 °C; T_{dec}: 218 °C; ^{1}H NMR (500 MHz; CD_3CN): δ 12.27 (s, 1H), 5.99 (s, 2H), 5.92 (s, 2H); ^{13}C NMR (CD_3CN, 125 MHz): δ 154.3, 153.6, 151.2, 149.8, 145.0, 144.0, 130.7, 129.8, 52.3, 51.6 ppm IR (cm^{-1}, KBr): 3373, 1548, 1477, 1420, 1338, 1287, 1233, 1175, 1009, 909, 885, 847, 806, 777, 670, 575; Anal. calcd for $C_{10}H_5N_{13}O_{12}$ (%): C 24.06, H 1.01, N 36.47; found: C 24.23, H 1.18, N 34.29; IS: 10 J; FS: 120 N.

3,5-Bis(3,4,5-trinitro-1H-pyrazol-1yl)methyl)-1-nitro-1,2,4-triazole (**3b-25**): Dropwise addition of 100% nitric acid (1.5 mL) to acetic anhydride (3 mL) was carried out at 0–5 °C. Subsequently, compound **3b-24** weighing 0.26 g (0.5 mmol) was slowly added at the same temperature range. After complete addition, the reaction mixture was stirred for an additional 30 min and then poured into ice water, resulting in the formation of a white precipitate. The precipitate was collected by filtration, washed with water, and dried in air. This process yielded 3b-25 as a white solid, weighing 0.18 g with a yield of 66%. T_{dec}: 134 °C; ^{1}H NMR (500 MHz, DMSO-d_6): δ 6.37 (s, 4H), 5.95 ppm (s, 4H); ^{13}C NMR (125 MHz, DMSO-d_6): δ 158.7, 157.2, 153.8, 149.5, 145.7, 144.2, 123.4, 122.0, 54.5, 53.4 ppm; IR (cm^{-1}, KBr): 3028, 2977, 2896, 1656, 1552, 1545, 1476, 1446, 1416, 1337, 1314, 1265, 1176, 1120, 1081, 1008, 968, 913, 858, 845, 808, 785, 736, 672, 600, 541; Anal. calcd for $C_{10}H_4N_{14}O_{14}$ (%): C 22.07, H 0.74, N 36.03; found: C 22.24, H 0.73, N 34.75; IS: 4 J; FS: 80 N.

3,5-Bis(3,5-dinitro-4-nitroamino-1H-pyrazol-1yl)methyl)-1H-1,2,4-triazole (**3b-26**): Compound 4-amino-3,5-bis(4-amino-3,5-dinitro-pyrazol-1yl)methyl)-1,2,4-triazole (**3b-22**) weighing 0.45 g (1 mmol) was slowly added to stirring 100% nitric acid at 0–5 °C. After 30 min, the resulting clear solution was poured into 30 g of ice water. The formed white solid was collected by filtration and washed with cold water. This process yielded pure compound **3b-26** weighing 0.35 g with a yield of 66%. T_{dec}: 122 °C; ^{1}H NMR (500 MHz, DMSO-d_6): δ 6.03 (s, 2H), 5.97 ppm (s, 2H); ^{13}C NMR (125 MHz, DMSO-d_6): δ 160.5, 158.2, 153.4, 152.4, 145.7, 144.8, 125.3, 124.8, 54.2, 53.8 ppm; IR (cm^{-1}, KBr): 3274, 1652, 1579, 1511, 1428, 1388, 1342, 1284, 1197, 1105, 1046, 988, 964, 879, 801, 756, 698, 624; Anal. calcd for $C_{10}H_7N_{15}O_{12}\cdot H_2O$ (%): C 21.95, H 1.66, N 38.39; found: C 22.19, H 1.59, N 37.47; IS: 9 J; FS: 120 N.

4-Amino-3,5-bis(3-nitro-{[4,5-d][1,2,3]triazole-3-oxide}-1H-pyrazol-1yl)-methyl)-1,2,4-triazole hydrate (**3b-28**): To a solution of **3b-27** weighing 0.49 g

(1.0 mmol) in methanol (10 mL), hydrazine monohydrate was added (0.2 mL, 4 mmol). The resulting mixture was stirred overnight at room temperature. The formed precipitate was filtered and then treated with water (10 mL). The pH of the mixture was adjusted to 1 by adding 10% HCl. The resulting precipitate was collected by filtration, yielding a gray solid. Yield 0.35 g (75%); T_{dec}: 308 °C; ^{1}H NMR (500 MHz, CD_3CN): δ 6.02 (s, 4H), 5.22 ppm (s, 2H); ^{13}C NMR (CD_3CN, 125 MHz): δ 162.3, 154.4, 147.2, 128.7, 50.6 ppm; IR (cm^{-1}, KBr): 3353, 1638, 151, 1482, 1426, 1318, 1078, 950, 831, 799, 624; Anal. calcd for $C_{10}H_8N_{16}O_6$ H_2O (%): C 25.76, H 2.16, N 48.06; found: C 25.77, H 1.91, N 48.94; IS: 18 J; FS: 240 N.

4-Amino-3,5-bis(3,4-dinitro-1H-pyrazol-1yl)methyl)-1,2,4-triazole (**3b-29**): A solution containing 3,5-dichloromethyl-1,2,4-triazole (0.36 g, 2.0 mmol) and ammonium 3,4-dinitropyrazolate (at a molar equivalent of 1.1; 2.2 mmol) in dimethyl formamide (10 mL) was subjected to overnight heating at 80 °C. The resulting reaction mixture was then poured into 75 g of ice water, leading to the formation of a solution. This solution was subsequently extracted with ethyl acetate. The solvent was removed under reduced pressure, and the remaining product was purified using column chromatography with a hexane/ethyl acetate eluent mixture in a ratio of 2:1. As a result of this purification process, compound **3b-29** was obtained as a white solid, weighing 0.25 g with a yield of 29%. T_m: 181 °C; T_{dec}: 299 °C; ^{1}H NMR (500 MHz, DMSO-d_6): δ 9.26 (s, 2H), 6.10 (s, 2H), 5.73 ppm (s, 4H); ^{13}C NMR (125 MHz, DMSO-d_6): δ 150.7, 148.3, 138.1, 125.4, 49.6 ppm; IR (cm^{-1}, KBr): 3358, 3138, 1663, 1545, 1520, 1465, 1423, 1362, 1339, 1258, 1147, 1119, 1061, 1010, 949, 865, 809, 747, 661; Anal. calcd for $C_{10}H_8N_{12}O_8$ (%): C 28.31, H 1.90, N 39.62; found: C 28.53, H 1.62, N 38.09; IS: 20 J; FS: 240 N.

Synthesis of 5,5′-(4-nitro-1H-pyrazole-3,5-diyl)bis(1H-1,2,4-triazol-3-amine) (**3b-30**): Dimethyl 4-nitro-1H-pyrazole-3,5-dicarboxylate weighing 4.60 g (20.0 mmol) was added gradually to a vigorously stirring mixture of aminoguanidine sulfate weighing 41.2 g (168.02 mmol) in freshly prepared NaOMe solution. The NaOMe solution was prepared using 1.980 g (86.08 mmol) of sodium metal in 50 mL of dry methanol (MeOH). This addition was performed at 0 °C. The resulting mixture was then heated to reflux for a duration of 20 h. After the reflux period, the mixture was cooled to room temperature (RT). Subsequently, it was poured into an ice water mixture consisting of 50 mL water and acidified to pH = 5 using concentrated hydrochloric acid (HCl). After allowing the reaction mixture to react for 20 min, the formed precipitate was collected by filtration. It was washed with cold water and dried in air. This process yielded compound **3b-30** as an orange solid, weighing 4.75 g with a yield of 85.7%. T_{dec}: 353.6 °C; ^{1}H NMR (500 MHz, DMSO-d_6): δ 6.26(s, 4H), 3.59(br) ppm; ^{13}C NMR (125 MHz, DMSO-d_6): δ 161.65, 154.85, 146.67, 136.79 ppm; IR (cm^{-1}, KBr): 3166, 1628, 1509, 1429, 1384, 1344, 1206, 1122, 1048, 953, 807, 703, 620, 542; Anal. calcd for $C_7H_7N_{11}O_2$ (%): C 30.33, H 2.55, N 55.58; found: C 31.29, H 2.75, N 56.37.

Synthesis of 5,5′-(4-nitro-1H-pyrazole-3,5-diyl)bis(3-nitro-1H-1,2,4-triazole) (**3b-31**): A solution of compound **3b-30** weighing 1.4 g (5.0 mmol) in 10 mL of 20% H_2SO_4 was added dropwise to a solution of sodium nitrite weighing 6.90 g

(0.1 mol, 10 equiv) in 40 mL of water at 40 °C. The resulting mixture was stirred at 50 °C for 2 h. After cooling the mixture to room temperature, it was acidified with 20% H_2SO_4 until no further evolution of nitrogen dioxide was observed. The resulting mixture was then extracted with ethyl acetate using 4 portions of 30 mL each. The combined organic phases were dried over magnesium sulfate, and the solvent was removed under vacuum. This process yielded a yellow solid weighing 1.05 g, with a calculated yield of 3.13 mmol, 62.5%; T_{dec}: 238.2 °C; ^{1}H NMR (500 MHz, DMSO-d_6): δ 5.55(br) ppm; ^{13}C NMR(125 MHz, DMSO-d_6): δ 162.88, 146.42, 132.47, 131.56 ppm; IR (cm^{-1}, KBr): 3152, 2146, 1697, 1619, 1521, 1423, 1378, 941, 825, 736, 569; Anal. calcd for $C_7H_3N_{11}O_6$ (%): C 24.94, H 0.9, N 45.70; found: C 25.24, H 1.26, N 45.25.

Synthesis of *N,N′*-((4-nitro-1*H*-pyrazole-3,5-diyl)bis(1*H*-1,2,4-triazole-5,3-diyl)) dinitramide (**3b-32**): To a solution of **3b-30** weighing 1.4 g (5.0 mmol) in concentrated H_2SO_4 (10.0 mL), 100% HNO_3 was added dropwise (3.0 mL) at 0 °C. The resulting mixture was allowed to warm to room temperature and stirred for an additional two hours. The clear solution obtained was then poured into 50 g of ice, resulting in the formation of a precipitate. This precipitate was collected by filtration as a white solid weighing 1.35 g, with a yield of 73.6%. It was further washed with water and dried in air. T_{dec}: 134.44 °C; ^{1}H NMR: δ 4.48(br, 2H) ppm; ^{13}C NMR (125 MHz, DMSO-d_6): δ 157.17, 152.78, 141.41, 131.68 ppm; IR (cm^{-1}, KBr): 3468, 3241, 1628, 1584, 1533, 1509, 1474, 1390, 1322, 1227, 1078, 1033, 932, 774; Anal. calcd for $C_7H_5N_{13}O_6$ (%): C 22.90, H 1.37, N 49.59%; found C 22.53, H 1.66, N 48.72.

Tris[2-(5-amino-1H-tetrazol-1-yl)ethyl]amine (**3b-33**): Dissolve 0.030 mol of cyanogen bromide in 50 ml of dry acetonitrile at 0 °C. Then, add 0.143 mol of sodium azide to the resulting solution. Stir the reaction mixture at a temperature range of 0–15 °C for a duration of 4 h. After the reaction is complete, filter out the inorganic salt from the reaction mixture. Next, add the solution to a suspension containing 5 mmol of triamine in 15 ml of water at 0 °C. Stir the reaction mixture for 1–2 days at ambient temperature. Once the reaction period is finished, dry the solvent by letting it evaporate in air. Wash the product with water and then with acetonitrile to remove any impurities. T_m: 123 °C; IR (cm^{-1}, KBr): 3341, 3175, 2952, 2840, 1647, 1589; ^{1}H NMR (500 MHz, CD_3CN-d_6): δ 6.66 (s, 6H), 4.06 (t, 6H), 2.90 (t, 6H); ^{13}C NMR (CD_3CN, 125 MHz): δ 155.58, 51.27, 42.83; ^{15}N NMR (CD_3CN): δ 6.87, − 20.52, − 89.30, − 173.92, − 334.39, − 351.98; Anal. calcd for $C_9H_{18}N_{16}O$ (350.35): C, 30.85; H, 5.18; N, 63.97; found C, 30.92; H, 5.29; N, 62.83.

1-(2-(Bis(2-(5-((2′,2′,2′-trinitroethyl)amino)-tetrazol-1-yl)ethyl)amino)eth-yl)-N-(2″,2″,2″-trinitroethyl)-tetrazol-5-amine (**3b-34**): Formalin (2 mmol of a 37% solution) was added slowly to an aqueous suspension containing **3b-33** weighing 0.35 g (1 mmol) in 2 mL of water. The resulting solution was then heated to 70 °C and stirred until it became clear. Afterward, it was cooled to room temperature (RT), and the water was removed by drying in air. Next, methanol (10 mL) was added to dissolve the resulting solid. To this solution, trinitromethane weighing 0.17 g (1.1 mmol) was added. The mixture was heated to 65 °C and stirred for 4 h. Following the reaction time, water (10 mL) was added, resulting in the formation of a precipitate. This precipitate was collected by filtration, washed with water, and finally dried to obtain the pure product, **3b-34**, as a pale yellow solid. The yield of the product was 0.34 g, corresponding to a yield of 41%. T_{dec}: 141 °C; ^{1}H NMR (500 MHz, CD_3CN-d_6): δ 6.18 (t, 3H), 5.28–5.25 (d, 6H), 4.09 (t, 6H), 2.90 ppm; ^{13}C NMR (CD_3CN, 125 MHz): δ 150.6, 127.3, 78.5, 48.9 ppm; IR (cm^{-1}, KBr): 3390, 3272, 3000, 2954, 1589, 1459, 1420, 1304, 1110, 891 955, 809, 637, 540; Anal. calcd for $C_{15}H_{21}N_{25}O_{18}$ (%): C 21.46, H 2.52, N 41.71; found C 21.45, H 2.50, N 40.71.

1-(2-(Bis(2-(5-((2′,2′,2′-trinitroethyl)amino)-tetrazol-1-yl)ethyl)amino)ethyl)-N-(2″,2″,2″-trinitroethyl)-tetrazol-5-amine (**3b-35**): Aminotetrazole compounds, obtained through the synthesis of primary amines and cyanogen azide, underwent treatment with excess 100% nitric acid. This treatment resulted in the formation of white solid nitroiminotetrazoles **3b-35** with good yields. IR (cm^{-1}, KBr): 3431, 3258, 3075, 2843, 1583, 1492, 1449, 1308, 1256, 1040, 719; ^{1}H NMR (500 MHz, DMSO-d_6): δ 2.99 (t, 6H), 4.12 (t, 6H), 12.71 ppm (br, 3H); ^{13}C NMR (125 MHz, DMSO-d_6): δ 44.7, 49.5, 149.8 ppm; ^{15}N NMR (DMSO-d_6): δ 351.7, 168.9, 156.3, 155.1, 26.1, 25.0, 15.1 ppm.

Tris(5-tetrazolo)methane (**3b-36**): A solution containing 3.88 g of potassium tricyanomethide (30 mmol), 7.81 g of sodium azide (120 mmol), and 20.27 g of zinc bromide (90 mmol) in 150 mL of water was prepared. To this solution, 8 mL of 4N HCl was added, and the mixture was refluxed overnight. The resulting mixture was then cooled and filtered. The obtained solids were washed with water and then dissolved in 200 mL of 1 M NaOH. After stirring at room temperature for 6 h, the solution was filtered, and the filtrate was neutralized with 75 mL of 4N HCl. The solids formed were filtered off, washed with water, and dried, resulting in an off-white solid weighing 2.52 g. The yield of the product was 38%. ^{13}C NMR (125 MHz, DMSO-d_6): 154.46, 29.37.

References

1. T.K. Kim, J.H. Choe, B.W. Lee, K. Chung, Bull. Korean Chem. Soc. **33**, 2765–2768 (2012)
2. H. Huang, Y. Shi, Y. Liu, J. Yang, Dalton Trans. **45**, 15382–15389 (2016)
3. Y. Zhang, C. Zhou, B. Wang, Y. Zhou, K. Xu, S. Jia, F. Zhao, Propellants Explos. Pyrotech. **39**, 809–814 (2014)
4. Y. Tang, C. He, L.A. Mitchell, D.A. Parrish, J.M. Shreeve, Chem. Eur. J. **22**, 11846–11853 (2016)
5. Q. Yu, H. Yang, X. Ju, C. Lu, G. Cheng, ChemistrySelect **2**, 688–696 (2017)
6. C. Yan, K. Wang, T. Liu, H. Yang, G. Cheng, Q. Zhang, Dalton Trans. **46**, 14210–14218 (2017)
7. L.L. Fershtat, I.V. Ananyev, N.N. Makhova, RSC Adv. **5**, 47248–47260 (2015)
8. E.C. Johnson, E.J. Bukowski, J.J. Sabatini, R.C. Sausa, E.F.C. Byrd, M.A. Garner, D.E. Chavez, ChemPlusChem **84**, 319–322 (2019)
9. T.S. Hermann, T.M. Klapotke, B. Krumm, J. Stierstorfer, J. Heterocycl. Chem. **55**, 852–862 (2018)
10. H. Wei, C. He, J. Zhang, J.M. Shreeve, Angew. Chem. Int. Ed. **54**, 9367–9371 (2015)
11. C. He, H. Gao, G.H. Imler, D.A. Parrishc, J.M. Shreeve, J. Mater. Chem. A **6**, 9391–9396 (2018)

12. L. Zhai, F. Bi, Y. Luo, N. Wang, J. Zhang, B. Wang, Sci. Rep. **9**, 4321–4329 (2019)
13. T.M. Klapotke, D.G. Piercey, J. Stierstorfer, Propellants Explos. Pyrotech. **36**, 160–167 (2011)
14. L. Zhai, X. Fan, B. Wang, F. Bi, Y. Lia, Y. Zhu, RSC Adv. **5**, 57833–57841 (2015)
15. A.B. Sheremetev, V.O. Kulagina, N.S. Aleksandrova, D.E. Dmitriev, Y.A. Strelenko, Propellants Explos. Pyrotech. **23**,142–149 (1998)
16. B. Wu, H. Yang, Q. Lin, Z. Wang, C. Lu, G. Cheng, New J. Chem. **39**, 179–186 (2015)
17. D.E. Chavez, M.A. Hiskey, J. Heterocycl. Chem. **35**, 1329–1332 (1998)
18. Z. Fu, R. Su, Y. Wang, Y. Wang, W. Zeng, N. Xiao, Y. Wu, Z. Zhou, J. Chen, F. Chen, Chem. Eur. J. **18**, 1886–1889 (2012)
19. Z. Fu, C. He, F. Chen, J. Mater. Chem. **22**, 60–63 (2012)
20. Z. Fu, Y. Wang, L. Yang, R. Su, J. Chen, F. Nie, J. Huang, F. Chen, RSC Adv. **4**, 11859–11861 (2014)
21. D.E. Chavez, M.A. Hiskey, D.L. Naud, Propellants Explos. Pyrotech. **29**, 209–215 (2004)
22. A. Saikia, R. Sivabalan, B.G. Polke, G.M. Gore, A. Singh, A.S, Rao, A.K. Sikder, J. Hazard. Mater. **170**, 306–313 (2009)
23. Y. Joo, B. Twamley, S. Garg, J.M. Shreeve, Angew. Chem. Int. Ed. **47**, 6236–6239 (2008)
24. F. Boneberg, A. Kirchner, T.M. Klapotke, D.G. Piercey, M.J. Poller, J. Stierstorfer, Chem. Asian J. **8**, 148–159 (2013)
25. D. Chen, H.W. Yang, Z.X. Yi, H.L. Xiong, L. Zhang, S.G. Zhu, G.B. Cheng, Angew. Chem. Int. Ed. **57**, 2081–2084 (2018)
26. T.K. Kim, B.W. Lee, H.W. Lee, K. Chung, J.S. Kim, Bull. Korean Chem. Soc. **34**, 1864–1866 (2013)
27. Y. Li, Z. Zhang, Z. Ge, B. Wang, W. Lai, Y. Luo, Chin. J. Chem. **31**, 520–524 (2013)
28. D. Fischer, T.M. Klapotke, M. Reymann, J. Stierstorfer, M.B.R. Volkl, New J. Chem. **39**, 1619–1627 (2015)
29. H. Huang, Z. Zhou, L. Liang, J. Song, K. Wang, D. Cao, C. Bian, W. Sun, M. Xue, Z. Anorg, Allg. Chem. **638**, 392–400 (2012)
30. T.I. Godovikova, S.K. Vorontsova, L.D. Konyushkin, S.I. Firgang, O.A. Rakitin, Russ. Chem. Bull. Int. Ed., **58**, 406–409 (2009)
31. Y. Liu, C. He, Y. Tang, G.H. Imler, D.A. Parrishc, J.M. Shreeve, Dalton Trans. **47**, 16558–16566 (2018)
32. A.A. Dippold, D. Izsak, T.M. Klapotke, C. Pfluger, Chem. Eur. J. **22**, 1768–1778 (2016)
33. G. Zhao, C. He, H. Gao, G.H. Imler, D.A. Parrish, J.M. Shreeve, New J. Chem. **42**, 16162–16166 (2018)
34. D. Srinivas, V.D. Ghule, K. Muralidharan, RSC Adv. **4**, 7041–7051 (2014)
35. Y. Tang, H. Gao, D.A. Parrish, J.M. Shreeve, Chem. Eur. J. **21**, 11401–11407 (2015)
36. M. Xu, G. Cheng, H. Xiong, B. Wang, X. Ju, H. Yang, New J. Chem. **43**, 11157–11163 (2019)
37. Q. Zhang, J. Zhang, D.A. Parrish, J.M. Shreeve, Chem. Eur. J. **19**, 11000–11006 (2013)
38. Y. Joo, J.M. Shreeve, Angew. Chem. Int. Ed. **121**, 572–575 (2009)
39. A. Chafin, D.J. Irvin, M.H. Mason, S.L. Mason, Tetrahedron Lett. **49**, 3823–3826 (2008)

Chapter 4
Tetracyclic-Based Energetic Materials

The more N-heterocycle structures present in a molecule, the higher the heat of formation. It can be easily attributed to the presence of a larger number of nitrogen atoms and –C=N– bonds in the four or more ring systems. Although there are many studies on bicyclic and tricyclic nitrogen-rich compounds, tetracyclic nitrogen-rich compounds are relatively less reported. Herein, the four or more ring systems recently reported as well as their synthesis and properties are presented.

4.1 Tetracyclic Oxadiazole-Based Energetic Compounds

Among those heterocyclic skeletons, oxadiazole has been paid considerable attention due to its high density and oxygen content. A series of energetic compounds containing multiple oxadiazole rings were reported. As shown in Scheme 4.1, many bicyclic oxadiazole were synthesized from 4-amino-N-hydroxyfurazan-3-carboximide (**AAOF**) under different reaction condition. The oxidative coupling reaction of **2a-44** above mentioned can give the tetracyclic compound 3,3′-azobis[4-(5-amino-1,2,4-oxadiazol-3-yl)-1,2,5-oxadiazole] (**4a-1**) using potassium permanganate in concentrated hydrochloric acid [1]. The further nitration of **4a-1** in a mixture of 100% HNO_3 and Ac_2O at 0 °C was reported by M. Lu's group. One amino group on **4a-1** was converted to a carbonyl group to give **4a-2** instead of nitramine group. Other nitration reagents were attempted to nitrify **4a-1** into nitramine compound such as 100% HNO_3 or a mixture of HNO_3 and H_2SO_4, but failed. The nitramine product **4a-3** can be obtained only using a mixture of 100% HNO_3 and Ac_2O in acetic acid [2]. The carbonyl derivative **4a-1** was transformed to dicarbonyl derivative **4a-4** by increasing the reaction temperature to 25 °C. Then the reduction of **4a-4** was proceeded using hydrazine monohydrate and followed by the addition of HCl (10%) result in the formation of the hydrazine **4a-5**. Furthermore, **4a-6** was prepared from the oxidation reaction of **4a-1** in the mixture of CH_3SO_3H, Na_2WO_4 and H_2O_2

G. Cheng et al., *Nitrogen-Rich Polycyclic-Based Energetic Materials*,
https://doi.org/10.1007/978-981-97-4569-2_4

(30%). The oxidation of **2a-44** in the mixture of H_2SO_4 and H_2O_2 can also give **4a-6**. Similar to the synthetic route to **4a-4**, the dicarbonyl derivative **4a-7** was obtained by nitration of **4a-6** in a mixture of 100% HNO_3 and Ac_2O at 25 °C [3]. Another tetracyclic compound **4a-8** was formed through a direct oxidative coupling reaction of **2a-46** using potassium permanganate in hydrochloric acid [4]. Similar to the oxidative coupling reaction of **4a-8**, tetracyclic compounds **4a-9**, **4a-10** and **4a-11** were synthesized through the oxidative coupling reaction of bicyclic intermediates, which were obtained by the treatment of **AAOF** with acetic anhydride, trifluoroacetic anhydride and the mixture of triethyl orthoformate and carbohydrazide, respectively [2]. As shown in Table 4.1, tetracyclic compounds **4a-(1–11)** containing azo-bridged furazan-oxadiazole reach a good balance between energy and safety. They exhibit relatively high detonation velocity (D > 6602 m s^{-1}) and detonation pressure (P > 22.6 GPa), especially for compound **4a-3** (D = 10,114 m s^{-1}, P = 46.1 GPa). And they all possess an acceptable sensitivity to impact and friction (IS > 13 J; FS > 80 N). It is noteworthy that the decomposition temperatures of these compounds are pretty high with values between 240 and 322 °C.

Apart from the azo-bridged tetracyclic compounds, the synthesis of bisnitramide-bridged oxadiazoles were also reported. When 4-(1,2,4-oxadiazol-3-yl)furazan-3-amine was treated with 37% aqueous formaldehyde solution, the methylene-bridged product **4a-12** was obtained. Furthermore, the ethylene-bridged **4a-14** was also synthesized by double nucleophilic substitution reaction of 3-nitro-4-(1,2,4-oxadiazol-3-yl) furazan and ethylene diamine starting from **AAOF** (Scheme 4.1). Further nitration of **4a-12** and **4a-14** was proceeded successfully in a mixture of acetic anhydride and 100% HNO_3 to obtain the bisnitramide tetracyclic products N,N′-dinitro-N,N′-bis[3-(1,2,4-oxadiazol)furazan-4-yl]ethylenediamine (**4a-13**) and N,N′-dinitro-N,N′-bis[3-(1,2,4-oxadiazol)furazan-4-yl]ethylenediamine (**4a-15**) [5].

The tetracyclic bis-1,2,4-oxadiazole (**4a-16**) can also be prepared through the cyclization reaction of **AAOF** and diethyl oxalate. Firstly, oxidation of **4a-16** using TFA and 70% aq H_2O_2 resulted in a complex mixture of target dinitro **4a-17** and mono-nitro product. And further attempt using a mixture of TFA, 90% aq H_2O_2 and $MeNO_2$ as oxidation reagent was proceeded to produce **4a-17** in a 73% yield [6]. And the nitramine product **4a-18** was obtained through the nitration reaction of **4a-16** using fuming HNO_3 [7]. The methylene- and ethylene-bridged tetracyclic compounds have comparable detonation performance (**4a-13**, D = 8182 m s^{-1}, P = 27.7 GPa; **4a-15**, D = 8185 m s^{-1}, P = 27.9 GPa) and sensitivity (**4a-13**, IS = 27 J, FS = 240 N; **4a-15**, IS = 31 J, FS = 240 N) compared with these azo-linkage compounds but a lower thermal stability (**4a-13**, T_{dec} = 149 °C; **4a-15**, T_{dec} = 175 °C). The nitro group and nitramino group substituted derivatives of **4a-16** have a significantly difference in thermal stability (**4a-17**, T_d = 295 °C; **4a-18**, T_d = 84 °C). Furthermore, the tetracyclic compound **4a-19** with high detonation performance (D = 9666 m s^{-1}, P = 42.8 GPa) was also synthesized by coupling reactions of **2a-46** using potassium permanganate in hydrochloric acid [8].

The synthesis of tetracyclic-based compounds containing 1,2,5-oxadiazole and 1,3,4-oxadiazole were shown in Scheme 4.2. Unlike the synthetic route to **2a-59**

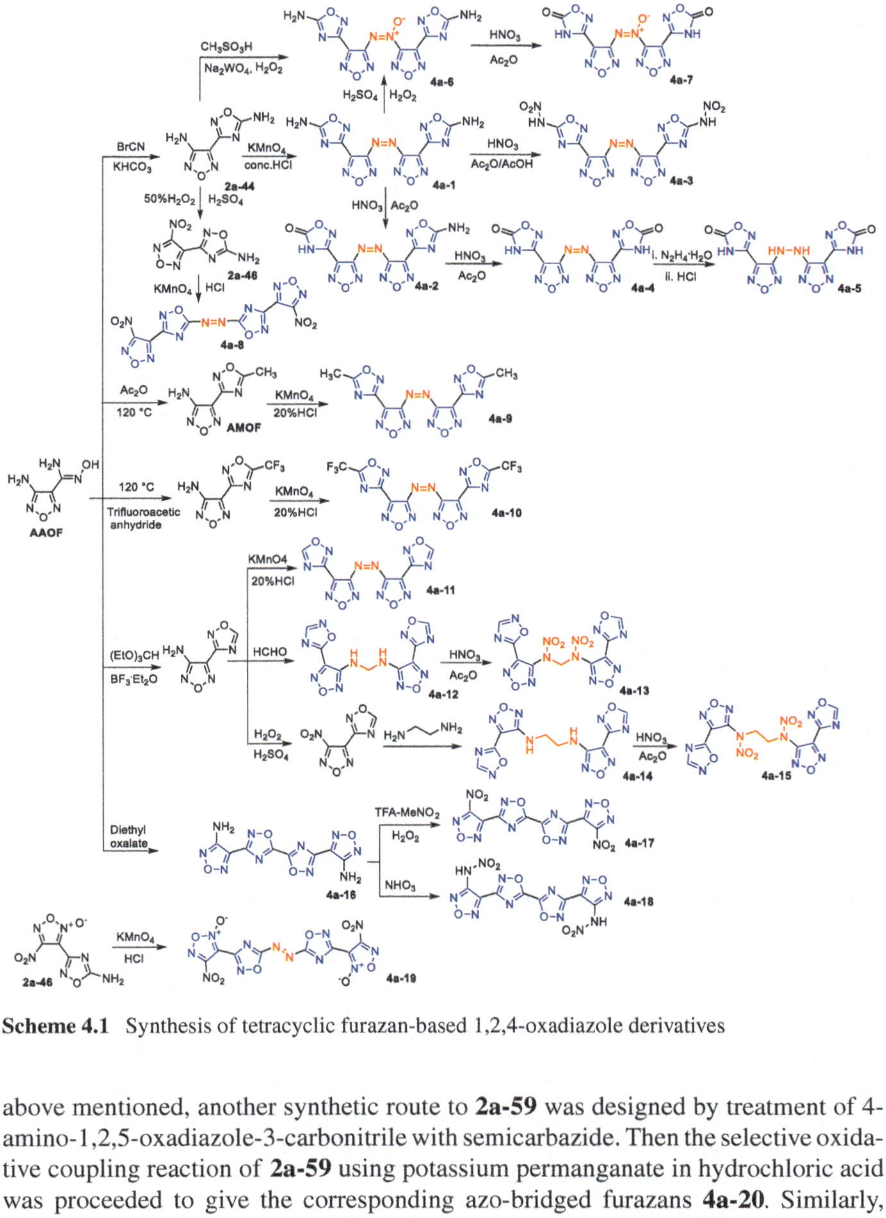

Scheme 4.1 Synthesis of tetracyclic furazan-based 1,2,4-oxadiazole derivatives

above mentioned, another synthetic route to **2a-59** was designed by treatment of 4-amino-1,2,5-oxadiazole-3-carbonitrile with semicarbazide. Then the selective oxidative coupling reaction of **2a-59** using potassium permanganate in hydrochloric acid was proceeded to give the corresponding azo-bridged furazans **4a-20**. Similarly, the tetracyclic nitro-based and nitramino-based azo compounds **4a-21** and **4a-22** were easily oxidative coupled from **2a-60** and **2a-61**, respectively [9]. As shown in Table 4.1, azo compounds 4a-(**20–21**) exhibit a good thermal stability ($T_d > 233$ °C) and detonation performance ($D > 8323$ m s^{-1}, $P > 27.3$ GPa). And the sensitivity to

Table 4.1 Energetic properties of tetracyclic oxadiazole-based energetic compound

Compound	T_{dec} (°C)	D (m s^{-1})	P (GPa)	$\Delta_f H$ (kJ mol^{-1})	IS (J)	FS (N)	d (g cm^{-3})
Tetracyclic furazan-based 1,2,4-oxadiazole derivatives							
4a-1	267	8445	28.3	711	> 80	> 360	1.85
4a-2	270	8419	29.6	943	26	240	1.80
4a-3	317	10,114	46.1	1038	60	80	2.12
4a-4	322	8507	30.7	530	16	160	1.89
4a-5	271	8209	27.9	289	30	240	1.86
4a-6	264	8568	30.8	1108	24	220	1.80
4a-7	307	8844	34.4	493	13	160	1.95
4a-8	256	9240	37.5	962	18	220	1.92
4a-9	271	7781	22.6	697	> 80	80	1.75
4a-10	240	6602	24.3	− 489	> 80	> 360	1.94
4a-11	259	7685	22.7	833	13	> 360	1.68
4a-13	149	8182	27.7	586	27	240	1.77
4a-15	175	8185	27.9	786	31	240	1.75
4a-17	295	/	/	/	16	250	/
4a-18	84	8793	33.3	786	16	180	1.85
4a-19	180	9666	42.8	1189	12	180	1.92
Tetracyclic furazan-based 1,3,4-oxadiazole derivatives							
4a-20	233	8323	27.3	684	> 80	> 360	1.83
4a-21	248	8409	29.3	786	3	78	1.77
4a-22	278	8800	35.1	854	15	220	1.83

impact of nitramino compound (**4a-21**, IS = 3 J) is higher than that of nitro compound (**4a-22**, IS = 15 J).

The synthetic route of a series of polycyclic furazan compounds containing an azo bridge is shown in Scheme 4.2. The oxidative coupling reaction of **BAFF** using potassium permanganate in hydrochloric acid can give the amino polyfurazan compound **4a-23** directly [10]. And the dinitro compound **4a-24** can also be obtained through the same coupling reaction of 4"-nitro-[3,3′,4′,3"-ter(1,2,5-oxadiazol)]-4-amine, which was synthesized by selective oxidation of **BAFF** using hydrogen peroxide in sulfuric acid or ammoniation of **BNFF** [11]. However, the detonation performance of **4a-23** and **4a-24** has not been reported.

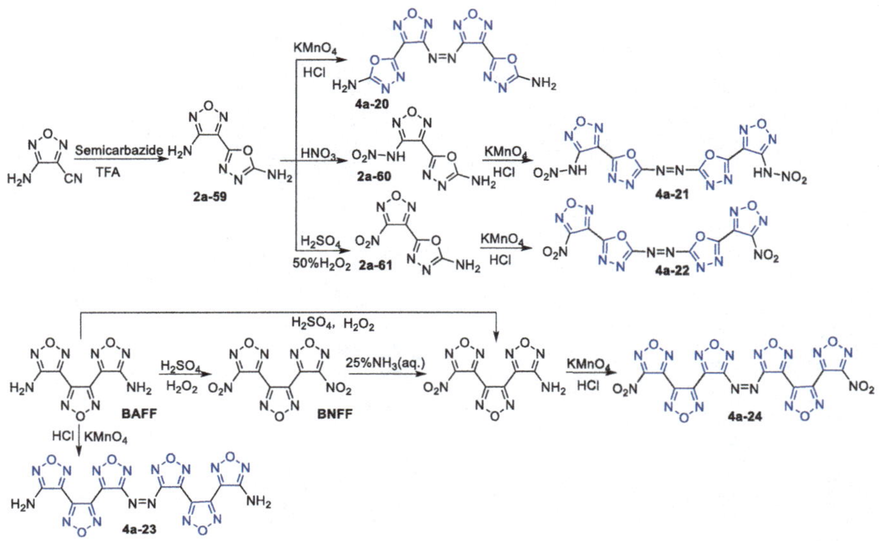

Scheme 4.2 Synthesis of polycyclic furazan-based 1,3,4-oxadiazole derivatives

4.2 Tetracyclic Mixed Heterocyclic-Based Energetic Compounds

Apart from tetracyclic oxadiazole-based energetic compounds, triazole and tetrazole rings are also introduced to tetracyclic compounds. As shown in Scheme 4.3, 5-methly-triazole and 5-amino-triazole were introduced to furazan ring starting from 3-amino-4-cyanofurazan in two steps through intermediate amidrazone. Then tetracyclic compounds **4b-1** and **4b-2** can be prepared by oxidative coupling reaction using potassium permanganate in 20% hydrochloric acid and concentrated hydrochloric acid, respectively. Another route to **4b-2** is the cyclization of diamidrazone, which was obtained from the substitution reaction of **2a-20** or **4a-11** [2]. The nitramino compound **4b-3** was also synthesized through two different routes. One synthetic route is the oxidative coupling reaction of **2f-7** and acidic potassium permanganate. Additionally, similar to the synthetic route to nitramino compound **2f-7**, the reaction of carboxylic acid hydrazides and MNNG also provides a feasible route to **4b-3** [12]. Another tetracyclic **4b-4** containing furazan and triazole was synthesized by a two-step reaction involved the cyclization of **AAOF** and 1,3-diaminourea and subsequently oxidative coupling reaction using potassium permanganate.

Moreover, dinitromethyl and trinitromethyl groups were also introduced to the tetracyclic backbone to boost the energy. As shown in Scheme 4.3, the synthesis process of intermediate dinitromethyl ester comprises four stages. The first is the substitution reaction of 3-carbohydrazide-4-aminofurazan and ethyl 3-ethoxy-3-iminopropionate. Subsequent cyclization in AcOH gives the triazole ring. Then the product was treated with potassium permanganate in hydrochloric acid solution to

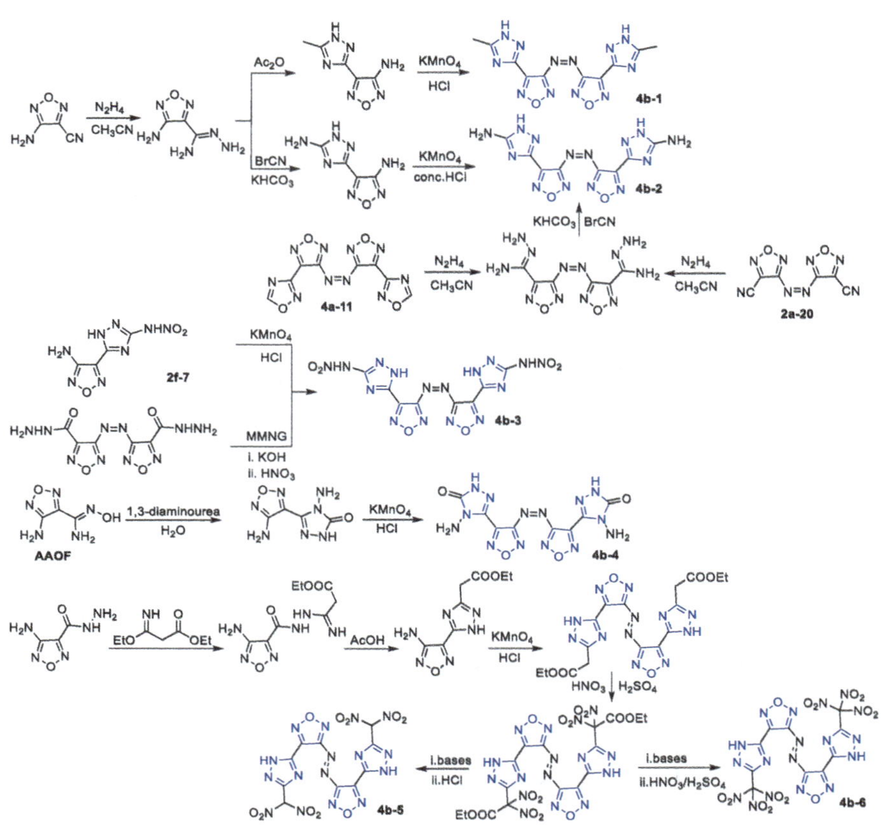

Scheme 4.3 Synthesis of tetracyclic furazan-triazole derivatives

give tetracyclic backbone containing furazan and triazole. Finally, this intermediate was obtained by the nitration reaction using the mixture of nitric acid and sulfuric acid. Then acidification of salt which was directly prepared by acid–base reaction of this intermediate using concentrated HCl in acetone resulted in the neutral dinitromethyl compound **4b-5** [13]. And trinitro compound **4b-6** was obtained by treating energetic salt with mixed nitric acid and sulfuric acid at room temperature [14]. In Table 4.2, compounds **4b-(1-6)** possess a relatively high heat of formation with $\Delta_f H$ values above 770 kJ mol^{-1}. And the decomposition temperatures are between 120 and 350 °C. The trinitro tetracyclic compound **4b-6** exhibits a high detonation performance ($D = 9354$ m s^{-1}, $P = 39.2$ GPa) but high sensitivity to impact (IS = 5 J).

Moreover, tetrazole and oxadiazole backbones have also been introduced to tetracyclic nitrogen-rich compound. As shown in Scheme 4.4, different from the synthesis route to the 3-amino-4-tetrazolefurazan (**HATF**) above mentioned, the treatment of amidrazone with sodium nitrite in HCl can also produce **HATF**. Then the azo

Table 4.2 Energetic properties of tetracyclic mixed heterocyclic-based energetic compound

Compound	T_d (°C)	D (m s^{-1})	P (GPa)	$\Delta_f H$ (kJ mol^{-1})	IS (J)	FS (N)	d (g cm^{-3})
Tetracyclic furazan-triazole derivatives							
4b-1	350	8482	26.4	864	40	80	1.81
4b-2	309	8458	26.2	921	40	> 360	1.79
4b-3	173	8973	34.9	1274	28	280	1.87
4b-4	241	7774	22.4	770	> 80	252	1.68
4b-5	120	8363	29.2	905	5	80	1.76
4b-6	144	9354	39.2	1253	5	160	1.90
Tetracyclic furazan-tetrazole derivatives							
4b-7	258	8477	28.0	1224	47	96	1.69
4b-9	154	9043	35.6	1426	16	180	1.83
4b-10	240	7977	26.5	859	2	/	1.74
Tetracyclic triazole-tetrazole and pyrazole-tetrazole derivatives							
4b-12	251	8190	26.4	474	3	240	1.82
4b-13	248	8884	33.1	396	16	324	1.76
4b-14	179	6810	20.0	750	/	/	1.65
TTP	276	6630	18.4	1045	/	/	1.64

product **4b-7** was acquired through the oxidative coupling reaction of **HATF** using acid $KMnO_4$. Also the oxidative coupling reaction of **2f-1** was performed to give hydroxytetrazole azo compound **4b-8** [15].

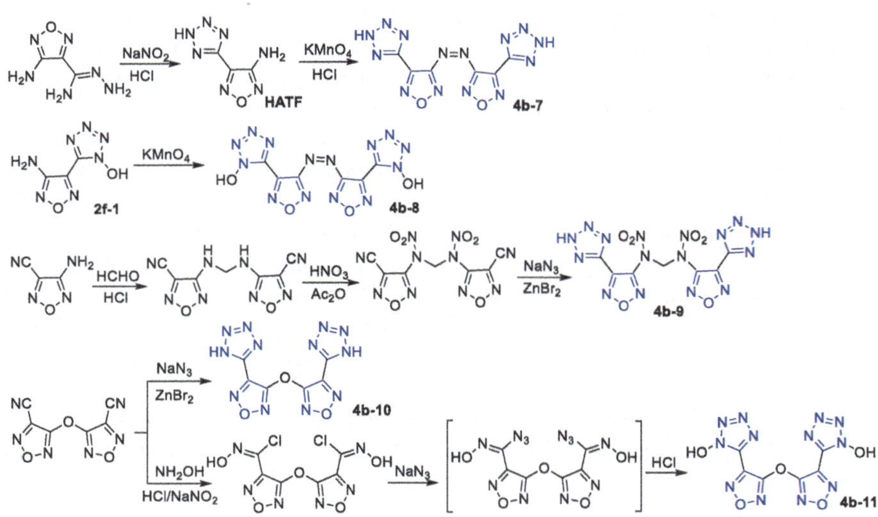

Scheme 4.4 Synthesis of tetracyclic furazan-tetrazole derivative

Other bridges such as ether bridge and methylene bridge were also introduced to link tetrazole and oxadiazole rings. The methylene-bridged compound 4,4′-(methylenebis(azanediyl))bis(1,2,5-oxadiazole-3-carbonitrile) was obtained by the condensation reaction of 3-amino-4-cyanofurazan and 37% aqueous formaldehyde solution under acidic condition. Further nitration of methylene-linked compound with a mixture of 100% nitric acid and acetic anhydride was proceeded successfully to give bisnitramide product. Using zinc bromide as a catalyst, this bisnitramide product was treated with sodium azide in a mixture of CH_3OH and H_2O to form **4b-9**. It is noteworthy that **4b-9** exhibits an interesting "3D-cube layer stacking", which breaks through the limitations of the planar molecular structures in 2D-plane layer stacking due to unusual U-shaped molecular structure and H-bonding interactions reported by M. Lu's group [16].

The synthesis of ether-linked tetracyclic compound 4,4′-oxybis[3,3′-(1H-5-tetrazol)] furazan (**4b-10**) was reported from 3,3′-dicyanodifurazanyl ether. Initial synthetic attempts to **4b-10** by the treatment of 3,3′-dicyanodifurazanyl ether with sodium azide using NH_4Cl as a catalyst was failed. While using zinc salt as a catalyst, the tetrazole was successfully formed from nitriles in water to give **4b-10** [17]. The chlorine oxime compound was reacted with sodium azide in a mixture of ethanol and water to form a azide intermediate, which was not isolated due to its sensitivity. And the reaction mixture was extracted with ethyl ether and then bubbled gaseous HCl into the solution to give the tetrazole-ol compound **4b-11** [18]. The detonation velocities and pressures of compounds **4b-(7–11)** are between 7977 and 9043 m s^{-1} and 28.0 to 35.6 Gpa, respectively. Although **4b-11** possesses an acceptable thermal stability ($T_d = 240$ °C) but a very high sensitivity to impact (IS = 2 J).

Among these N-bridge such as –NH–NH–, –N=N–, –N=N–N–, –N–N=N–N– and –N=N–N–N=N–, the hydrazine (–NH–NH–) is advantageous over the corresponding azo group (–N=N–) both in heat of formation and the formation of hydrogen bonds subsequently leading to higher energy and stability. The most common method to introduce hydrazine group is the reductive reaction of corresponding azo parent using magnesim or zinc powders. Z. Zhou group found that the hydrazine was directly formed during the ion exchange process between the azo silver salt and organic chloride salts without addition of any other reductive agents [19]. As shown in Scheme 4.5, 1,2-bis(3-nitro-1-(1H-tetrazol-5-yl)-1H-1,2,4-triazol-5-yl)diazene (**4b-12**) was obtained through the oxidation of **2f-18** using potassium permanganate. Then many attempts to obtain salts of **4b-12** by ionic metathesis reaction between silver salt and hydroxyl amine hydrochloride was unsuccessful but resulted in corresponding salts of hydrazine compound. Then the netural hydrazine compound **4b-13** was formed by the treatment of corresponding salts with concentrated hydrochloric acid.

And the imidazole-tetrazole based compound 1-[4,5-di(1H-tetrazol-5-yl)-1H-imidazol-2-yl]-1H-tetrazole (**4b-14**) was obtained by the cyclization reaction

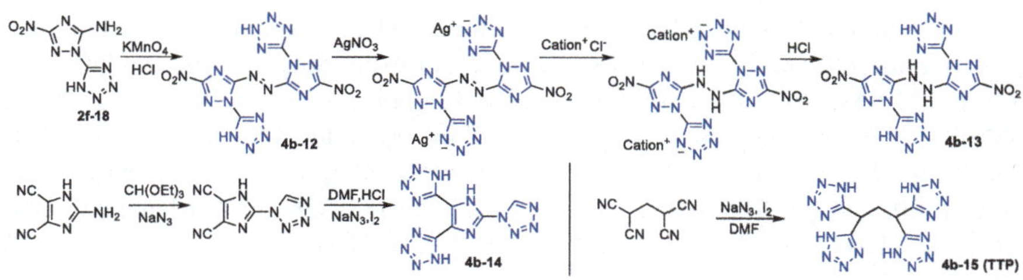

Scheme 4.5 Synthesis of tetracyclic triazole-tetrazole and pyrazole-tetrazole derivatives

using sodium azide and iodine in DMF starting from 2-(1H-tetrazol-1-yl)-1H-imidazole-4,5-dicarbonitrile, which was prepared through the reaction of 2-amino-1H-imidazole-4,5-dicarbonitrile and sodium azide in the presence of triethylorthoformate [20]. Similar to the synthetic route to tetrazole, 1,1,3,3-tetra(1H-tetrazol-5-yl)propane (**TTP**) can be synthesized by the reaction of propane-1,1,3,3-tetracarbonitrile and sodium azide in the presence of iodine [21]. In Table 4.2 compounds **4b-(12-15)**, both azo compounds **4b-12** and **4b-13** exhibit the higher detonation velocity with a value of 8190 and 8884 m s^{-1} than pyrazole-bridged **4b-14** and propane-bridged **4b-15** with a value of 6810 and 6630 m s^{-1}, respectively. And all these compounds possess a high decomposition temperature above 179 °C.

4.3 Experimental Section

3,3′-Azobis[4-(5-amino-1,2,4-oxadiazol-3-yl)-1,2,5-oxadiazole] (**4a-1**): Amino furazans (10 mmol) were dispersed in 50 mL of 20% hydrochloric acid with stirring at an ice-salt bath. A solution of potassium permanganate ($KMnO_4$, 10 mmol) in a minimal amount of water was then added dropwise to the amino furazans solution. The reaction mixture was stirred for 120 min at 30 °C after the addition. Subsequently, the reaction was quenched by using 5% hydrogen peroxide (H_2O_2) until the solution became colorless. To isolate the desired product, the precipitate formed during the reaction was filtered and washed with cold water. Finally, the precipitate was dried under vacuum to obtain a yellow solid with a yield of 76%. T_d (onset): 265 °C; T_d (peak): 267 °C; ^1H NMR (DMSO-d_6, 300 MHz) δ: 8.38(s, 4H) ppm; ^{13}C NMR (DMSO-d_6, 300 MHz) δ: 171.07, 160.11, 155.88, 140.27 ppm; IR (KBr): ν 3577, 3445, 3393, 3319, 3271, 1652, 1599, 1502, 1489, 1374, 1348, 1173, 995, 879, 778, 603 cm^{-1}; HRMS (ESI): calcd for $C_8H_4N_{12}O_4^+$ [M+Na^+] 355.1852, found 355.0363; Anal. calcd for $C_8H_4N_{12}O_4$ (332.20): C 28.92, H 1.21, N 50.60; found C 28.73, H 1.31, N 49.65.

3,3′-Azobis[4-(1,2,4-oxadiazol-5-one-5′-amino-3-yl)-1,2,5-oxadiazole] (**4a-2**): To prepare **4a-2**, 0.33 g (1 mmol) of **4a-1** was added in portions to a mixture of 2 mL of Ac_2O and 1 mL of 100% HNO_3 at 0 °C. The reaction mixture was stirred for 0.5 h. After the completion of the reaction, the mixture was poured into ice-water. A yellow precipitate formed, which was then filtered. The precipitate was washed with water and dried in air to obtain 0.18 g of **4a-2**, corresponding to a yield of 54%. T_d: 269 °C. ^{1}H NMR (500 MHz, DMSO-d_6): d 8.42 ppm (s, 2H), ^{13}C NMR (500 MHz, DMSO-d_6): due to the poor solubility of **4a-2** in deuterated solvents, the ^{13}C NMR data for **4a-2** were not collected. IR (KBr): n 3450, 3280, 1770, 1670, 1610, 1560, 1490, 1340, 1240, 1170, 1030, 973, 895, 811, 764, 615, 526 cm^{-1}. Elemental analysis for $C_8H_3N_{11}O_5$ (332.20): calcd C 28.84, H 0.91, N 46.24%. Found: C 28.83, H 0.92, N 46.26%.

3,3′-Azobis[4-(5-nitramino-1,2,4-oxadiazol-3-yl)-1,2,5-oxadiazole] (**4a-3**): To obtain **4a-3**, 0.66 g (2 mmol) of compound **4a-1** was added in small portions to 10 mL of ice-cold 100% HNO_3 at 0 °C. The mixture was stirred for 30 min at 10 °C until a clear solution formed. The resulting solution was then poured into 50 mL of ice-water. Upon this addition, a precipitate formed. The precipitate was collected by filtration, resulting in a solidified residue. The solid residue was then dried under vacuum conditions (vacuo) to yield **4a-3** as a yellow solid. The yield of this process was found to be 89%. T_d (onset): 304 °C; T_d (peak): 317 °C; ^{1}H NMR (THF-d_6, 300 MHz) δ: 4.29 (s, 2H) ppm; ^{13}C NMR (THF-d_6, 300 MHz) δ: 160.74, 157.15, 146.17, 137.84 ppm; IR (KBr): ν 3435, 3004, 2824, 2708, 1817, 1766, 1616, 1547, 1432, 1384, 1313, 1247, 1130, 1025, 946, 908, 757, 699, 663, 608 cm^{-1}; HRMS (ESI): calcd for $C_8H_2N_{14}O_8^-$ [M-H$^+$] 421.1825, found 421.0005; Anal. calcd for $C_8H_4N_{16}O_6$ (420.05): C 22.87, H 0.96, N 53.33; found C 23.02, H 0.95, N 51.87.

3,3′-Azobis[4-(1,2,4-oxadiazol-5(4H)-one-3-yl)-1,2,5-oxadiazole] (**4a-4**): After obtaining **4a-2**, the mixture was further stirred at room temperature for 4 h. Subsequently, the mixture was poured into ice-water. This resulted in the formation of a yellow precipitate. The yellow precipitate was then filtered and washed with water. Finally, the precipitate was dried in air to obtain compound **4a-4**. The yield of this process was determined to be 0.20 g, corresponding to a yield of 61%. T_d: 322 °C. ^{1}H NMR (500 MHz, DMSO-d_6): no peak. ^{13}C NMR (500 MHz, DMSO-d_6): d 162.18, 159.52, 148.26, 139.89 ppm. IR (KBr): n 2990, 2820, 2700, 1800, 1760, 1620, 1540, 1430, 1310, 1250, 1140, 1020, 991, 943, 906, 873, 795, 749, 698, 618, 526 cm^{-1}. Elemental analysis for $C_8H_2N_{10}O_6$ (334.17): calcd C 28.75, H 0.60, N 41.92%. Found: C 28.73, H 0.61, N 41.93%.

3,3′-Hydrazine-1,2-diylbis[4-(1,2,4-oxadiazol-5(4H)-one-3-yl)-1,2,5-oxadiazole] (**4a-5**): To prepare **4a-5**, a solution of **4a-4** (0.33 g, 1 mmol) in 5 mL of methanol was prepared. Hydrazine (0.12 g, 2 mmol) was then added to the solution, and the mixture was stirred at 25 °C for 0.5 h. Following this, 5 mL of concentrated HCl was added to the mixture, and stirring was continued for another 0.5 h. Methanol was subsequently evaporated under reduced pressure. The resulting precipitate was filtered and washed with water. This yielded 0.25 g of **4a-5** with a yield of approximately 74.9%. T_d: 271 °C. ^{1}H NMR (500 MHz, DMSO-d_6): d 9.16 ppm (s, 2H). ^{13}C NMR (500 MHz, DMSO-d_6): d 159.62, 159.36, 149.35, 134.72 ppm. IR (KBr): n 3359, 1762, 1591, 1496, 1438, 1345, 1287, 1245, 1142, 1209, 1096, 1028, 992, 903, 848, 749, 671, 581 cm^{-1}. Elemental analysis for $C_8H_4N_{10}O_6$ (336.18): calcd C 28.58, H 1.20, N 41.66%. Found: C 28.56, H 1.21, N 41.67%.

3,3′-Azoxybis[4-(5-amino-1,2,4-oxadiazol-3-yl)-1,2,5-oxadiazole] (**4a-6**): Method 1: To obtain **4a-6**, 0.33 g (1 mmol) of compound **4a-1** was added to a mixture of 10 mL of 30% H_2O_2 and 10 mL of concentrated H_2SO_4 at 0 °C. The resulting mixture was stirred for 1 h at 55 °C. After the reaction, the mixture was poured into 20 mL of ice water. To extract the product, the mixture was then treated with ethyl acetate in three portions of 10 mL each. The organic layer containing the product was separated, and the ethyl acetate was removed by vacuum concentration. As a result, 0.15 g of **4a-6** was obtained with a yield of approximately 43.1%.

Method 2: 0.34 g (2 mmol) of compound **2a-44** was added to a mixture containing 10 mL of CH_3SO_3H and a solution of Na_2WO_4 (1.65 g, 5 mmol) in 10 mL of 30% H_2O_2. The reaction was carried out at 0 °C and stirred for 2 h at room temperature (RT). After the reaction, the mixture was poured into 50 mL of ice-water. As a result, a white solid precipitated from the solution.

The white solid precipitate obtained from the mixture was filtered and washed with water. Afterward, it was dried under ambient conditions to obtain 0.17 g of **4a-6**. The yield of this process was approximately 48.9%. T_d: 264 °C. ^1H NMR (500 MHz, DMSO-d_6): d 8.44 ppm (s, 2H) 8.39 ppm (s, 2H). ^{13}C NMR (500 MHz, DMSO-d_6): d 173.42, 173.32, 158.20, 157.02, 153.66, 143.12, 141.67 ppm. IR (KBr): n 3430, 3390, 3150, 1660, 1590, 1570, 1510, 1490, 1470, 1400, 1340, 1280, 1170, 1130, 1040, 913, 876, 764, 652 cm^{-1}. Elemental analysis for $C_8H_4N_{12}O_5$ (336.18): calcd C 27.60, H 1.16, N 48.27%. Found: C 27.61, H 1.17, N 48.26%.

3,3′-Azoxybis[4-(1,2,4-oxadiazol-5(4H)-one-3-yl)-1,2,5-oxadiazole] (**4a-7**): To prepare **4a-7**, 0.35 g (1 mmol) of **4a-6** was added in portions to a mixture of 2 mL of Ac_2O and 1 mL of 100% HNO_3 at 0 °C. The reaction mixture was then warmed to 25 °C and stirred for 4 h. After the completion of the reaction, 0.23 g of **4a-7** was obtained with a yield of approximately 66.1%. T_d: 306 °C. ^1H NMR (500 MHz, DMSO-d_6): no peak. ^{13}C NMR (500 MHz, DMSO-d_6): d 159.61, 159.39, 159.12, 153.07, 148.34, 147.21, 141.61, 139.37 ppm. IR (KBr): n 3000, 1760, 1620, 1530, 1420, 1310, 1250, 1140, 1020, 990, 930, 884, 749, 682, 638, 573, 527 cm^{-1}. Elemental analysis for $C_8H_2N_{10}O_7$ (336.18): calcd C 27.44, H 0.58, N 40.00%. Found: C 27.45, H 0.57, N 40.02%.

3-(4-Nitro-1,2,5-oxadiazol-3-yl)-5-((3-(4-nitroisoxazol-3-yl)-1,2,4-oxadiazol-5-yl)diazenyl)-1,2,4-oxadiazole (**4a-8**): To prepare compound **4a-8**, a solution of **2a-46** (0.594 g, 3.0 mmol) in 10 mL of 37% hydrochloric acid and 3 mL of acetonitrile was prepared. A separate solution of $KMnO_4$ (0.553 g, 3.5 mmol) in 20 mL of water was also prepared. The solution of $KMnO_4$ was added to the solution of **2a-46**, and the resulting reaction mixture was kept at 25 °C for 6 h. Afterward, the reaction mixture was treated with 30% hydrogen peroxide. The resulting precipitate was filtered off, washed with water, and dried in air. This yielded compound **4a-8** with a mass of 0.488 g and a yield of approximately 86%. T_m: 157 °C. T_d: 256 °C. ^{13}C NMR (500 MHz, DMSO-d_6): δ 138.76, 153.19, 159.80, 168.57 ppm. IR (KBr): n 1599, 1551, 1397, 1368, 1287, 1271, 1124, 1042, 1020, 965, 894, 775, 762 cm^{-1}. Elemental analysis for $C_8N_{12}O_8$ (392.16): calcd C 24.50, N 42.86%. Found: C 24.46, N 42.81%.

3,3′-Bis(4-(5-methyl-1,2,4-oxadiazol-3-yl)-4,4′-azofurazan (**4a-9**): To carry out the reaction, 10 mmol of **AMOF** was dispersed in 50 mL of 20% hydrochloric acid with stirring at an ice-salt bath. A solution of potassium permanganate ($KMnO_4$, 10 mmol) in a minimal amount of water was added dropwise to this solution. The reaction mixture was then stirred for 120 min at 30 °C. After the completion of the reaction, it was quenched by adding 5% hydrogen peroxide (H_2O_2) until the solution turned colorless. The precipitate formed during the reaction was isolated by filtration and washed with cold water. It was then dried under vacuum conditions, resulting in the desired product as a yellow solid. The yield of the reaction was found to be 86%. T_m: 142.5 °C; T_d: 271 °C; ^1H NMR (DMSO-d_6, 300 MHz) δ: 2.78 (s, 6H) ppm; ^{13}C NMR (DMSO-d_6, 300 MHz) δ: 179.48, 162.39, 159.98, 157.35, 140.56 ppm; IR (KBr): ν 3466, 2930, 2360, 1713, 1626, 1587, 1459, 1254, 1149, 1024, 979, 927, 902, 865, 710, 596; HRMS (ESI): calcd for $C_{10}H_6N_{10}O_4^+$ [M+Na$^+$] 353.2091, found 353.0443; Anal. calcd for $C_{10}H_6N_{10}O_4$ (330.22): C 36.37, H 1.83, N 42.42; found C 35.86, H 1.82, N 41.25.

3,3′-bis(4-(5-(trifluoromethyl)-1,2,4-oxadiazol-3-yl)-4,4′-azofurazan (**4a-10**):10 mmol of amino furazans was dispersed in 50 mL of 20% hydrochloric acid with stirring at an ice-salt bath. A solution of potassium permanganate ($KMnO_4$, 10 mmol) in a minimal amount of water was then added dropwise to the solution. The reaction mixture was stirred for 120 min at 30 °C. After the reaction time, the reaction was quenched by adding 5% hydrogen peroxide (H_2O_2) until the solution turned colorless. The resulting precipitate was isolated by filtration and washed with cold water. It was then dried under vacuum conditions, resulting in the desired product as a light yellow solid. The yield of the reaction was determined to be 82%. T_d (onset): 224 °C; T_d (peak): 239 °C; ^{13}C NMR (CD_3CN, 300 MHz) δ: 167.03, 166.73, 160.97, 155.98, 136.86 ppm; IR (KBr): ν 3468, 3338, 3329, 2925, 1687, 1632, 1556, 1462, 1334, 1179, 997, 966, 929, 864, 766, 572, 530 cm^{-1}; HRMS (ESI): calcd for $C_{10}F_6N_{10}O_4^+$ [M+H$^+$] 438.1621, found 438.0343; [M+Na$^+$] 461.1915, found 461.1038; Anal. calcd for $C_{10}F_6N_{10}O_4$ (438.16): C 27.41, N 31.97; found C 29.51, N 31.19.

3,3′-Bis (5-amine-1,3,4-oxadiazol-yl)-4,4′-azofurazn (**4a-11**): 10 mmol of amino furazans was dispersed in 50 mL of 20% hydrochloric acid with stirring at an ice-salt bath. A solution of potassium permanganate ($KMnO_4$, 10 mmol) in a minimal amount of water was added dropwise to the solution. The reaction mixture was then stirred for 120 min at 30 °C. After the reaction time, the reaction was quenched by adding 5% hydrogen peroxide (H_2O_2) until the solution turned colorless. The resulting precipitate was isolated by filtration and washed with cold water. It was then dried under vacuum conditions to obtain the desired product as a light yellow solid. The yield of the reaction was determined to be 72%. T_d (onset): 176 °C; T_d (peak): 233 °C; ^1H NMR (DMSO-d_6, 300 MHz) δ: 8.54 (d, J = 6.0 Hz, 2H), 8.40 (d, J = 6.0 Hz, 2H) ppm; ^{13}C NMR (DMSO-d_6, 300 MHz) δ: 162.26, 160.10, 146.93, 140.01 ppm; IR (KBr): ν 3434, 1687, 1636, 1385, 1098, 1037, 805, 705, 469 cm^{-1}; HRMS (ESI): calcd for $C_8H_4N_{12}O_4^+$ [M+H$^+$] 333.2035, found 333.0076; Anal. calcd for $C_8H_4N_{12}O_4$ (332.20): C 28.92, H 1.21, N 50.60; found C 28. 73, H 1.31, N 49.65.

N,N′-Bis[3-(1,2,4-oxadiazol)furazan-4-yl]methylenediamine (**4a-12**): To prepare **4a-12**, 1.53 g (10 mmol) of 4-(1,2,4-oxadiazol-3-yl)furazan-3-amine was suspended in a minimum amount of water at 60 °C. Aqueous formaldehyde solution (37%, 0.81 g) and six drops of concentrated hydrochloric acid were added to the suspension with stirring. The reaction mixture was stirred at 80 °C for 1 h and then cooled to room temperature. As a result, a white precipitate formed. The white precipitate was collected by filtration, washed with water, and dried to obtain **4a-12**. The yield of this process was determined to be 1.51 g, corresponding to a yield of approximately 95%. ^{1}H NMR (DMSO-d_6, 300 MHz) d: 10.02 (s, 2H, CH), 7.01 (t, 2H, NH), 4.97 (t, 2H, CH_2).

N,N′-Dinitro-N,N′-bis[3-(1,2,4-oxadiazol)furazan-4-yl]methylenediamine (**4a-13**): To synthesize **4a-13**, 0.64 g (2 mmol) of compound **4a-12** was added in portions to a mixture of 6 mL of acetic anhydride and 4 mL of 100% nitric acid at 0 °C. The resulting mixture was then stirred at room temperature for 6 h. After the reaction, the mixture was poured into ice water. This caused a white precipitate to form. The white precipitate was collected by filtration, washed with ethanol, and air-dried. This process yielded 0.61 g of **4a-13** with a yield of approximately 75%. IR (KBr): n 3140, 1590, 1416, 1284, 1175, 1097, 1069, 955, 912 cm^{-1}. ^{1}H NMR d: 10.05 (s, 2H, CH), 6.74 (s, 6H, CH_2); ^{13}C NMR d: 169.22, 156.61, 150.38, 143.12, 65.90; anal. calcd for $C_9H_4N_{12}O_8$ (408.21): C 26.48, H 0.99, N 41.18; found C 26.62, H 0.94, N 40.65.

N,N′-Bis[4-(1,2,4-oxadiazol-3-yl)furazan-3-yl]ethane-1,2-diamine (**4a-14**): A solution of ethylenediamine (6.5 g, 0.11 mol) in 30 mL of acetonitrile (MeCN) was added dropwise over 1 h to a vigorously stirred and cooled solution of 4-(1,2,4-oxadiazol-3-yl)-3-nitro-furazan (18.3 g, 0.10 mol) in 100 mL of MeCN. After the addition was complete, the mixture was stirred at room temperature for 1–2 h. Then, it was acidified with acetic acid (AcOH) to adjust the pH to 6. The solvent was removed under vacuum, and the resulting residue was stirred with cold water (100 mL). This led to the formation of a precipitate. The formed precipitate was filtered off and then recrystallized from a 1:1 mixture of methanol (MeOH) and acetic acid (AcOH). The yield of the process was 10.6 g (64%), resulting in a light-gray fine crystalline powder. T_d, 235–236 °C. IR spectrum, ν, cm^{-1}: 3383, 3107, 3036, 2991, 2945, 1624, 1586, 1524, 1499, 1454, 1330, 1284, 1229, 1139, 1096, 1020, 963, 901, 872, 765, 714, 612, 584, 575, 522, 421. Found, %: C 36.43; H 2.66; N 42.02. $C_{10}H_8N_{10}O_4$. Calculated, %: C 36.15; H 2.43; N 42.16.

N,N′-Dinitro-N,N′-bis[3-(1,2,4-oxadiazol)furazan-4-yl]ethylenediamine (**4a-15**): To synthesize **4a-15**, 0.66 g (2 mmol) of N,N′-bis[4-(1,2,4-oxadiazol-3-yl)furazan-3-yl]ethane-1,2-diamine was added in portions to a mixture of 6 mL of acetic anhydride and 4 mL of 100% nitric acid at 0 °C. The resulting mixture was then stirred at room temperature for 6 h. After the reaction, the mixture was poured into ice-water, causing a white precipitate to form. The white precipitate was collected by filtration, washed with ethanol, and air-dried. This process yielded 0.66 g of **4a-15** with a yield of approximately 78%. IR (KBr): n 3127, 1586, 1521, 1441, 1338, 1281, 1178, 1099, 1026, 916, 812 cm^{-1}. ^{1}H NMR d: 10.06 (s, 2H, CH), 4.76 (s, 4H, CH_2); ^{13}C NMR d: 169.02, 156.82, 151.35, 142.91, 50.33; anal.

calcd for $C_{10}H_6N_{12}O_8$ (422.04): C 28.45, H 1.43, N 39.81; found C 28.09, H 1.32, N 39.04.

4,4′-([5,5′-bi(1,2,4-oxadiazole)]-3,3′-diyl)bis(1,2,5-oxadiazol-3-amine) (**4a-16**): A round-bottom flask equipped with a magnetic stir bar, Dean-Stark trap/condenser, thermometer, and liquid addition funnel was prepared. Amidoxime **AAOF** (100 g, 0.700 mol) was added to the flask. Solution A (70 mL) was prepared. This solution consisted of diethyl oxalate (120 g, 0.820 mol), 2,4,6-collidine (18.0 g, 0.150 mol), anisole (300 mL), and PhMe (40 mL). Solution A was added to the flask containing the amidoxime AAOF. The mixture in the flask was heated at 150 °C while collecting the resulting azeotrope. After heating for 2 h, approximately 30 mL of solution A was added. After another 1.5 h, approximately 35 mL of solution A was added. Once the complete addition of solution A was done, the reaction mixture was heated at 150 °C for 1.5 h. The reaction mixture was then cooled to room temperature. Acetone (400 mL) was added to the cooled reaction mixture while stirring. The resulting precipitate was filtered off and washed with acetone (100 mL). The solid was further stirred with water (800 mL), collected by filtration, and washed with water (100 mL) and acetone (100 mL). The collected solid was dried under suction to obtain product 4a-16 as a colorless solid. The yield of the reaction was found to be 82.8 g (78%). ^1H NMR spectrum: (90 MHz, DMSO-d_6), δ, ppm: 6.59 (2H, s, NH_2). ^1H NMR spectrum (90 MHz, DMF-d_7), δ, ppm: 6.60 (2H, s, NH_2). ^{13}C NMR spectrum (22.5 MHz, DMF-d_7), δ, ppm: 165.1; 161.9; 157.0; 137.6.

3,3′-Bis(4-nitro-1,2,5-oxadiazol-3-yl)-5,5′-bi(1,2,4-oxadiazole) (**4a-17**): A round-bottom flask was prepared with a thermometer, magnetic stir bar, and liquid addition funnel. Into the flask, oxadiazole **4a-16** (50 g, 164 mmol), TFA (200 mL), and $MeNO_2$ (750 mL) were added. The resulting mixture was stirred at 50 °C. To the stirred mixture at 50 °C, H_2O_2 (90%, 75 mL, 95 g, 2.8 mol) was added dropwise over a period of 18 h. After the complete addition of H_2O_2, if the reaction mixture was not clear, the heating was extended until a clear solution was obtained. Once the reaction was complete, the reaction mixture was cooled to room temperature and then poured into a mixture of ice water (400 mL) and EtOH (100 mL). The resulting mixture was stirred at room temperature for 1 h. The precipitate formed in the mixture was filtered off, washed with water (200 mL), and dried in air. The crude material obtained was further recrystallized from a mixture of EtOAc-EtOH to yield product **4a-17** as a colorless crystalline solid. The yield of the reaction was determined to be 44 g (73%). T_d, 182 °C (DSC peak endotherm). ^{13}C NMR spectrum (22.5 MHz, DMSO-d_6), δ, ppm: 164.5; 160.1; 158.1; 140.5.

N,N′-([5,5′-Bi(1,2,4-oxadiazole)]-3,3′-diylbis(1,2,5-oxadiazole-4,3-diyl))dinitramide (**4a-18**): To prepare compound **4a-18**, a solution of compound **4a-16** (3.04 g, 10 mmol) was slowly added to 95% HNO_3 (6 mL) at 0 °C. The resulting mixture was then allowed to react for 1 h. After the reaction time, the mixture was poured into ice water (200 mL), causing the insoluble matter to precipitate. The precipitate was collected by filtration to obtain compound **4a-18** as a light-yellow solid. The yield of this process was determined to be 2.88 g, corresponding to a yield of 73.0%. T_d: 84 °C. ^1H NMR (500 MHz, DMSO-d_6): δ 7.08 (s) ppm. ^{13}C NMR (500 MHz, DMSO-d_6): δ 163.84, 159.74, 152.87, 141.94 ppm. IR

(KBr): n 3471, 2728, 1979, 1690, 1621, 1595, 1514, 1460, 1321, 1237, 1167, 1038, 977, 872, 800, 761, 679 cm^{-1}. Elemental analysis for $C_8H_2N_{12}O_8$ (394.18): calcd. C 24.38, H 0.51, N 42.64%. Found: C 24.53, H 0.57, N 42.42%.

Bis(4-nitro-1,2,5-oxadiazole-2-oxid-3-yl)-azo-1,2,4-oxadiazole (**4a-19**): A solution of compound **2a-46** (0.50 g, 2.3 mmol) in 5 mL of 37% hydrochloric acid was prepared. To this solution, a solution of $KMnO_4$ (0.55 g, 3.45 mmol) in 20 mL water was added. The reaction mixture was kept at 25 °C for 2 h. After the designated reaction time, the mixture was treated with 30% hydrogen peroxide. This caused a precipitate to form. The precipitate was then filtered off, washed with water, and dried in air. As a result, an orange solid weighing 0.43 g was obtained. The yield of this process was determined to be 89.1%. ^{13}C NMR (50.69 MHz, DMSO-d_6): δ 100.73, 153.96, 159.62, 171.46 ppm. IR (KBr): n 3440.72, 3137.88, 1678.53, 1632.32, 1572.95, 1535.87, 1500.76, 1389.36, 1312.77, 1227.10, 1095.32, 1070.17, 1048.46, 991.25, 958.39, 907.97, 846.16, 789.99, 758.17, 663.44 cm^{-1}. Elemental analysis for $C_8N_{12}O_{10}$ (424.16): calcd C 22.65, N 39.63%. Found: C 22.62, N 39.59%.

3,3′-Bis (5-amine-1,3,4-oxadiazol-yl)-4,4′-azofurazn (**4a-20**): **2a-59** (10 mmol) was dispersed in 50 mL of 20% hydrochloric acid with stirring at an ice-salt bath. A solution of potassium permanganate ($KMnO_4$, 10 mmol) in a minimal amount of water was prepared separately. This solution was then added dropwise to the dispersion of 2a-59 in hydrochloric acid. After the complete addition of the $KMnO_4$ solution, the reaction mixture was stirred for 120 min at 30 °C. To quench the reaction, 5% hydrogen peroxide (H_2O_2) was added until the solution turned colorless. The resulting precipitate was isolated by filtering and washed with cold water. The isolated solid was dried under vacuum to obtain the desired product as a light yellow solid. The yield of the process was determined to be 72%. T_d (onset): 176 °C; T_d (peak): 233 °C; ^1H NMR (DMSO-d_6, 300 MHz) δ: 8.54 (d, J = 6.0 Hz, 2H), 8.40 (d, J = 6.0 Hz, 2H) ppm; ^{13}C NMR (DMSO-d_6, 300 MHz) δ: 162.26, 160.10, 146.93, 140.01 ppm; IR (KBr): n 3434, 1687, 1636, 1385, 1098, 1037, 805, 705, 469 cm^{-1}; HRMS (ESI): calcd for $C_8H_4N_{12}O_4^+$ [M+H$^+$] 333.2035, found 333.0076; Anal. calcd for $C_8H_4N_{12}O_4$ (332.20): C 28.92, H 1.21, N 50.60; found C 28.73, H 1.31, N 49.65.

1,2-Bis(5-(4-nitramino-1,2,5-oxadiazol-3-yl)-1,3,4-oxadiazol-2-yl)diazene (**4a-21**): A solution of compound **2a-60** (0.64 g, 3.0 mmol) in 10 mL of 37% hydrochloric acid and 3 mL of acetonitrile was prepared. To this solution, a solution of $KMnO_4$ (0.553 g, 3.5 mmol) in 20 mL of water was added. The reaction mixture was kept at 25 °C for 4 h. After the designated reaction time, the mixture was treated with 30% hydrogen peroxide. This caused a precipitate to form. The precipitate was then filtered off, washed with water, and dried in air. As a result, compound **4a-21** weighing 0.53 g and corresponding to 1.25 mmol was obtained. The yield of this process was determined to be 83%. T_m: 181 °C. T_d: 248 °C. ^1H NMR (500 MHz, DMSO-d_6): δ 7.33 (s, 2H) ppm. ^{13}C NMR (500 MHz, DMSO-d_6): δ 139.73, 154.35, 155.32, 168.06 ppm. IR (KBr): n 3196, 1621, 1606, 1470, 1370, 1271, 1126, 1019, 969, 912, 862, 747 cm^{-1}. Elemental analysis for $C_8H_2N_{14}O_8$ (422.24): calcd C 22.76, H 0.48, N 46.45%. Found: C 22.73, H 0.49, N 46.43%.

1,2-Bis(5-(4-nitro-1,2,5-oxadiazol-3-yl)-1,3,4-oxadiazol-2-yl)diazene (**4a-22**): A solution of compound **2a-61** (0.59 g, 3.0 mmol) in 10 mL of 37% hydrochloric acid and 3 mL of acetonitrile was prepared. To this solution, a solution of $KMnO_4$ (0.553 g, 3.5 mmol) in 20 mL of water was added. The reaction mixture was kept at 25 °C for 4 h. After the designated reaction time, the mixture was treated with 30% hydrogen peroxide. This caused a precipitate to form. The precipitate was then filtered off, washed with water, and dried in air. As a result, compound **4a-22** weighing 0.49 g and corresponding to 1.25 mmol was obtained. The yield of this process was determined to be 83%. T_m: 157 °C. T_d: 278 °C. ^{13}C NMR (500 MHz, DMSO-d_6): δ 142.68, 148.38, 157.23, 172.04 ppm. IR (KBr): n 1609, 1564, 1458, 1402, 1377, 1324, 1231, 1100, 1019, 983, 965, 894, 767 cm^{-1}. Elemental analysis for $C_8N_{12}O_8$ (392.20): calcd C 24.50, N 42.86%. Found: C 24.47, N 42.82%.

Bis(4″-amino[3,3′:4′,3″]terfurazan-4-yl)diazene (**4a-23**): A solution of compound **BAFF** (0.64 g, 3.0 mmol) in 10 mL of 37% hydrochloric acid and 3 mL of acetonitrile was prepared. To this solution, a solution of $KMnO_4$ (0.553 g, 3.5 mmol) in 20 mL of water was added. The reaction mixture was kept at 25 °C for 4 h. After the designated reaction time, the mixture was treated with 30% hydrogen peroxide. This caused a precipitate to form. The precipitate was then filtered off, washed with water, and dried in air. As a result, compound **4a-23** weighing 0.53 g and corresponding to 1.25 mmol was obtained. The yield of this process was determined to be 83%.

Bis(4″-nitro[3,3′:4′,3″]terfurazan-4-yl)diazene (**4a-24**): An initial quantity of 8 g (0.03 mol) 4″-nitro-[3,3′:4′,3″-ter(1,2,5-oxadiazol)]-4-amine was added to a mixture consisting of acetic acid (50 mL) and 35% aqueous hydrochloric acid (50 mL). The reaction mixture was then heated to a temperature range of 30–35 °C. Next, $KMnO_4$ (2.76 g, 0.01 mol) was incrementally added to the mixture while vigorous stirring and cooling were employed to maintain the reaction mixture's temperature below 50 °C. Following this addition, the reaction mixture was stirred for 30 min at 50 °C before being cooled to 20 °C. To neutralize any unreacted $KMnO_4$, a small amount of a 30% aqueous solution of H_2O_2 was introduced into the mixture. Subsequently, the resulting mixture was diluted with cold water (100 mL), leading to the formation of a precipitate. This precipitate was separated and subjected to recrystallization from a mixture of acetic acid and water (in a 2:1, v/v ratio). The overall yield of the process was determined to be 7.13 g (74%). T_d, 116–117 °C. Calculated (%): C, 27.29; H, 0.00; N, 42.43. $C_{12}N_{16}O_{10}$. Found (%): C, 27.42; H, 0.02; N, 42.29.

3,3′-Bis(4-(5-methyl-1H-1,2,4-triazol-3-yl)-4,4′-azofurazan (**4b-1**): 4-(5-Methyl-1H-1,2,4-triazol-3-yl)-1,2,5-oxadiazol-3-amine (10 mmol) was dispersed in 50 mL of 20% hydrochloric acid with stirring at an ice-salt bath. A solution of potassium permanganate ($KMnO_4$, 10 mmol) in a minimal amount of water was prepared separately. This $KMnO_4$ solution was then added dropwise to the dispersion of 4-(5-methyl-1H-1,2,4-triazol-3-yl)-1,2,5-oxadiazol-3-amine in hydrochloric acid. After the complete addition of the $KMnO_4$ solution, the reaction mixture was stirred for 120 min at 30 °C. To quench the reaction, 5% hydrogen peroxide (H_2O_2) was added until the solution turned colorless. The resulting precipitate was isolated by filtration and washed with cold water. The isolated solid was dried under vacuum to obtain the desired product as a light yellow solid. The yield of the process was determined to be 72%. T_m: 289.7 °C; T_d (onset): 327.6 °C; T_d (peak): 349.7 °C;

^{1}H NMR (C_3D_6O, 300 MHz) δ: 14.41 (s, 1H), 10.33 (s, 1H), 2.48 (s, 3H), 2.19 (s, 3H) ppm; ^{13}C NMR (CD_3CN, 300 MHz) δ: 168.94, 155.03, 150.42, 149.76, 23.80 ppm; IR (KBr): n 3434, 3285, 3209, 1703, 1562, 1550, 1441, 1376, 1322, 1280, 1242, 1172, 1049, 980, 873, 847, 679, 592, 525 cm^{-1}; HRMS (ESI): calcd for $C_{10}H_8N_{12}O_2^-$ [M-H$^-$] 327.2417, found 327.0777; Anal. calcd for $C_{10}H_8N_{12}O_2$ (328.25): C 36.59, H 2.46, N 51.21; found C 36.51, H 2.40, N 51.19.

4-(5-Amino-1H-1,2,4-triazol-3-yl)-1,2,5-oxadiazol-3-amine (**4b-2**): 4-(5-Methyl-1H-1,2,4-triazol-3-yl)-1,2,5-oxadiazol-3-amine, a starting material of 10 mmol was dispersed in 50 mL of 20% hydrochloric acid with stirring at an ice-salt bath. A separate solution of potassium permanganate ($KMnO_4$, 10 mmol) in a small amount of water was prepared. This $KMnO_4$ solution was then added dropwise to the dispersion of the starting material in hydrochloric acid. After completing the addition of $KMnO_4$, the reaction mixture was stirred for 120 min at 30 °C. To stop the reaction, 5% hydrogen peroxide (H_2O_2) was added gradually until the solution became colorless. The resulting precipitate was isolated by filtration, thoroughly washed with cold water, and then dried under vacuum. As a result, an orange solid corresponding to 67% yield of the desired product was obtained. T_d (onset): 279 °C; T_d (peak): 308 °C; ^{1}H NMR (DMSO-d_6, 300 MHz) δ: 8.53 (s, 2H), 5.96 (s, 4H), ppm; ^{13}C NMR (DMSO-d_6, 300 MHz) δ: 155.88, 148.58, 140.99, 136.11 ppm; IR (KBr): n 3373, 3250, 3148, 2193, 1653, 1556, 1400, 1323, 1274, 1169, 1136, 1035, 995, 910, 799, 573 cm^{-1}; HRMS (ESI): calcd for $C_8H_6N_{14}O_2^+$ [M+H$^+$] 331.2337, found 331.0881; Anal. calcd for $C_8H_6N_{14}O_2$ (330.23): C 29.10, H 1.83, N 59.38; found C 29.07, H 1.86, N 59.30.

Di(5-nitramino-1,2,4-triazoly)-azo-furazan (**4b-3**): Method 1: A solution of compound **2f-7** (2.30 g, 10.0 mmol) in 25 mL of 37% hydrochloric acid was prepared. To this solution, a solution of $KMnO_4$ (1.98 g, 12.5 mmol) in 30 mL of water was added. The reaction mixture was then kept at 50 °C for 4 h. After the designated reaction time, the mixture was treated with 30% hydrogen peroxide. This led to the formation of a precipitate, which was subsequently filtered off, washed with water, and dried in air. As a result, compound **4b-3** weighing 1.39 g was obtained. The yield of this process was determined to be 66%. Method 2: A solution of N-methyl-N-nitroso-N′-nitroguanidine (0.53 g, 3.6 mmol) in 20 mL of methanol was added to a suspension of dicarbohydrazide-azo-furazan (0.42 g, 1.5 mmol) in 30 mL of water under stirring. The resulting mixture was heated and maintained at the boiling point for 2 h before being cooled to room temperature. Subsequently, a solution of potassium hydroxide (0.28 g, 5.0 mmol) in 15 mL of water was added to the mixture. The mixture was stirred for an additional 2 h at the boiling point and then cooled to 50 °C. To achieve acidification to pH = 3, 68% nitric acid was added to the mixture. This caused a precipitate to form. The precipitate was filtered, washed with cold water, and dried in air, resulting in compound **4b-3** weighing 0.41 g and corresponding to 1.0 mmol. The yield of this process was determined to be 67%. ^{1}H NMR (300 MHz, DMSO-d_6): δ 6.88 (br) ppm; ^{13}C NMR (75 MHz, DMSO-d_6): δ 161.84, 152.64, 142.00, 141.48 ppm. IR (KBr): n 3445, 2768, 1708, 1630, 1548, 1432, 1385, 1312, 1184, 1155, 1035, 980, 616 cm^{-1}. Elemental analysis for

$C_8H_4N_{16}O_6$ (420.23): calcd. C 22.87, H 0.96, N 53.33%; Found: C 22.81, H 1.03, N 53.77%.

3,3′-bis(4-amino-1H-1,2,4-triazol-5(4H)-one)-4,4′-azofurazan (**4b-4**): 4-Amino-5-(4-amino-1,2,5-oxadiazol-3-yl)-2,4-dihydro-3H-1,2,4-triazol-3-one, a starting material of 10 mmol was dispersed in 50 mL of 20% hydrochloric acid with stirring at an ice-salt bath. A solution of potassium permanganate ($KMnO_4$, 10 mmol) in a minimal amount of water was prepared separately. This $KMnO_4$ solution was then added dropwise to the dispersion of the starting material in hydrochloric acid. After completing the addition of $KMnO_4$, the reaction mixture was stirred for 120 min at 30 °C. To stop the reaction, 5% hydrogen peroxide (H_2O_2) was added gradually until the solution became colorless. The resulting precipitate was isolated by filtration, thoroughly washed with cold water, and then dried under vacuum. As a result, a light yellow solid corresponding to a yield of 92% of the desired product was obtained. T_d (onset): 232 °C; T_d (peak): 241 °C; ^1H NMR (DMSO-d_6, 300 MHz) δ: 8.54 (s, 1H), 8.43 (s, 1H), 4.20 (s, 4H) ppm; ^{13}C NMR (DMSO-d_6, 300 MHz) δ: 162.11, 157.56, 144.76, 144.10 ppm; IR (KBr): n 3337, 3184, 1685, 1625, 1543, 1438, 1355, 1175, 1110, 1033, 824, 703, 659, 544 cm^{-1}; HRMS (ESI): calcd for $C_8H_6N_{14}O_4^+$ [M+Na^+] 385.2134, found 385.0475; [M+K^+] 401.3228, found 401.0384; Anal. calcd for $C_8H_6N_{14}O_4$ (362.22): C 26.53, H 1.67, N 54.14; found C 26.17, H 1.56, N 54.11.

3,3′-Bis(4-dinitromethyl-1H-1,2,4-triazol)-4,4′-azofurazan (**4b-5**): A suspension was prepared by combining 4-amino-1,2,5-oxadiazole-3-carboximihydrazide (2.84 g, 20 mmol), ethyl 3-ethoxy-3-iminopropanoate hydrochloride (4.3 g, 22 mmol), and 20 drops of boron trifluoride diethyl etherate in acetic acid (50 mL). The resulting suspension was heated to 50 °C for 3 h. Subsequently, the reaction mixture was refluxed for 18 h. After cooling the solution to ambient temperature, insoluble solids were removed by filtration, and the filtrate was concentrated under vacuum using a rotary evaporator. The resulting residue was then suspended in water (200 mL) and neutralized with solid sodium bicarbonate to achieve a pH of 5–6. Pale white solids were obtained by filtration and dried in air, yielding 3.57 g (75%) of product 2-[5-(4-amino-1,2,5-oxadiazol-3-yl)-1H-1,2,4-triazol-3-yl)]acetate. T_m: 165 °C. T_d (onset): 270 °C. 1H NMR: 14.63 (br), 6.41 (s, 2H), 4.13 (q, 2H), 4.05 (s, 2H), 1.21 (t, 2H) ppm. ^{13}C NMR: 167.9, 155.2, 152.1, 149.3, 139.1, 61.1, 32.3, 13.9 ppm. IR (KBr): 3508, 3467, 3405, 3168, 1742, 1730, 1705, 1625, 1561, 1549, 1464, 1396, 1370, 1323, 1220, 1183, 1097, 1063, 1026, 1008, 981, 899, 874, 862, 809, 767, 734, 663 cm^{-1}. Elemental analysis for $C_8H_{10}N_6O_3$ (238.20): Calcd C 40.34, H 4.23, N 35.28%. Found: C 40.64, H 4.78, N 35.03%.

Compound 2-[5-(4-amino-1,2,5-oxadiazol-3-yl)-1H-1,2,4-triazol-3-yl)]acetate (2.38 g, 10.0 mmol) was dissolved in concentrated HCl (20 mL) with the addition of water (40 mL). The resulting solution was cooled to below 10 °C using an ice bath. A suspension of potassium permanganate (1.60 g, 10.0 mmol) in a minimal amount of water was prepared separately. This suspension was then added dropwise to the reaction mixture while maintaining the temperature below 10 °C. The solution was warmed to room temperature and stirred for 24 h. Dilute hydrogen peroxide (5%) was added until the solution became colorless. The resulting precipitate was collected by filtration, washed with water (200 mL), and dried, resulting in diethyl 2,2′-((diazene-1,2-diylbis(1,2,5-oxadiazole-4,3-diyl))bis(1H-1,2,4-triazole-5,3-diyl))(E)-diacetate weighing 1.33 g. The yield of this process was determined to be 56%. The product obtained was in the form of yellow solid. T_d (onset): 194 °C. ^{1}H NMR: δ 14.67 (s, 2H), 4.12 (q, 4H), 4.03 (s, 4H), 1.18 (t, 6H) ppm. ^{13}C NMR: δ 167.8, 161.8, 152.0, 148.9, 144.3, 61.1, 40.0, 13.9 ppm. IR (KBr): 3132, 2991, 1727, 1556, 1497, 1476, 1439, 1373, 1336, 1299, 1270, 1215, 1138, 1059, 1032, 990, 933, 886, 844 cm^{-1}. Elemental analysis for $C_{16}H_{16}N_{12}O_6$ (472.38): Calcd C 40.68, H 3.41, N 35.58%. Found: C 40.38, H 3.34, N 35.41%.

Compound diethyl 2,2′-((diazene-1,2-diylbis(1,2,5-oxadiazole-4,3-diyl))bis(1H-1,2,4-triazole-5,3-diyl))(E)-diacetate (0.52 g, 1.1 mmol) was slowly added to a cooled and well-stirred mixture of concentrated sulfuric acid (10 mL) and 100% nitric acid (3 mL) at 0 °C. The temperature of the reaction mixture was allowed to slowly warm to room temperature over a period of 1 h, and the mixture was then stirred for an additional 24 h. The resulting solution was poured into ice water (100 g), leading to the formation of a yellow precipitate. The yellow precipitate was collected by filtration and washed with water (30 mL × 2). As a result, 3,3′-bis(4-dinitroacetate-1H-1,2,4-triazol)-4,4′-azofurazan•H_2O weighing 0.50 g was obtained, corresponding to a yield of 68%. The product obtained was in the form of a yellow solid. T_d (onset): 190 °C. ^{1}H NMR: 4.54 (q, 4H), 1.29 (t, 6H) ppm. ^{13}C NMR: 161.8, 156.0, 151.3, 144.7, 141.2, 113.4, 66.7, 13.3 ppm. IR (KBr): 3182, 3089, 2974, 1756, 1662, 1583, 1501, 1456, 1437, 1399, 1185, 1038, 996, 924, 883, 816, 753, 692 cm^{-1}. Elemental analysis for $C_{16}H_{14}N_{16}O_{15}$ (670.38): Calcd C 28.67, H 2.10, N 33.43%. Found: C 28.23, H 1.76, N 33.58%.

A suspension of 3,3′-bis(4-dinitroacetate-1H-1,2,4-triazol)-4,4′-azofurazan (0.34 g, 0.50 mmol) in ethanol (20 mL) was prepared. Methanolic ammonia (4 mL, 2.0 mol L^{-1}) was then added to the suspension. The resulting reaction mixture was stirred for 6 h at room temperature. After completion of the reaction, the precipitate was collected by filtration. The obtained precipitate weighed 0.27 g, corresponding to a yield of 94%. To purify the product, it was washed with ethanol (10 mL) and diethyl ether (10 mL). The final product was obtained as a yellow solid. T_d (onset): 165 °C. ^{1}H NMR: δ 7.12 (br) ppm. ^{13}C NMR: 156.5, 152.6, 150.1, 139.8, 126.0 ppm. IR (KBr): 1636, 1584, 1529, 1503, 1440, 1325, 1245, 1195, 1081, 1035, 1007, 986, 925, 880, 861, 776, 743, 698, 617, 607 cm^{-1}. Elemental analysis for $C_{10}H_{14}N_{18}O_{12}$ (578.33): Calcd C 20.77, H 2.44, N 43.59%. Found: C 20.56, H 2.66, N 43.05%.

A suspension was formed by combining the compound (0.29 g, 0.50 mmol) with acetone (20 mL) at 25 °C. Concentrated HCl (0.5 mL) was then added to the suspension. The resulting mixture was stirred for 1 h. A white solid formed and was separated from the solution through filtration. The filtrate was dried using magnesium sulfate as a drying agent, and the solvent was subsequently removed. The remaining residue was solidified by adding dichloromethane (5 mL). The resulting precipitate, identified as **4b-5**, was filtered and dried in air. T_d (onset): 120 °C. ^{1}H NMR (CD_3CN): δ 8.24 (s, 2H) ppm. ^{13}C NMR (CD_3CN): δ 162.9, 153.1, 145.5, 141.4, 106.3 ppm. IR (KBr): 1577, 1445, 1400, 1328, 1239, 1192, 1145, 1059, 995, 873, 815, 785,

718 cm^{-1}. Elemental analysis for $C_{10}H_4N_{16}O_{10}$ (508.24): Calcd C 23.63, H 0.79, N 44.09%. Found: C 23.95, H 1.13, N 44.83%.

3,3′-bis(4-trinitro-1H-1,2,4-triazol)-4,4′-azofurazan (**4b-6**): Compound (0.5 g, 0.92 mmol) was added to 3 mL of H_2SO_4 and stirred for 0.5 h. Then, 1.5 mL of 100% HNO_3 was added to this mixture at 0 °C and stirred at room temperature for 3 h. After the reaction period, the mixture was quenched with ice water and extracted with diethyl ether. The organic layer was dried in air, resulting in the formation of a white solid identified as **4b-6**. The white solid weighed 0.44 g, corresponding to a yield of 81%, and the amount obtained was 0.75 mmol. ^1H NMR (500 MHz, DMSO-d_6) $\delta = 6.80$(s), 6.73(s) ppm. ^{13}C NMR (126 MHz, DMSO-d_6) $\delta = 161.94$, 148.76, 146.67, 141.66, 127.14 ppm. IR: 3191.43, 1631.71, 1610.43, 1585.92, 1436.49, 1280.15, 1195.81, 1147.99, 1080.95, 1053.26, 990.02, 970.05, 911.44, 898.81, 868.46, 844.24, 796.87, 766.09, 751.72, 733.98, 709.66, 695.67, 643.48, 620.85, 604.51 cm^{-1}. Elemental analysis for **4b-6**, $C_{10}H_2N_{18}O_{14}$ (598.00): calcd C 20.08, H 0.34, N 42.14%; found C 20.16, H 0.30, N 42.33%.

3,3′-bis(4-amino-1H-1,2,4-triazol-5(4H)-one)-4,4′-azofurazan (**4b-7**): **HATF** (10 mmol) was dispersed in 50 mL of 20% hydrochloric acid with stirring at an ice-salt bath to maintain a low temperature. A solution of potassium permanganate ($KMnO_4$, 10 mmol) in a minimal amount of water was prepared separately and added dropwise to the HATF solution. After the addition, the reaction mixture was stirred for 120 min at 30 °C. To stop the reaction, 5% hydrogen peroxide (H_2O_2) was slowly added until the solution turned colorless. The resulting precipitate was isolated by filtration, washed with cold water, and then dried under vacuum conditions. This process yielded the desired product as a yellow solid with a yield of 76%. T_d (onset): 238 °C; T_d (peak): 251 °C; ^{13}C NMR (DMSO-d_6, 300 MHz) δ: 162.16, 159.98, 139.80 ppm; HRMS (ESI): calcd for $C_6H_2N_{14}O_2^-$ [M-H$^-$] 301.1647, found 301.0410; Anal. calcd for $C_6H_2N_{14}O_2$ (302.17): C 23.85, H 0.67, N 64.89; found C 23.88, H 0.65, N 63.06.

3,3′-Bis(tetrazol-1-oxhydryl)-4,4′-azofurazan (**4b-8**): 5-(4-Amino-1,2,5-oxadiazol-3-yl)-1-hydroxytetrazole (1 g, 5.9 mmol) was added to acetonitrile (15 mL) in a 100 mL round-bottomed flask equipped with a liquid addition funnel and a magnetic stir bar. Concentrated HCl (35%, 15 mL) was then added, resulting in the dissolution of most of the 5-(4-amino-1,2,5-oxadiazol-3-yl)-1-hydroxytetrazole. With rapid stirring, a solution of $KMnO_4$ (1 g, 6.4 mmol) in water (15 mL) was added dropwise to the flask. This caused the color of the solution to change from yellow to green, and finally to purple/brown. After completing the addition, the reaction mixture was stirred for 20 min. Excess $KMnO_4$ was quenched by adding hydrogen peroxide (5%) dropwise until the mixture turned yellow. The reaction mixture was then poured into ice water (50 mL) and extracted with diethyl ether (25 mL × 4). The organic phases were combined and dried using anhydrous $MgSO_4$. Finally, the mixture was filtered to obtain the desired product.

N,N′-Methylenebis(N-(4-(2H-tetrazol-5-yl)-1,2,5-oxadiazol-3-yl)-nitramide) (**4b-9**): A mixture containing the compound (3.22 g, 10 mmol), $ZnBr_2$ (2.25 g, 10 mmol), and NaN_3 (1.63 g, 25 mmol) was prepared in water (15 mL) and methanol (25 mL). The mixture was then stirred at 40 °C for 2 h. Afterward, it was cooled, and concentrated hydrochloric acid was added to adjust the pH to 2. Stirring was continued for an additional 0.5 h. The resulting mixture was then concentrated under vacuum to remove methanol. The obtained precipitate was collected and washed with hydrochloric acid (2%) and water. As a result, **4b-9** was obtained with a yield of 2.49 g (69%). T_d: 154 °C. ^{1}H NMR (DMSO-d_6, 500 MHz): δ 6.65 (s, 2H, CH_2), 4.98 (2H, NH) ppm. ^{13}C NMR (DMSO-d_6, 500 MHz): δ 150.35 148.97 144.23 66.60 ppm. IR (KBr): 3376 2988 1585 1405 1276 1190 1078 1023 1001 881 750 730 642 608 cm^{-1}. Elemental analysis calcd for $C_7H_4N_{16}O_6$ (408.22): C 20.60, H 0.99, N 54.90%; found: C 20.96, H 0.87, N 54.67%.

4,4′-Oxybis[3,3′-(1H-5-tetrazol)]furazan (**4b-10**): A mixture containing the compound (1.000 g, 4.9 mmol), NaN_3 (0.700 g, 9.8 mmol), and $ZnBr_2$ (1.103 g, 4.9 mmol) in water (10 mL) was prepared in a 25 mL round-bottomed flask. The reaction mixture was heated at 100 °C for 4 h with vigorous stirring. After completion of the reaction, the mixture was cooled to room temperature and acidified with 2% HCl. The resulting mixture was then filtered to obtain a solid product. To purify the crude product, it was recrystallized using H_2O. This process yielded white crystals weighing 0.800 g, corresponding to a yield of 56%. ^{1}H NMR (DMSO-d_6, 400 MHz): 12.92 (br. s, 2H, NH) ppm. ^{13}C NMR (DMSO-d_6, 100 MHz): 160.4, 147.5, 138.7 ppm. IR (neat): 3131, 3073, 2910, 2443, 1636, 1589, 1487, 1452, 1373, 1233, 1196, 1168, 1101, 1077, 1044, 1016, 987, 915, 889, 864, 844, 753, 584, 510, 430, 412, 388 cm^{-1}. MS (ESI): m/z (%): 291.0. Elemental analysis for $C_6H_2N_{12}O_3$ (290.16): calcd. C 24.84, H 0.69, N 57.93%; found: C 24.87, H 0.76, N 58.08%.

4,4′-Oxybis[3,3′-(1-hydroxytetrazolyl)]furazan (**4b-11**): 4,4′-Oxybis(N-hydroxy-1,2,5-oxadiazole-3-carbimidoyl chloride) (0.93 g, 3.0 mmol) was dissolved in ethanol (10 mL) and the solution was cooled to 0 °C. A solution of sodium azide (0.49 g, 7.5 mmol) in water (5 mL) was then added slowly to the cooled solution. After complete addition, the reaction mixture was stirred for 3 h at 0–5 °C. The mixture was then poured into ice-water (30 g), resulting in the formation of a white precipitate known as 4,4′-oxybis(N′-hydroxy-1,2,5-oxadiazole-3-carbimidoyl azide). The suspension was extracted with ethyl ether (3 × 25 mL), and the organic phases were combined and dried using anhydrous magnesium sulfate. After filtration, gaseous HCl was bubbled into a solution of the obtained product in diethyl ether below 0 °C. The reaction mixture was then allowed to reach room temperature and stirred overnight. The progress of the reaction was monitored using thin-layer chromatography (TLC) until no spot corresponding to 4,4′-oxybis(N′-hydroxy-1,2,5-oxadiazole-3-carbimidoyl azide) was observed. The solvent was removed with air, and the resulting residue was dried under reduced pressure. Compound **4b-11** was obtained as a colorless oily liquid weighing 0.73 g, corresponding to a yield of 2.2 mmol. T_d = 187 °C; IR (KBr): n = 3284, 1632, 1586, 1438, 1356, 1271, 1053, 982, 883, 752, 567 cm^{-1}; ^{1}H NMR (CD_3CN): d = 7.74 ppm (s, 2H); ^{13}C NMR (CD_3CN): d = 161.1, 137.9, 136.5 ppm; ^{15}N NMR (CD_3CN): 46.7, 5.5, − 1.0, − 15.0, − 53.9, − 117.7 ppm; elemental analysis calcd (%) for $C_6H_4N_{12}O_6$ (340.17): C 21.18, H 1.19, N 49.41; found: C 21.30, H 1.42, N 50.49.

1,2-Bis(3-nitro-1-(1H-tetrazol-5-yl)-1H-1,2,4-triazol-5-yl)diazene (**4b-12**): **2f-18** (10 mmol) was dispersed in 50 mL of 20% hydrochloric acid with stirring at an ice-salt bath to maintain a low temperature. A solution of potassium permanganate ($KMnO_4$, 10 mmol) in a minimal amount of water was prepared separately and added dropwise to the **2f-18** solution. After the addition, the reaction mixture was stirred for 120 min at 30 °C. To stop the reaction, 5% hydrogen peroxide (H_2O_2) was slowly added until the solution turned colorless. The resulting precipitate was isolated by filtration, washed with cold water, and then dried under vacuum conditions. The obtained product was a bright yellow solid weighing 304 mg, corresponding to a yield of 78%. ^{1}H NMR (400 MHz, DMSO-d_6): δ = 8.22 (s, 3H); ^{13}C NMR (176 MHz, DMSO-d_6): δ = 165.2, 158.2, 156.1 ppm. IR (neat): 3064, 2905, 2791, 1566, 1434, 1321, 1015, 841 cm^{-1}. Anal. calcd for $C_4H_4N_8O_7$: C 18.47, H 0.52, N 64.61; found: C 18.33, H 0.50, N 64.09. HRMS: calc. for $C_6H_2N_8O_4$ $[M+H]^+$: 391.0579; found: 391.0587.

1,2-Bis(3-nitro-1-(1H-tetrazol-5-yl)-1H-1,2,4-triazol-5-yl)hydrazine (**4b-13**): **4b-12** (390 mg, 1 mmol) was added to a solution containing the corresponding hydrazine hydrate (1 mmol). The resulting mixture was stirred at 80 °C for 10 h. Afterwards, the hot solution was filtered and the solvent was evaporated under vacuum. The obtained residue was subjected to recrystallization from an ethanol–water solution to obtain the desired product. To convert the product into its neutral hydrazine form, the corresponding salts were treated with concentrated hydrochloric acid. The final product, **4b-13**, was obtained as a pale yellow solid weighing 357 mg, with a yield of 91%. ^{1}H NMR (400 MHz, DMSO-d_6): δ = 10.36 (s, 2H), 7.09 (s, 2H) ppm; ^{13}C NMR (176 MHz, DMSO-d_6): δ = 161.3, 157, 155.3 ppm. IR (neat): 3590, 3265, 2940, 2640, 1990, 1630, 1605, 1380, 1308, 1036 cm^{-1}. Anal. calcd for $C_7H_7N_{15}O_6$: C 18.37, H 1.03, N 64.28; found: C 18.03, H 1.01, N 64.01. HRMS: calc. for $C_6H_4N_{18}O_4^-$ $[M]^-$: 391.0590; found: 391.0595.

5,5′-(2-(1H-Tetrazol-1-yl)-1H-imidazole-4,5-diyl)bis(1Htetrazole) (**4b-14**): A solution of 2-(1H-tetrazol-1-yl)-1H imidazole-4,5-dicarbonitrile (0.215 g, 1.1 mmol) in DMF (10 mL) was prepared. Sodium azide (0.172 g, 2.65 mmol) and iodine (0.120 g, 47.2 mmol) were added to the solution. The reaction mixture was then refluxed for 6 h with stirring. After the reflux, the mixture was cooled to room temperature and hydrochloric acid (20 mL, 1 M) was added. The resulting solution was extracted with ethyl acetate. The crude product was further purified by column chromatography using silica gel as the stationary phase and n-hexane–EtOAc as the eluent. The purified product was dried over sodium sulfate. The solvent was removed under reduced pressure, resulting in the isolation of the desired product as a white solid. The yield of the isolated product was 0.060 g (15%), and it exhibited satisfactory purity. DSC-TGA (10 °C min^{-1}): 179 °C (T_d). IR (KBr): 3324, 3136, 2934, 2241, 2150, 1651, 1523, 1437, 1388, 1344, 1307, 1251, 1197, 667, 642, 505 cm^{-1}. ^{1}H NMR (400 MHz, DMSO): d (ppm) 9.89 (s, 1H). ^{13}C NMR (100 MHz, DMSO): d (ppm) 160.5, 153.9, 146.3, 142.1. HRMS (ESI) for $C_6H_4N_{14}$ (M+H): calcd 273.0743, found 273.0707. LC–MS (ES, m/z): 273 [M+H]+. C, H, N analysis (%): $C_6H_4N_{14}$ (272), calculated result: C, 26.48; H, 1.48; N, 72.04; found: C, 26.35; H, 1.41; N, 72.15%.

1,1,3,3-Tetra(1H-tetrazol-5-yl)propane (**TTP**) (**4b-15**): Method A: a solution of propane-1,1,3,3-tetracarbonitrile (0.100 g, 0.69 mmol) in DMF (10 mL) was prepared. Sodium azide (0.202 g, 3.1 mmol) and iodine (0.240 g, 3.1 mmol) were added to the solution. The resulting mixture was heated at reflux for 6 h with stirring. After completion of the reaction, the reaction mixture was cooled to room temperature and 1 M HCl (10 mL) was added. The resulting solution was then extracted with ethyl acetate. The organic layer was dried over sodium sulfate. The solvent was removed under reduced pressure, yielding the product as a white powder. The isolated product had a yield of 0.160 g (73.3%) and exhibited satisfactory purity. Method B: A mixture of propane-1,1,3,3-tetra-carbonitrile (5.0 g, 34.6 mmol) and zinc bromide (1.562 g, 6.93 mmol) was prepared in water (50 mL). To this mixture, sodium azide (1.124 g, 17.3 mmol) was added, and the resulting solution was heated at reflux. After stirring overnight, the mixture was cooled and concentrated HCl (10 mL) was added. Upon addition of HCl, the initially dissolved solids reprecipitated and formed a new solid after approximately 10 min. The mixture was further stirred for an additional 50 min. The resulting solid product was filtered off and washed with water, yielding an orange solid. This solid could be recrystallized from water (70 mL) using decolorizing carbon as an aid to obtain the desired product as white crystals. The isolated product weighed 7.1 g, corresponding to a yield of 63%. ^{1}H NMR (400 MHz, DMSO): d = 5.03 (t, 2H), 2.49 ppm (t, 2H); ^{13}C NMR (100 MHz, DMSO): d = 34.4, 30.0 ppm; IR (KBr): n = 3144, 3069, 2962, 2910, 2729, 2613, 2465, 1747, 1674, 1556, 1442, 1383, 1298, 1277, 1248, 1168, 1076, 1055 cm^{-1}; DSC-TGA (heating rate: 10 °C min^{-1}): 276 °C (T_d); LCMS (EI): m/z: 317 $[M]^+$; elemental analysis calcd (%) for $C_7H_8N_{16}$ (316.11): C 26.59, H 2.55, N 70.86; found: C 26.61, H 2.51, N 70.68.

References

1. S.D. Shaposhnikov, I.V. Tselinskii, Russ. J. Org. Chem. **38**, 1351–1355 (2002)
2. Y. Qu, Q. Zeng, J. Wang, Q. Ma, H. Li, H. Li, G. Yang, Chem. Eur. J. **22**, 12527–12532 (2016)
3. Q. Sun, C. Shen, X. Li, Q. Lin, M. Lu, J. Mater. Chem. A **5**, 11063–11070 (2017)
4. Q. Wang, Y. Shao, M. Lu, Cryst. Growth Des. **18**, 6150–6154 (2018)
5. J. Zhang, S. Dharavath, L.A. Mitchell, D.A. Parrish, J.M. Shreeve, J. Mater. Chem. A **4**, 16961–16967 (2016)
6. P.F. Pagoria, M. Zhang, N.B. Zuckerman, A.J. DeHope, D.A. Parrish, Chem. Heterocycl. Compd **53**, 760–778 (2017)
7. Q. Sun, Q. Lin, M. Lu, CrystEngComm **20**, 4321–4328 (2018)
8. H. Xiong, G. Cheng, Z. Zhang, H. Yang, New J. Chem. **43**, 7784–7789 (2019)
9. Q. Wang, Y. Shao, M. Lu, Cryst. Growth Des. **19**, 839–844 (2019)
10. S. Jia, H. Zhang, B. Wang, H. Huo, C. Zhou, Chin. J. Energ. Mater. **23**, 882–886 (2015)
11. A.I. Stepanov, V.S. Sannikov, D.V. Dashko, A.G. Roslyakov, A.A. Astratev, Z.G. Aliev, T.K. Goncharov, S.M. Aldoshin, Russ. Chem. Bull. Int. Ed. **65**, 2063–2067 (2016)
12. Z. Xu, G. Cheng, H. Yang, X. Ju, P. Yin, J. Zhang, J.M. Shreeve, Angew. Chem. Int. Ed. **56**, 5877–5881 (2017)
13. Y. Tang, C. He, G.H. Imler, D.A. Parrishc, J.M. Shreeve, Dalton Trans. **48**, 7677–7684 (2019)
14. J. Ma, G. Zhang, H. Gu, J. Tang, B. Wang, H. Yang, G. Cheng, New J. Chem. **43**, 8370–8375 (2019)
15. H. Wei, J. Zhang, C. He, J.M. Shreeve, Chem. Eur. J. **21**, 8607–8612 (2015)

16. Q. Sun, C. Shen, X. Li, Q. Lin, M. Lu, Cryst. Growth Des. **17**, 6105–6110 (2017)
17. L. Liang, H. Huang, K. Wang, C. Bian, J. Song, L. Ling, F. Zhao, Z. Zhou, J. Mater. Chem. **22**, 21954–21964 (2012)
18. Y. Tang, C. He, G.H. Imler, D.A. Parrish, J.M. Shreeve, Chem. Asian J. **11**, 3113–3117 (2016)
19. B. Zhao, X. Li, P. Wang, Y. Ding, Z. Zhou, New J. Chem. **42**, 14087–14090 (2018)
20. D. Srinivas, V.D. Ghule, K. Muralidharan, H.D.B. Jenkins, Chem. Asian J. **8**, 1023–1028 (2013)
21. P. Yin, Q.H. Zhang, J.M. Shreeve, Acc. Chem. Res. **49**, 4–16 (2016)

Chapter 5
Conclusions and Perspectives

As we known, it is attractive and challenging to achieve a fine balance between energy and safety in the field of HEDMs. Therefore, the in-depth study in the design and syntheses of new HEDMs especially for nitrogen-rich polycyclic-based energetic materials is growing in interest among chemical researchers. Up to now, there are four promising routes to modern HEDMs: crystal engineering, constructing energetic metal organic frameworks, designing polynitrogen compounds and forming energetic salts. In fact, molecular design is the most commonly used method to achieve the balance through the combination of stable skeletons with high energy units through suitable linkages. In this book, we outline the major synthetic strategies, energetic properties and potential applications of a series of nitrogen-rich heterocycles. By summarizing these researches, the main heterocyclization routes are as follows (Fig. 5.1): (1) cyclization reactions of oxime or hydrazide to construct different oxadiazole forms; (2) amino or cyano compounds react with azides to form tetrazole; (3) reactions of ester to give triazole; (4) reactions of the hydrazine compound and carbonyl compound to prepare tetrazine, pyrazole and imidazole. In addition, the increase of the ring system and the change of performance can be realized by introducing a bridge such as imine, hydrazine, azo, triazene, tetrazene, azoxy, oxy, alkoxy, alkyl and bisnitramino-methylene linkage. In order to replace traditional explosives (**TNT**, **RDX** and **HMX**), the next generation of energetic materials should exhibit a good performance both in energy but also in safety such as the detonation velocity above 8500 m s^{-1}, the thermal stability above 200 °C, impact sensitivity above 7 J, friction sensitivity above 120 N, long-term storage above 15 years. Moreover, the products after combustion or detonation and the synthesis process should be environmentally friendly. Among these nitrogen-rich heterocycles in this book, several compounds have better thermal stability, detonation properties and sensitivity compared to **RDX** (**2a-11**, **2c-9**, **2c-17**, **2c-19**, **2c-23**, **2e-16**, **2e-32**, **2e-33**, **4a-7**, **4a-22**, **4b-13**) and **HMX** (**TKX-50**, **H_2bta**, **2f-19**, **2f-33**, **4a-3**, **4a-8**). Some of them with high oxygen content can be a suitable candidate as an oxidizer in solid rocket propellants due to high specific impulse such as **2a-12**, **2a-16**, **2a-51**, **2a-62** and so on.

G. Cheng et al., *Nitrogen-Rich Polycyclic-Based Energetic Materials*,
https://doi.org/10.1007/978-981-97-4569-2_5

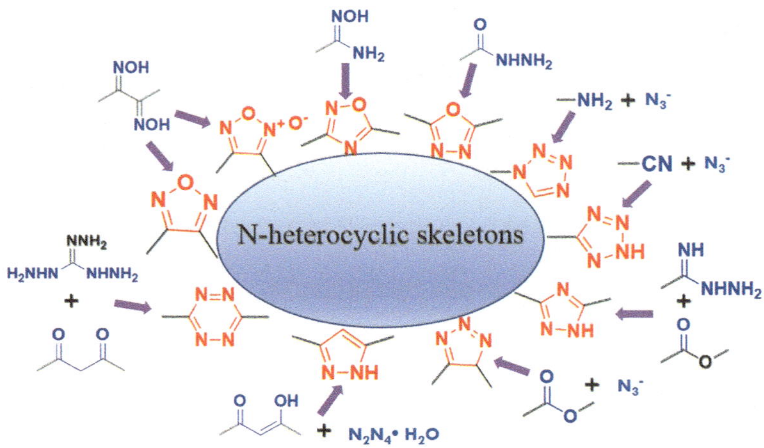

Fig. 5.1 Heterocyclization routes

And the introduction of alkyl or alkoxy groups makes these heterocyclic compounds as possible candidates as melt-cast explosives due to low melting points. In addition, the pyrazole or imidazole analogues with energetic groups usually possess a high thermal stability to apply in heat resistant explosives such as **2a-56**, **2e-18**, **2e-23**, **3b-30**, **4b-1** (D > 8000 m s^{-1}, T_d > 350 °C). Also, the nitrogen-rich heterocycle with poly-azide groups often can be used as a potential environmentally friendly candidate for lead-based primary explosives represented by **3b-7**.

The numerous works on the synthesis and study of polycyclic-based energetic materials in recent years reveal that this field is of major interest to the chemical community. We believe that the widely use of ecofriendly polycyclic-based energetic materials can help to achieve a positive impact of chemistry on our society. And authors hope that, more and more new powerful and strategic energetic materials may appear in literature in the future.

Caution: Readers are reminded that the information given in this book is intended to cover the progress of recent research on nitrogen-rich polycyclic-based energetic materials. Most of the molecules (salts) collected in this book are nevertheless energetic materials that may be explosive under certain conditions. Their syntheses should be carried out by experienced personnel and handled with caution. In any case, carefully planned safety protocols and proper protective equipment, such as Kevlar gloves, ear protection, safety shoes and plastic spatulas, should be utilized at all times especially when working on a large scale (> 1 g). The authors strongly suggest that the original references should be consulted for detailed safety information.

The manufacturer's authorised representative in the EU is Springer Nature Customer Service Centre GmbH, Europaplatz 3, 69115 Heidelberg, Germany. If you have any concerns regarding our products, please contact ProductSafety@springernature.com

Printed and bound by CPI Group (UK) Ltd, Croydon, CR0 4YY
30/09/2026
02225300-0006